Progress in Mathematics
Volume 226

Arithmetic of Higher-Dimensional Algebraic Varieties

Bjorn Poonen
Yuri Tschinkel
Editors

Birkhäuser
Boston • Basel • Berlin

Bjorn Poonen
Department of Mathematics
University of California, Berkeley
Berkeley, CA 94720
U.S.A.

Yuri Tschinkel
Mathematisches Institut
Bunsenstr. 3-5
D-37073 Göttingen
Germany

Library of Congress Cataloging-in-Publication Data

A CIP catalogue record for this book is available from the Library of Congress, Washington D.C., USA.

AMS Subject Classifications: 11D45, 11D72, 11F72, 11G35, 11G50, 11Y50, 14D72, 14E30, 14G05, 14G25

ISBN 0-8176-3259-X
ISBN 3-7643-3259-X

Printed on acid-free paper.

Birkhäuser

Printed in the United States of America. (MV)

9 8 7 6 5 4 3 2 1 SPIN 10936412

Birkhäuser is part of *Springer Science+Business Media*

birkhauser.com

CONTENTS

Part II. Research Articles

ABSTRACTS

We survey some of the outstanding problems concerning rational points on curves and surfaces.

We discuss connections between analytic number theory and geometry of higher-dimensional algebraic varieties.

We give an introduction to the study of weak approximation on algebraic varieties.

Around 1989, Manin initiated a program to understand the asymptotic behaviour of rational points of bounded height on Fano varieties. This program led to the search of new methods to estimate the number of points of bounded height on various classes of varieties. Methods based on harmonic analysis were successful for compactifications of homogeneous spaces. However, they do not apply to other types of varieties.

Universal torsors, introduced by Colliot-Thélène and Sansuc in connection with the Hasse principle and weak approximation, turned out to be a useful tool in the treatment of other varieties. The aim of this short survey is to describe the use of torsors in various representative examples.

Let X_r be a smooth Del Pezzo surface obtained from $\mathbb{P}^2$ by blowing up $r \leqslant 8$ points in general position. It is well known that for $r \in \{3,4,5,6,7,8\}$ the Picard group $\mathrm{Pic}(X_r)$ contains a canonical root system $R_r \in \{A_2 \times A_1, A_4, D_5, E_6, E_7, E_8\}$. We prove some general properties of the Cox ring of X_r $(r \geqslant 4)$ and show its similarity to the homogeneous coordinate ring of the orbit of the highest weight vector in some irreducible representation of the algebraic group G associated with the root system R_r.

Let $X \subset \mathbb{P}^4$ be an irreducible hypersurface and $\varepsilon > 0$ be given. We show that there are $O(B^{3+\varepsilon})$, resp. $O(B^{55/18+\varepsilon})$, rational points on $\mathbb{P}^4$ lying on X when X is of degree $d \geqslant 4$, resp. $d = 3$. The implied constants depend only on d and ε.

On établit l'approximation faible pour les espaces homogènes de groupes linéaires connexes définis sur le corps des fonctions d'une courbe complexe. On en déduit la même propriété pour les variétés qui se ramènent à de tels espaces par fibrations successives. Ainsi l'approximation faible vaut pour une surface fibrée en coniques au-dessus d'une droite. On montre par contre que l'approximation faible peut être en défaut pour une surface d'Enriques.

If d is an even positive integer, the generic complex K3 surface of degree d has Picard number one. However, because $\bar{\mathbb{Q}}$ is countable it is not a priori obvious that there exists such a K3 surface defined over a number field. We prove that for every d, there exists a number field K and a K3 surface X/K which has geometric Picard number one.

We ask the question: If a pencil of curves of genus one defined over $\mathbf{Q}$ admits no section, can we find a number field $L/\mathbf{Q}$ and a member of that pencil defined over L having no L-rational points?

We study the equations of universal torsors on rational surfaces.

Consider hypersurfaces of fixed degree d in a fixed projective space $\mathbf{P}^n$ over $\mathbb{Q}$. We present a conjecture about the fraction of these that have rational points, and present evidence for the conjecture, including a proof that a positive fraction of the hypersurfaces have points over every completion of $\mathbb{Q}$, provided that $n, d \geq 2$ and $(n, d) \neq (2, 2)$. Generalizations to number fields are discussed. One of our proofs uses a result of Colliot-Thélène, proved in an appendix, that there is no Brauer–Manin obstruction to the Hasse principle for smooth complete intersections of dimension ≥ 3 in projective space over number fields. Colliot-Thélène's proof, in turn, uses a consequence of the Weak Lefschetz Theorem proved in an appendix by Katz.

We study descent and obstructions to the Hasse principle on simply connected surfaces over number fields.

We explain our approach to the problem of counting rational points of bounded height on equivariant compactifications of semi-simple groups.

INTRODUCTION

The study of diophantine equations has a long history. Recently, it has become clear that a multitude of modern branches of mathematics can be successfully applied in the study of such equations, and that sometimes it is only by combining techniques that progress can be made. Examples of such branches include

- complex algebraic geometry
- Galois and étale cohomology (including class field theory and its higher-dimensional generalizations)
- harmonic analysis on algebraic groups, automorphic forms
- analytic number theory (circle method).

And conversely, insights about rational points lead to advances in some of these areas. For example, even if one is interested solely in complex algebraic geometry, one is inevitably led to consider families of varieties, and the generic member of such a family is a variety over a function field, which is not algebraically closed, so that arithmetic questions become relevant.

During December 11–20, 2002, the American Institute of Mathematics hosted a workshop on the subject of

"Rational and integral points on higher-dimensional varieties"

in Palo Alto, California. One of the main objectives of the workshop was to bring together key researchers in these various areas to facilitate the exchange of ideas across the boundaries of their fields and to stimulate new research at their intersections.

This volume contains the proceedings of this meeting, including expository articles based on talks given by leading experts in their respective fields, as well as research contributions. An extensive glossary, developed in collaboration with the workshop participants, explains concepts and notions that are most important to the field. A carefully prepared index, covering the entire volume, is also included.

We thank the American Institute of Mathematics for sponsoring and hosting the workshop which led to these proceedings.

Bjorn Poonen
Yuri Tschinkel

August 12, 2003

Part I

Expository Articles

Arithmetic of Higher-dimensional Algebraic Varieties
(B. Poonen, Yu. Tschinkel, eds.), p. 3–35
Progress in Mathematics, Vol. 226, © 2004 Birkhäuser Boston, Cambridge, MA

DIOPHANTINE EQUATIONS: PROGRESS AND PROBLEMS

Sir Peter Swinnerton-Dyer
DPMMS, Centre for Mathematical Sciences, University of Cambridge,
Wilberforce Road, Cambridge, CB3 0WB, UK
E-mail : H.P.F.Swinnerton-Dyer@dpmms.cam.ac.uk

Abstract. We survey some of the outstanding problems concerning rational points on curves and surfaces.

1. Introduction

A *diophantine problem* over **Q** is concerned with the solutions either in **Q** or in **Z** of a finite system of polynomial equations

$$F_i(X_1, \ldots, X_n) = 0 \quad (1 \leqslant i \leqslant m) \tag{1}$$

with coefficients in **Q**. Without loss of generality we can obviously require the coefficients to be in **Z**. A system (1) is also called a system of *diophantine equations.* Often one will be interested in a family of such problems rather than a single one; in this case one requires the coefficients of the F_i to lie in some $\mathbf{Q}(c_1, \ldots, c_r)$, and one obtains an individual problem by giving the c_j values in **Q**. Again one can get rid of denominators. Some of the most obvious questions to ask about such a family are:

(A) Is there an algorithm which will determine, for each assigned set of values of the c_j, whether the corresponding diophantine problem has solutions, either in **Z** or in **Q**?

(B) For values of the c_j for which the system is soluble, is there an algorithm for exhibiting a solution?

For individual members of such a family, it is also natural to ask:

(C) Can we describe the set of all solutions, or even its structure?
(D) Is the phrase 'density of solutions' meaningful, and if so, what can we say about it?

The attempts to answer these questions have led to the introduction of new ideas and these have generated new questions. On some of them I expect progress within the next decade, and I have restricted myself to these in the text below. Progress in mathematics usually means proven results; but there are cases where even a well justified conjecture throws new light on the structure of the subject. (For similar reasons, well motivated computations can be helpful; but computations not based on a deep feeling for the structure of the subject have generally turned out to be a waste of time.) But I have not included those problems (such as the Riemann Hypothesis and the Birch–Swinnerton-Dyer conjecture) on which I do not expect further progress within so short a timescale. The reader may also find it useful to read Silverberg [**38**], which has the same purpose as this article but rather little overlap with it.

Though the study of solutions in $\mathbf{Z}$ and in $\mathbf{Q}$ may look very similar (and indeed were believed for a long time to be so), it now appears that they are actually very different and that the theory for solutions in $\mathbf{Q}$ has much more structure than that for solutions in $\mathbf{Z}$. The main reason for this seems to be that in the rational case the system (1) defines a variety in the sense of algebraic geometry, and many of the tools of that discipline can be used. Despite the advent of Arakelov geometry, this is much less true of integral problems. However, for varieties of degree greater than 2 it is only in low dimension that we yet know enough of the geometry for it to be useful.

Uniquely, the Hardy–Littlewood method is useful both for integral and for rational problems; it was designed for integral problems but can also be applied to rational problems in projective space, because then the F_i in (1) are homogeneous and it does not matter whether we treat the variables X_ν as integral or rational. There is a brief discussion of this method in Section 8, and a comprehensive survey in [**48**].

Denote by V the variety defined by the equations (1) and let V' be any variety birationally equivalent to V over $\mathbf{Q}$. The problem of finding solutions of (1) in $\mathbf{Q}$ is the same as that of finding rational points on V, which is almost the same as that of finding rational points on V'. Hence (except possibly for Question (D) above) one expects the properties of the rational solutions of (1) to be essentially determined by the birational equivalence class of V; and the way in which algebraic geometers classify varieties should provide at least a first rough guide to the classification of diophantine problems — though they mainly study birational equivalence over **C** rather than over **Q**. But it does at the moment seem that even for surfaces the geometric classification needs some refinements if it is to fit the number-theoretic results and conjectures.

Without loss of generality we can assume that V is absolutely irreducible. For if V has proper components defined over **Q** it is enough to ask the questions above for each of the proper components; and if V is the union of varieties conjugate over **Q**, then any rational point on V lies on the intersection of these conjugates, which is a proper subvariety of V. Since we can desingularize V by a birational transformation defined over **Q**, it is natural to concentrate on the case when V is projective and nonsingular.

The definitions and the questions above can be generalized to an arbitrary algebraic number field and the ring of integers in it; the answers are usually known or conjectured to be essentially the same as over **Q** or **Z**, though the proofs can be very much harder. (But there are exceptions; for example, the modularity of elliptic curves only holds over **Q**.) The questions above can also be posed for other fields of number-theoretic interest — in particular for finite fields and for completions of algebraic number fields — and when one studies diophantine problems it is essential to consider these other fields also. If V is defined over a field K, the set of points on V defined over K will always be denoted by $V(K)$. If $V(K)$ is not empty we say that V is *soluble in* K. In the special case where $K = k_v$, the completion of an algebraic number field k at the place v, we also say that V is *locally soluble at* v. From now on we denote by $\mathbf{Q}_v$ any completion of **Q**; thus $\mathbf{Q}_v$ means **R** or some $\mathbf{Q}_p$.

One major reason for considering solubility in complete fields and in finite fields is that a necessary condition for (1) to be soluble in **Q**, for example, is that it is soluble in every $\mathbf{Q}_v$. The condition of solubility in every $\mathbf{Q}_v$ is computationally decidable; see Section 2. Moreover, at least for primes p for which the system (1) has good reduction mod p, the first step in deciding solubility in $\mathbf{Q}_p$ is to decide whether the reduced system is soluble in the finite field $\mathrm{GF}(p)$ of p elements.

Some geometers are already used to studying varieties over fields k which are not algebraically closed; what makes diophantine problems special is the number-theoretic nature of the fields k. But it seems that only a few of the properties peculiar to such fields are useful in this context, so that a geometer need not learn much number theory in order to work on diophantine problems. On the other hand, a number theorist would be wise to learn quite a lot of geometry.

Diophantine problems were first introduced by Diophantus of Alexandria, the last of the great Greek mathematicians, who lived at some time between 300 B.C. and 300 A.D.; but he was handicapped by having only one letter available to represent variables, all the others being used in classical Greek to represent specific numbers. Individual diophantine problems were studied by such great mathematicians as Fermat, Euler and Gauss. But it was Hilbert's address to the International Congress in 1900 which started the development of a systematic theory. His tenth problem asked:

> Given a diophantine equation with any number of unknown quantities and with rational integral numerical coefficients: to devise a process according to which it can be determined by a finite number of operations whether the equation is soluble in rational integers.

Most of the early work on diophantine equations was concerned with rational rather than integral solutions; presumably Hilbert posed this problem in terms of integral solutions because such a process for integral solutions would automatically provide the corresponding process for rational solutions also, by restricting to the special case when the equations are homogeneous. In those confident days before the First World War, it was assumed that such a process must exist; but in 1970 Matijasevič showed that this was impossible. Indeed he exhibited a polynomial $F(c; x_1, \dots, x_n)$ such that there cannot exist an algorithm which will decide for every given c whether $F = 0$ is soluble in integers. His proof is part of the great program on decidability initiated by Gödel; good accounts of it can be found in **[9]**, pp. 323–378 or **[14]**. The corresponding question for rational solutions is still open; I am among the very few who believe that it may have a positive answer.

But even if the answer to the analogue of Hilbert's tenth problem for rational solutions is positive, one must expect that a separate algorithm will be needed for each kind of variety. Thus we shall need not one algorithm but an infinity of them. So number theorists depend on the development by geometers of an adequate classification of varieties. At the moment, such a classification is reasonably complete for curves and surfaces, but it is still fragmentary even in dimension 3; so number theorists have to concentrate on curves and surfaces, and on certain particularly simple kinds of variety in higher dimension.

2. The Hasse Principle and the Brauer–Manin obstruction

Let V be a variety defined over $\mathbf{Q}$. If V is locally soluble at every place of $\mathbf{Q}$, we say that it satisfies the *Hasse condition*. If $V(\mathbf{Q})$ is not empty, then V certainly satisfies the Hasse condition. What makes this remark valuable is that the Hasse condition is computable — that is, one can decide in finitely many steps whether a given V satisfies the Hasse condition. This follows from the next two lemmas.

Lemma 1. *Let W be an absolutely irreducible variety of dimension n defined over the finite field $k = \mathrm{GF}(q)$. Then $N(q)$, the number of points on W defined over k, satisfies*

$$|N(q) - q^n| < Cq^{n-1/2}$$

where the constant C depends only on the degree and dimension of W and is computable.

This follows from the Weil conjectures, for which see Section 3; but weaker results which are adequate for the present application were known much earlier. Since the singular points of W lie on a proper subvariety, there are at most $C_1 q^{n-1}$ of them, where C_1 is also computable. It follows that if q exceeds a computable bound depending only on the degree and dimension of W, then W contains a nonsingular point defined over k.

Now let V be an absolutely irreducible variety defined over $\mathbf{Q}$. If V has good reduction at p, which happens for all but a finite computable set of primes p, denote that reduction by $\tilde{V}_p$. If p is large enough, it follows from the remarks above that $\tilde{V}_p$ contains a nonsingular point Q_p defined over $\mathrm{GF}(p)$. The result which follows, which is known as Hensel's Lemma though the idea of the proof goes back to Sir Isaac Newton, now shows that V contains a point P_p defined over $\mathbf{Q}_p$.

Lemma 2. *Let V be an absolutely irreducible variety defined over $\mathbf{Q}$ which has a well-defined reduction* mod p, *denoted by $\widetilde{V_p}$. If $\tilde{V}_p$ contains a nonsingular point Q_p defined over* $\mathrm{GF}(p)$, *then V contains a nonsingular point P_p defined over $\mathbf{Q}_p$ whose reduction* mod p *is Q_p.*

In view of this, to decide whether V satisfies the Hasse condition one only has to check individually solubility in $\mathbf{R}$ and in finitely many $\mathbf{Q}_p$. Each of these checks can be shown to be a finite process, using ideas similar to those in the proof of Lemma 1.

A family $\mathscr{F}$ of varieties is said to satisfy the *Hasse Principle* if every V contained in $\mathscr{F}$ and defined over $\mathbf{Q}$ which satisfies the Hasse condition actually contains at least one point defined over $\mathbf{Q}$. Again, a family $\mathscr{F}$ is said to admit *weak approximation* if every V contained in $\mathscr{F}$ and defined over $\mathbf{Q}$, and such that $V(\mathbf{Q})$ is not empty, has the following property: given any finite set of places v and corresponding non-empty sets $\mathscr{N}_v \subset V(\mathbf{Q}_v)$ open in the v-adic topology, there is a point P in $V(\mathbf{Q})$ which lies in each of the $\mathscr{N}_v$. In the special case when $\mathscr{F}$ consists of a single variety V, and $V(\mathbf{Q})$ is not empty, we simply say that V admits weak approximation. Whether V admits weak approximation is in general not computable; for an important exception, see **[46]**.

The most important families which are known to have either of these properties (and which actually have both) are the families of quadrics of any given dimension; this was proved by Minkowski for quadrics over $\mathbf{Q}$ and by Hasse for quadrics over an arbitrary algebraic number field. But many families, even of very simple varieties, do not satisfy either the Hasse Principle or weak approximation. (For example, neither of them holds for nonsingular cubic surfaces.) It is therefore natural to ask

Question 1. *For a given family $\mathscr{F}$, what are the obstructions to the Hasse Principle and to weak approximation?*

For weak approximation there is a variant of this question which may be both more interesting and easier to answer. Another way of stating weak approximation on V is to say that if $V(\mathbf{Q})$ is not empty, then it is dense in the adelic space $V(\mathbf{A}) = \prod_v V(\mathbf{Q}_v)$. This suggests the following:

Question 2. *For a given V, or family $\mathscr{F}$, what can be said about the closure of $V(\mathbf{Q})$ in the adelic space $V(\mathbf{A})$?*

However, there are families for which Question 1 does not seem to be a sensible question to ask; these probably include for example all families of varieties of general type. So one should back up Question 1 with

Question 3. *For what kinds of families is either part of Question* 1 *a sensible question to ask?*

The only systematic obstruction to the Hasse Principle which is known is the Brauer–Manin obstruction, though obstructions can be found in the literature which are not Brauer–Manin. Let A be a *central simple algebra* — that is, a simple algebra which is finite dimensional over a field K which is its centre. Each such algebra consists, for fixed D and n, of all $n \times n$ matrices with elements in a division algebra D with centre K. Two central simple algebras over K are *equivalent* if they have the same underlying division algebra. Formation of tensor products over K gives the set of equivalence classes the structure of a commutative group, called the *Brauer group* of K and written $\mathrm{Br}(K)$. There is a canonical isomorphism $\imath_p : \mathrm{Br}(\mathbf{Q}_p) \simeq \mathbf{Q}/\mathbf{Z}$ for each p; and there is a canonical isomorphism $\imath_\infty : \mathrm{Br}(\mathbf{R}) \simeq \{0, \frac{1}{2}\}$, the nontrivial division algebra over $\mathbf{R}$ being the classical quaternions.

Let B be an element of $\mathrm{Br}(\mathbf{Q})$; tensoring B with any $\mathbf{Q}_v$ gives rise to an element of $\mathrm{Br}(\mathbf{Q}_v)$, and this element is trivial for almost all v. There is an exact sequence

$$0 \to \mathrm{Br}(\mathbf{Q}) \to \bigoplus \mathrm{Br}(\mathbf{Q}_v) \to \mathbf{Q}/\mathbf{Z} \to 0,$$

due to Hasse, in which the third map is the sum of the $\imath_v$; it tells us when a set of elements, one in each $\mathrm{Br}(\mathbf{Q}_v)$ and almost all trivial, can be generated from some element of $\mathrm{Br}(\mathbf{Q})$.

Now let V be a complete nonsingular variety defined over $\mathbf{Q}$ and A an *Azumaya algebra* on V — that is, a simple algebra with centre $\mathbf{Q}(V)$ which has a good specialization at every point of V. The group of equivalence classes of Azumaya algebras on V is denoted by $\mathrm{Br}(V)$. If P is any point of V, with field of definition $\mathbf{Q}(P)$, we obtain a simple algebra $A(P)$ with centre $\mathbf{Q}(P)$ by specializing at P. For all but finitely many p, we have $\imath_p(A(P_p)) = 0$ for all p-adic points P_p on V. Thus a necessary condition for the existence of a rational point P on V is that for every v there should be a v-adic point P_v on V such that

$$\sum \imath_v(A(P_v)) = 0 \quad \text{for all } A. \tag{2}$$

Similarly, a necessary condition for V with $V(\mathbf{Q})$ not empty to admit weak approximation is that (2) should hold for all Azumaya algebras A and all adelic points $\prod_v P_v$. In each case this is the *Brauer–Manin condition*. It is clearly unaffected if we add to A a constant algebra — that is, an element of $\mathrm{Br}(\mathbf{Q})$. So what we are really interested in is $\mathrm{Br}(V)/\mathrm{Br}(\mathbf{Q})$.

All this can be put into highbrow language. Even without any hypotheses on V, there is an injection of $\mathrm{Br}(V)$ into the étale cohomology group $\mathrm{H}^2(V, \mathbf{G}_m)$; and if for example V is a complete nonsingular surface, this injection is an isomorphism. If we write

$$\mathrm{Br}_1(V) = \ker(\mathrm{Br}(V) \to \mathrm{Br}(\bar{V})) = \ker(\mathrm{H}^2(V, \mathbf{G}_m) \to \mathrm{H}^2(\bar{V}, \mathbf{G}_m)),$$

there is a filtration

$$\mathrm{Br}(\mathbf{Q}) \subset \mathrm{Br}_1(V) \subset \mathrm{Br}(V).$$

Here only the abstract structure of $\mathrm{Br}(V)/\mathrm{Br}_1(V)$ is known; and in general there is no known way of finding Azumaya algebras which represent nontrivial elements of this quotient, though in a particular case Harari [**20**] has exhibited a Brauer–Manin obstruction coming from such an algebra. In contrast, provided the Picard variety of V is trivial there is an isomorphism

$$\mathrm{Br}_1(V)/\mathrm{Br}(\mathbf{Q}) \simeq \mathrm{H}^1(\mathrm{Gal}(\bar{\mathbf{Q}}/\mathbf{Q}), \mathrm{Pic}(V \otimes \bar{\mathbf{Q}})),$$

and this is computable in both directions provided $\mathrm{Pic}(V \otimes \bar{\mathbf{Q}})$ is known. (For details of this, see [**8**].)

There is no known systematic way of determining $\mathrm{Pic}(V \otimes \bar{\mathbf{Q}})$ for arbitrary V, and there is strong reason to suppose that this is really a number-theoretic rather than a geometric problem. If V is defined over $\mathbf{Q}$ (rather than over an arbitrary algebraic number field) there is a tentative algorithm, depending on the Birch/Swinnerton-Dyer conjecture, for determining an algebraic number field K (depending on V) such that $\mathrm{Pic}(V \otimes \bar{\mathbf{Q}}) = \mathrm{Pic}(V \otimes K)$, and this may be the right first step towards determining $\mathrm{Pic}(V \otimes \bar{\mathbf{Q}})$; but one hopes not, because even for so elementary a variety as a cubic surface, we may need to have $[K : \mathbf{Q}] \geqslant 51840$. It seems to me likely that a better approach to this question will be through the Tate conjectures, for which see Section 3; but this is a very long term prospect. However, it is usually possible to determine $\mathrm{Pic}(V \otimes \bar{\mathbf{Q}})$ for any particular V that one is interested in.

Question 4. *Is there a general algorithm (even conjectural) for determining* $\mathrm{Pic}(V \otimes \bar{\mathbf{Q}})$ *for varieties V defined over an algebraic number field?*

Bombieri and Lang have independently conjectured that if V is a variety of general type defined over an algebraic number field K, then there is a finite union $\mathscr{S}$ of proper subvarieties of V such that every point of $V(K)$ lies in $\mathscr{S}$. (Faltings' theorem, for which see Section 4, is the special case of this for curves.) This raises another question, similar to Question 4 but probably easier:

Question 5. *Is there an algorithm for determining* $\mathrm{Pic}(V)$ *where V is a variety defined over an algebraic number field?*

The Brauer–Manin obstruction was introduced by Manin **[28]** in order to bring within a single framework various sporadic counterexamples to the Hasse principle for rational surfaces. The theory of this obstruction has been extensively developed by Colliot-Thélène and Sansuc. In particular, for rational varieties they showed how to go back and forth between the Brauer–Manin condition and the descent condition for torsors under tori. They also showed that if there is no Brauer–Manin obstruction to the Hasse principle on a variety V, then there exists a universal torsor over V which has points everywhere locally. This suggests that one should pay particular attention to diophantine problems on universal torsors. Unfortunately, it is usually not easy to exploit what is known about the geometric structure of universal torsors. Indeed there are very few families for which the Brauer–Manin obstruction can be nontrivial but for which it has been shown that it is the only obstruction to the Hasse principle. (See however **[12]** and, subject to Schinzel's hypothesis, **[43]** and **[13]**.) It is generally believed that the Brauer–Manin obstruction is the only obstruction to the Hasse principle for rational surfaces — that is, surfaces birationally equivalent to $\mathbf{P}^2$ over $\bar{\mathbf{Q}}$. On the other hand, Skorobogatov (**[40]**, and see also **[39]**) has exhibited an obstruction to the Hasse principle on a bielliptic surface which is definitely not Brauer–Manin.

Question 6. *Is the Brauer–Manin obstruction the only obstruction to the Hasse principle for all unirational (or all Fano) varieties?*

For the method of universal torsors, the immediate question to address must be the following:

Question 7. *Does the Hasse principle hold for universal torsors over a rational surface?*

We can of course ask similar questions for weak approximation. Both for the Hasse principle and for weak approximation one can alternatively ask what is the most general class of varieties for which the Brauer–Manin obstruction is the only one. Colliot-Thélène has suggested that this class probably includes, and may even be equal to, all rationally connected varieties.

There are families $\mathscr{F}$ whose universal torsors appear to be too complicated to be systematically investigated, but for which it is still possible to identify the obstruction to the Hasse principle. It is sometimes possible to start from the absence of a Brauer–Manin obstruction; but there are also alternative strategies. Implementing these falls naturally into two parts:

(i) Assuming that V in $\mathscr{F}$ satisfies the Hasse condition, one finds a necessary and sufficient condition for V to have a rational point, or to admit weak approximation.

(ii) One then shows that this necessary and sufficient condition is equivalent to the Brauer–Manin condition.

Both these strategies have been applied to pencils of conics, in each case assuming Schinzel's Hypothesis; the curious reader may wish to compare the approaches in **[43]** and in **[13]**. Except for Skorobogatov's example above, I know of no families for which it has been possible to carry out (i) but not (ii). But there are families for which it has been possible to find a sufficient condition for solubility (additional to the Hasse condition) which appears rather weak but which is definitely stronger than the Brauer–Manin condition. The obvious examples of such a condition are the various forms of what is called Condition D in **[11]**, **[44]**, **[2]** and **[45]**. However, in these cases it is not obvious that a condition stronger than the Brauer–Manin condition is actually necessary; and I am provisionally inclined to attribute the gap to clumsiness in the proofs.

Question 8. *When the Brauer–Manin condition is trivial, how can one make use of this fact?*

3. Zeta functions and L-series

Let $W \subset \mathbf{P}^n$ be a nonsingular and absolutely irreducible projective variety of dimension d defined over the finite field $k = \mathrm{GF}(q)$, and denote by $\varphi(q)$ the Frobenius automorphism of W given by

$$\varphi(q) : (x_0, x_1, \ldots, x_n) \mapsto (x_0^q, x_1^q, \ldots, x_n^q).$$

For any $r > 0$ the fixed points of $(\varphi(q))^r$ are precisely the points of W which are defined over $\mathrm{GF}(q^r)$; suppose that there are $N(q^r)$ of them. Although the context is totally different, this is almost the formalism of the Lefschetz Fixed Point theorem, since for geometric reasons each of these fixed points has multiplicity $+1$. This analogy led Weil to conjecture that there should be a cohomology theory applicable in this context. This would imply that there were finitely many complex numbers α_{ij} such that

$$N(q^r) = \sum_{i=0}^{2d} \sum_{j=1}^{B_i} (-1)^i \alpha_{ij}^r \quad \text{for all } r > 0, \tag{3}$$

where B_i is the dimension of the ith cohomology group of W and the α_{ij} are the characteristic roots of the map induced by $\varphi(q)$ on the ith cohomology. For each i duality implies that $B_i = B_{2d-i}$ and the $\alpha_{2d-i,j}$ are a permutation of the q^d/α_{ij}. If we define the local zeta function $Z(t, W)$ by either of the equivalent relations

$$\log Z(t) = \sum_{r=1}^{\infty} N(q^r)t^r/r \quad \text{or} \quad tZ'(t)/Z(t) = \sum_{r=1}^{\infty} N(q^r)t^r,$$

then (3) is equivalent to

$$Z(t) = \frac{P_1(t, W) \cdots P_{2d-1}(t, W)}{P_0(t, W)P_2(t, W) \cdots P_{2d}(t, W)}$$

where $P_i(t, W) = \prod_j (1 - \alpha_{ij}t)$. Each $P_i(t, W)$ must have coefficients in $\mathbf{Z}$, and the analogue of the Riemann hypothesis is that $|\alpha_{ij}| = q^{i/2}$. (For a fuller account of Weil's conjectures and their motivation, see the excellent survey **[26]**.) All this has now been proved, the main contributor being Deligne.

Now let V be a nonsingular and absolutely irreducible projective variety defined over an algebraic number field K. If V has good reduction at a prime $\mathfrak{p}$ of K we can form $\tilde{V}_\mathfrak{p}$, the reduction of V mod $\mathfrak{p}$, and hence form the $P_i(t, \tilde{V}_\mathfrak{p})$. For s in $\mathbf{C}$, we can now define the ith global L-series $L_i(s, V)$ of V as a product over all places of K, the factor at a prime $\mathfrak{p}$ of good reduction being $(P_i(q^{-s}, \tilde{V}_\mathfrak{p}))^{-1}$ where $q = \mathrm{Norm}_{K/\mathbf{Q}}\mathfrak{p}$. The rules for forming the factors at the primes of bad reduction and at the infinite places can be found in **[36]**. These L-series of course depend on K as well as on V. In particular, $L_0(s, V)$ is just the zeta-function of the algebraic number field K.

To call a function $F(s)$ a (global) zeta-function or L-series carries with it certain implications:

- $F(s)$ must be the product of a Dirichlet series and possibly some Gamma-functions, and the half-plane of absolute convergence for the Dirichlet series must have the form $\Re s > \sigma_0$ with $2\sigma_0$ in $\mathbf{Z}$.
- The Dirichlet series must be expressible as an Euler product $\prod_p f_p(p^{-s})$ where the f_p are rational functions.
- $F(s)$ must have an analytic continuation to the entire s-plane as a meromorphic function all of whose poles are in $\mathbf{Z}$.
- There must be a functional equation relating $F(s)$ and $F(2\sigma_0 - 1 - s)$.
- The zeroes of $F(s)$ in the critical strip $\sigma_0 - 1 < \Re s < \sigma_0$ must lie on $\Re s = \sigma_0 - \frac{1}{2}$.

In our case, the first two implications are trivial; and fortunately one is not expected to prove the last three, but only to state them as conjectures. The last one is the Riemann Hypothesis, which appears to be out of reach even in the simplest case, which is the classical Riemann zeta-function; and the third and fourth have so far only been proved in a few favourable cases.

Question 9. *Can one extend the list of V for which analytic continuation and the functional equation can be proved?*

It seems likely that any proof of analytic continuation will carry a proof of the functional equation with it.

It has been said about the zeta-functions of algebraic number fields that 'the zeta-function knows everything about the number field; we just have to prevail on it to tell us'. If this is so, we have not yet unlocked the treasure house. Apart from the classical formula which relates hR to $\zeta_K(0)$ all that has so far been proved are certain results of Borel **[6]** which relate the behaviour of $\zeta_K(s)$ near $s = 1 - m$ for integers $m > 1$ to the K-groups of $\mathfrak{O}_K$. I would be reluctant to claim that the L-series of a variety V contains all the information which one would like to have about the number-theoretical properties of V; but one might hope that when a mysterious number turns up in the study of diophantine problems on V, some L-series contains information about it.

Suppose for convenience that V is defined over $\mathbf{Q}$, and let its dimension be d. Even for varieties with $B_1 = 0$ we do not expect a product like

$$(4) \qquad \prod_p N(p)/p^d \quad \text{or} \quad \prod_p N(p) \Big/ \left(\frac{p^{d+1}-1}{p-1}\right)$$

to be necessarily absolutely convergent. But in some contexts there is a respectable expression which is formally equivalent to one of these, with appropriate modifications of the factors at the bad primes. The idea that such an expression should have number-theoretic significance goes back to Siegel (for genera of quadratic forms) and Hardy and Littlewood (for what they called the *singular series*). Using the ideas above, we are led to replace the study of the products (4) by a study of the behaviour of $L_{2d-1}(s, V)$ and $L_{2d-2}(s, V)$ near $s = d$. By duality, this is the same as studying $L_1(s, V)$ near $s = 1$ and $L_2(s, V)$ near $s = 2$. The information derived in this way appears to relate to the Picard group of V, defined as the group of divisors defined over $\mathbf{Q}$ modulo linear equivalence. By considering simultaneously both V and its Picard variety (the abelian variety which parametrises divisors algebraically equivalent to zero modulo linear equivalence), one concludes that $L_1(s, V)$ must be associated with the Picard variety and $L_2(s, V)$ with the group of divisors modulo algebraic equivalence — that is, with the Néron–Severi group of V. These remarks motivate the weak forms of the Birch–Swinnerton-Dyer conjecture (for which see Section 4) and the case $m = 1$ of the Tate conjecture below. For the strong forms (which give expressions for the leading coefficients of the relevent Laurent series expansions) heuristic arguments are less convincing; but one can formulate conjectures for these coefficients by asking what other mysterious numbers turn up in the same context and should therefore appear in the formulae for the leading coefficients.

The weak form of the Tate conjecture asserts that the order of the pole of $L_{2m}(s, V)$ at $s = m+1$ is equal to the rank of the group of classes of m-cycles on V defined over K, modulo algebraic equivalence; it is a natural generalization of the case $m = 1$ for which the heuristics have just been shown. For a more detailed account of both of these, including the conjectural formulae for the leading coefficients, see **[47]** or **[42]**.

Question 10. *What information about V is contained in its L-series?*

There is in the literature a beautiful edifice of conjecture, lightly supported by evidence, about the behaviour of the $L_i(s, V)$ at integral points. The principal architects of this edifice are Beilinson, Bloch and Kato. Beilinson's conjectures relate to the order and leading coefficients of the Laurent series expansions of the $L_i(s, V)$ about integer values of s; in them the leading coefficients are treated as elements of $\mathbf{C}^*/\mathbf{Q}^*$. (For a full account see **[32]** or **[25]**.) Bloch and Kato (**[5]** and **[4]**) have strengthened these conjectures by treating the leading coefficients as elements of $\mathbf{C}^*$. But I do not believe that anything like the full story has yet been revealed.

4. Curves

The most important invariant of a curve is its genus. In the language of algebraic geometry over **C**, curves of genus 0 are called *rational*, curves of genus 1 are called *elliptic* and curves of genus greater than 1 are *of general type*. But note that for a number theorist an elliptic curve is a curve of genus 1 with a distinguished point P_0 on it, both being defined over the ground field K. The effect of this is that the points on an elliptic curve form an abelian group with P_0 as its identity element, the sum of P_1 and P_2 being the other zero of the function (defined up to multiplication by a constant) with poles at P_1 and P_2 and a zero at P_0.

A canonical divisor on a curve Γ of genus 0 has degree -2; hence by the Riemann–Roch theorem Γ is birationally equivalent over the ground field to a conic. The Hasse principle holds for conics, and therefore for all curves of genus 0; this gives a complete answer to Question (A) at the beginning of these notes. But it does not give an answer to Question (B). Over **Q**, a very simple answer to Question (B) is as follows:

Theorem 1. *Let a_0, a_1, a_2 be nonzero elements of* **Z**. *If the equation*

$$a_0X_0^2 + a_1X_1^2 + a_2X_2^2 = 0$$

is soluble in **Z**, *then it has a solution for which each $a_iX_i^2$ is absolutely bounded by $|a_0a_1a_2|$.*

Siegel [**37**] has given an answer to Question (B) over arbitrary algebraic number fields, and Raghavan [**2**] has generalized Siegel's work to quadratic forms in more variables.

The knowledge of one rational point on Γ enables us to transform Γ birationally into a line; so there is a parametric solution which gives explicitly all the points on Γ defined over the ground field. This answers Question (C).

If Γ is a curve of general type defined over an algebraic number field K, Mordell conjectured and Faltings proved that $\Gamma(K)$ is finite; and a number of other proofs have appeared since then. But it does not seem that any of them enable one to compute $\Gamma(K)$, though some of them come tantalizingly close. For a survey of several such proofs, see [**15**].

Question 11. *Is there an algorithm for computing $\Gamma(K)$ when Γ is a curve of general type defined over an algebraic number field K?*

The study of rational points on elliptic curves is now a major industry, almost entirely separate from the study of other diophantine problems. If Γ is an elliptic curve defined over an algebraic number field K, the group $\Gamma(K)$ is called the *Mordell–Weil group.* Mordell proved that $\Gamma(K)$ is finitely generated; Weil's contribution was to extend this result to all abelian varieties. Thanks to Mazur (see **[29]**) and Merel (see **[30]**) the theory of the torsion part of the Mordell–Weil group is now reasonably complete; but for the non-torsion part all that was known before 1960 is that $\Gamma(K)$ can be embedded into a certain group which is finitely generated and computable. The process involved, which is known as the method of infinite descent, goes back to Fermat; for use in Section 6 I shall illustrate it below in a particularly simple case. By means of this process one can always compute an upper bound for the rank of the Mordell–Weil group of any particular Γ, and the upper bound thus obtained can frequently be shown to be equal to the actual rank by exhibiting enough elements of $\Gamma(K)$. It was also conjectured that the difference between the upper bound thus computed and the actual rank was always an even integer, but apart from this the actual rank was mysterious. This not wholly satisfactory state of affairs has been radically changed by the Birch–Swinnerton-Dyer conjecture, the weak form of which is described at the end of this section. A survey of what is currently known or conjectured about the ranks of Mordell–Weil groups can be found in **[34]**.

I now turn to the situation in which Γ is a curve of genus 1 defined over K but not necessarily containing a point defined over K. Let J be the Jacobian of Γ, defined as a curve whose points are in one-one correspondence with the divisors of degree 0 on Γ modulo linear equivalence. Then J is also a curve of genus 1 defined over K, and $J(K)$ contains the point which corresponds to the trivial divisor. So J is an elliptic curve in our sense.

Conversely, if we fix an elliptic curve J defined over K we can consider the equivalence classes (for birational equivalence over K) of curves Γ of genus 1 defined over K which have J as Jacobian. For number theory, the only ones of interest are those which contain points defined over each completion K_v. These form a commutative torsion group, called the *Tate–Shafarevich group* and usually denoted by Ш; the identity element of this group is the class which contains J itself, and it consists of those Γ which have J as Jacobian and which contain a point defined over K. (The simplest example of a nontrivial element of a Tate–Shafarevich group is the curve

$$3X_0^3 + 4X_1^3 + 5X_2^3 = 0 \quad \text{with Jacobian} \quad Y_0^3 + Y_1^3 + 60Y_2^3 = 0.)$$

Thus for curves of genus 1 the Tate–Shafarevich group is by definition the obstruction to the Hasse principle.

Suppose in particular that the elliptic curve J is defined over $\mathbf{Q}$ and has the form

$$Y^2 = (X - c_1)(X - c_2)(X - c_3)$$

where the distinguished point is taken to be the point at infinity. The three points $(c_i, 0)$ on J have order 2; they are called the 2-*division points.* To any rational point (x, y) on Γ there exist m_1, m_2, m_3 and y_1, y_2, y_3 such that

$$m_i(x - c_i) = y_i^2 \quad \text{for} \quad i = 1, 2, 3;$$

here the m_i are really elements of $\mathbf{Q}^*/\mathbf{Q}^{*2}$ but it is convenient to treat them as square-free integers. We must have $m_1m_2m_3 = m^2$ for some integer m, and $my = y_1y_2y_3$. Conversely the equations

$$Y_i^2 = m_i(X - c_i) \ \ (i = 1, 2, 3) \quad \text{and} \quad mY = Y_1Y_2Y_3$$

for any m, m_i with $m_1m_2m_3 = m^2$ define a curve $\mathscr{C} = \mathscr{C}(m_1, m_2, m_3)$ and a four-to-one map $\mathscr{C} \to J$. If $\mathscr{C}(\mathbf{Q})$ is not empty, its image under this map is a coset of $2J(\mathbf{Q})$ in $J(\mathbf{Q})$, and we obtain all such cosets in this way. Thus we could find $J(\mathbf{Q})$ if we could decide which $\mathscr{C}$ are soluble in $\mathbf{Q}$.

After a change of variables we can assume that the c_i are in $\mathbf{Z}$. Define the *good primes* for J as those which do not divide $2(c_1 - c_2)(c_2 - c_3)(c_3 - c_1)$; then it is not hard to show that $\mathscr{C}$ is locally soluble at all good primes if and only if all the m_i are units at all good primes. So there are only finitely many $\mathscr{C}$ whose solubility in $\mathbf{Q}$ is at all hard to decide.

The curves $\mathscr{C}$ obtained in this way are called 2-*coverings* of J, and the process of obtaining them is called a 2-*descent.* They form a group under multiplication of the corresponding triples (m_1, m_2, m_3). The finite subgroup consisting of those 2-coverings which are everywhere locally soluble is called the 2-*Selmer group.* It is easily computable; and since there is a canonical embedding of $J(\mathbf{Q})/2J(\mathbf{Q})$ into the 2-Selmer group, this provides an upper bound for the rank of $J(\mathbf{Q})$. The descent process can be continued, though with somewhat greater difficulty; for 4-descents see **[10]**. One can also carry out 2-descents for the more general elliptic curve

$$Y^2 = X^3 + aX^2 + bX + c;$$

but in order to do this one requires information about the splitting field of the right hand side.

The weak form of the Birch–Swinnerton-Dyer conjecture states that the rank of the Mordell–Weil group of an elliptic curve J is equal to the order of the zero of $L_1(s, J)$ at $s = 1$; the conjecture also gives an explicit formula for the leading coefficient of the power series expansion at that point. Note that this point is in the critical strip, so that the conjecture pre-supposes the analytic continuation of $L_1(s, J)$. At present there are two well-understood cases in which analytic continuation is known: when $K = \mathbf{Q}$, so that J can be parametrised by means of modular functions, and when J admits complex multiplication. As consequence, these two cases are likely to be easier than the general case; but even here I do not expect much further progress in the next decade. In each of these two cases, if one assumes the Birch–Swinnerton-Dyer conjecture one can derive an algorithm for finding the Mordell–Weil group and the order of the Tate–Shafarevich group; and in the first of the two cases this algorithm has been implemented by Gebel (see [**16**]). Without using the Birch–Swinnerton-Dyer conjecture, Heegner long ago produced a way of generating a point on J whenever $K = \mathbf{Q}$ and J is modular; and Gross and Zagier ([**17**] and [**19**]) have shown that this point has infinite order precisely when $L'(1, J) \neq 0$. Building on their work, Kolyvagin (see [**18**]) has shown the following.

Theorem 2. *Suppose that the Heegner point has infinite order; then the group* $J(\mathbf{Q})$ *has rank* 1 *and* $\text{Ш}(J)$ *is finite.*

Kolyvagin [**27**] has also obtained sufficient conditions for both $J(\mathbf{Q})$ and $\text{Ш}(J)$ to be finite. The following result is due to Nekovar and Plater.

Theorem 3. *If the order of* $L(s, J)$ *is odd, then either* $J(\mathbf{Q})$ *is infinite or the* p*-part of* $\text{Ш}(J)$ *is infinite for every good ordinary* p.

If J can be parametrized by modular functions for some arithmetic subgroup of $\text{SL}_2(\mathbf{R})$, then analytic continuation and the functional equation follow; but there is not even a plausible conjecture identifying the J which have this property, and there is no known analogue of Heegner's construction.

In the complex multiplication case, what is known is as follows.

Theorem 4. *Let* K *be an imaginary quadratic field and* J *an elliptic curve defined and admitting complex multiplication over* K. *If* $L(1, J) \neq 0$, *then*

(i) $J(K)$ *is finite;*
(ii) *for every prime* $p > 7$ *the* p*-part of* $\text{Ш}(J)$ *is finite and has the order predicted by the Birch–Swinnerton-Dyer conjecture.*

Here (i) is due to Coates and Wiles, and (ii) to Rubin. For an account of the proofs, see [**33**]. Katz has generalized (i) and part of (ii) to behaviour over an abelian extension of $\mathbf{Q}$, but with the same J as before.

In general we do not know how to compute Ш. It is conjectured that it is always finite; and indeed this assertion can be regarded as part of the Birch–Swinnerton-Dyer conjecture, for the formula for the leading coefficient of the power series for $L_1(s, J)$ at $s = 1$ contains the order of $\text{Ш}(J)$ as a factor. If indeed this order is finite, then it must be a square; for Cassels has proved the existence of a skew-symmetric bilinear form on Ш with values in $\mathbf{Q}/\mathbf{Z}$, which is nonsingular on the quotient of Ш by its maximal divisible subgroup. Thus finiteness implies that if Ш contains at most $p-1$ elements of order exactly p for some prime p, then it actually contains no such elements; hence an element which is killed by p is trivial, and the curves of genus 1 in that equivalence class contain points defined over K. For use later, we state the case $p=2$ as a lemma.

Lemma 3. *Suppose that* Ш(J) *is finite and the quotient of the* 2*-Selmer group of* J *by its soluble elements has order at most* 2*; then that quotient is actually trivial.*

This result will play a crucial role in Sections 6 and 7.

5. Generalities about surfaces

Over $\mathbf{C}$ a full classification of surfaces can be found in [**1**]. A first coarse classification is given by the *Kodaira dimension* κ, which for surfaces can take the values $-\infty, 0, 1$ or 2. What also seems to be significant for the number theory (and cuts across this classification) is whether the surface is *elliptic* — that is, whether over $\mathbf{C}$ there is a map $V \to C$ for some curve C whose general fibre is a curve of genus 1. The case when the map $V \to C$ is defined over the ground field K and C has genus 0 is discussed in Section 6; in this case the diophantine problems for V are only of interest when $C(K)$ is nonzero, in which case C can be identified with $\mathbf{P}^1$. When C has genus greater than 1, the map $V \to C$ is essentially unique and it and C are therefore both defined over K. By Faltings' theorem, $C(K)$ is then finite; thus each point of $V(K)$ lies on one of a finite set of fibres, and it is enough to study these. In contrast, we know nothing about the case when C is elliptic.

The surfaces with $\kappa = -\infty$ are precisely the *ruled surfaces* — that is, those which are birationally equivalent over $\mathbf{C}$ to $\mathbf{P}^1 \times C$ for some curve C. Among these, by far the most interesting are the *rational surfaces*, which are birationally equivalent to $\mathbf{P}^2$ over $\mathbf{C}$.

Surfaces with $\kappa = 0$ fall into four families:

- Abelian surfaces. These are the analogues in two dimensions of elliptic curves, and there is no reason to doubt that their number-theoretical properties largely generalize those of elliptic curves.
- K3 surfaces, including in particular Kummer surfaces. Some but not all K3 surfaces are elliptic.
- Enriques surfaces, whose number theory has been very little studied. They are necessarily elliptic.
- bielliptic surfaces.

Surfaces with $\kappa = 1$ are necessarily elliptic.

Surfaces with $\kappa = 2$ are called *surfaces of general type* — which in mathematics is generally a derogatory phrase. About them there is currently nothing to say beyond the Bombieri–Lang conjecture stated in Section 2.

In the next two sections I shall outline what can at present be said about rational surfaces and K3 surfaces respectively; these appear to be the two most interesting families of surfaces for the number theorist. In both cases many of the most recent results depend on one or both of two major conjectures. One of these (for the reason given near the end of Section 4) is the finiteness of the Tate–Shafarevich group; the other is Schinzel's Hypothesis, which we now describe. It gives a conjectural answer to the following question: given finitely many polynomials $F_1(X), \ldots, F_n(X)$ in $\mathbf{Z}[X]$ with positive leading coefficients, is there an arbitrarily large integer x at which they all take prime values? There are two obvious obstructions to this:

- One or more of the $F_i(X)$ may split in $\mathbf{Z}[X]$.
- There may be a prime p such that for any value of x mod p at least one of the $F_i(x)$ is divisible by p.

Clearly the second obstruction can only happen for $p \leqslant \sum \deg(F_i)$. Schinzel's Hypothesis is that these are the only obstructions: in other words, if neither of them happens then we can choose an arbitrarily large x so that every $F_i(x)$ is a prime. There are various more complicated variants of this hypothesis (including ones in other algebraic number fields), but they all follow fairly easily from the hypothesis in its original form.

No one in his right mind would attempt to prove Schinzel's Hypothesis; indeed one instance is the notoriously intractable twin primes problem, which is the special case when the F_i are the two polynomials $X+1$ and $X-1$. But probabilistic arguments suggest that the hypothesis is in fact true. At the very least it would be perverse to look for counterexamples to results which have been proved subject to Schinzel's Hypothesis.

6. Rational surfaces

From the number-theoretic point of view, there are two kinds of rational surface:

- Pencils of conics, given by an equation of the form

$$a_0(u,v)X_0^2 + a_1(u,v)X_1^2 + a_2(u,v)X_2^2 = 0 \tag{5}$$

where the $a_i(u,v)$ are homogeneous polynomials of the same degree. Pencils of conics can be classified in more detail according to the number of bad fibres.

- Del Pezzo surfaces of degree d, where $0 < d \leqslant 9$. Over $\mathbf{C}$, such a surface is obtained by blowing up $(9-d)$ points of $\mathbf{P}^2$ in general position. It is known that Del Pezzo surfaces of degree $d > 4$ satisfy the Hasse principle and weak approximation; indeed those of degree 5 or 7 necessarily contain rational points. Del Pezzo surfaces of degree 2 or 1 have no aesthetic merits and have attracted little attention; it seems sensible to ignore them until the problems coming from those of degrees 4 and 3 have been solved. The Del Pezzo surfaces of degree 3 are the nonsingular cubic surfaces, which have an enormous but largely irrelevent literature, and those of degree 4 are the nonsingular intersections of two quadrics in $\mathbf{P}^4$. For historical reasons, attention has been concentrated on the Del Pezzo surfaces of degree 3; but the problems presented by those of degree 4 are necessarily simpler.

We consider first pencils of conics, and assume that (5) is defined over $\mathbf{Q}$, the argument for an arbitrary algebraic number field not being essentially different. We can require the coefficients of the $a_i(u,v)$ to be in $\mathbf{Z}$. Since the Hasse principle holds for conics, it is enough to choose $u = u_0, v = v_0$ in such a way that (5) is locally soluble at $2, \infty$ and all the odd primes which divide any of the $a_i(u_0, v_0)$. As it stands, this appears to involve arguing in a circle; the way to make the argument respectable is as follows.

Assume that (5) is everywhere locally soluble. By absorbing suitable factors into the X_i, we can ensure that the $a_i(u,v)$ are square-free and coprime. To achieve this, we have to drop the condition that the $a_i(u,v)$ are all of the same degree; but it is still true that their degrees are all even or all odd. Denote by $\mathscr{B}$ the set of bad places, which turns out to consist of $2, \infty$, the primes which divide the discriminant of $a_0(u,v)a_1(u,v)a_2(u,v)$ and the primes which do not exceed the degree of that product. Let $\mathscr{S}$ be the space of all pairs of coprime integers u_0, v_0, with the topology induced by the places of $\mathscr{B}$; and let $\mathscr{S}_0$ be the subset of $\mathscr{S}$ consisting of the points at which (5) is locally soluble at every place in $\mathscr{B}$. By hypothesis, $\mathscr{S}_0$ is not empty; and it is open in $\mathscr{S}$. To obtain solubility in $\mathbf{Q}$, we have to choose u_0, v_0 in $\mathscr{S}_0$ so that (5) is locally soluble at each good prime p_0 which divides one of the $a_i(u_0, v_0)$; for solubility at the other good primes is trivial, and we have already taken care of the bad places. Let $c(u,v)$ be the irreducible factor of that one of the $a_i(u,v)$ for which $p_0 | c(u_0, v_0)$, and to fix ideas assume that $c(u,v)$ divides $a_2(u,v)$; here $c(u,v)$ is unique because p_0 does not divide the discriminant of $a_0a_1a_2$. The condition of local solubility at p_0 is

$$(a_0(u_0,v_0)a_1(u_0,v_0), c(u_0,v_0))_{p_0} = +1 \tag{6}$$

where the outer bracket is the Hilbert symbol. So a necessary condition for the solubility of (5) is that all the conditions like

$$\prod (a_0(u_0,v_0)a_1(u_0,v_0), c(u_0,v_0))_p = +1 \tag{7}$$

hold simultaneously for some (u_0, v_0) in $\mathscr{S}_0$, where the product is taken over all p which divide $c(u_0, v_0)$.

What is unexpected is that this turns out to be useful, because of the following lemma. The proof of the lemma is straightforward since the function Φ behaves like a quadratic residue symbol and can be evaluated by a Euclidean algorithm process very like that which is used for such symbols.

Lemma 4. *Let $F(u,v), G(u,v)$ be homogeneous polynomials in $\mathbf{Z}[u,v]$, with* $\deg(F)$ *even. Let $\mathscr{B}$ be a finite set of places of $\mathbf{Q}$ which contains $2, \infty$ and all the primes which divide the discriminant of FG. For any coprime u_0, v_0 in $\mathbf{Z}$, write*

$$\Phi(\mathscr{B}; F, G; u_0, v_0) = \prod (F(u_0,v_0), G(u_0,v_0))_p \tag{8}$$

where the outer bracket on the right is the Hilbert symbol and the product is taken over all primes p not in $\mathscr{B}$ such that $p|G(u_0,v_0)$. Then $\Phi(u_0,v_0)$ is continuous in the topology on $\mathscr{S}$, and computable.

In this result we take $F = a_0a_1, G = c$; we noted above that $\deg(a_0a_1)$ is necessarily even. It follows that a necessary condition for the solubility of (5) is that there is a point (u_0, v_0) in $\mathscr{S}_0$ such that $\Phi(u_0, v_0) = +1$ for all Φ which can be generated from (5) in this way. This condition is computable, and it is unsurprising (though not obvious) that it turns out to be equivalent to the Brauer–Manin condition for (5).

If one assumes Schinzel's Hypothesis, this condition is also sufficient. For suppose that u_0, v_0 have been so chosen that there is only one good prime p_0 which divides $c(u_0, v_0)$; then the product in (7) reduces to the left side of (6), and so (6) holds for this prime. Now choose an open set $\mathscr{N} \subset \mathscr{S}_0$ such that (7) holds throughout $\mathscr{N}$ for each $c(u,v)$; by a slightly modified version of Schinzel's Hypothesis we can choose (u_0, v_0) in $\mathscr{N}$ so that every $c(u_0, v_0)$ is the product of one good prime and possibly some factors in $\mathscr{B}$. As c runs through all irreducible factors of $a_1a_2a_3$, p_0 runs through all those primes for which we have to verify (6). Thus (5) is everywhere locally soluble for the pair u_0, v_0, and therefore globally soluble. With minor modifications, the same argument shows that (subject to Schinzel's Hypothesis) the Brauer–Manin obstruction is also the only obstruction to weak approximation.

If there is no Brauer–Manin obstruction, this construction finds infinitely many conics in the pencil which contain rational points. But, somewhat unexpectedly, even if we know some conics of the pencil which are soluble, without Schinzel's Hypothesis we do not know how to generate more such conics.

Question 12. *Given a pencil of conics and finitely many conics in the pencil each of which contains rational points, can we generate further conics of the pencil which contain rational points without using Schinzel's Hypothesis?*

If we know even one rational point on a Del Pezzo surface V of degree 3 or 4, we can obtain an infinity of curves of genus 0 each of which lies on V, though they will be singular and for degree 3 it will usually not be true that each point of $V(\mathbf{Q})$ lies on at least one curve of the family. But without such a point, the best we can do is to find on V a family of curves of genus 1. At first sight, it would seem that in these circumstances nothing resembling the argument above can be applied; for an essential component of that argument was that conics satisfy the Hasse principle, and this is not true for curves of genus 1. However Lemma 3 provides us with a way round this obstacle.

The arguments involved are applicable to some surfaces V which are not necessarily rational, but which are elliptic with a fibration $V \to \mathbf{P}^1$. Consider a pencil of curves $\mathscr{C}_\lambda$ of genus 1, each of which is a 2-covering of its Jacobian J_λ. If we can choose λ in such a way that $\mathscr{C}_\lambda$ is everywhere locally soluble and at least half the elements of the 2-Selmer group of J_λ are soluble (and if we assume the finiteness of the relevent Tate–Shafarevich group), then it will follow from Lemma 3 that $\mathscr{C}_\lambda$ is soluble. For this machinery to have any chance of working, we must be able to implement the 2-descent on J_λ in a manner which is uniform in λ. This more or less requires J_λ to have its 2-division points defined over $\mathbf{Q}(\lambda)$ and therefore to have the form

$$Y^2 = (X - c_1(\lambda))(X - c_2(\lambda))(X - c_3(\lambda)) \tag{9}$$

where the $c_i(\lambda)$ are in $\mathbf{Q}(\lambda)$; but an additional trick, given in [**2**], enables the method to be used even if J_λ has just one 2-division point in $\mathbf{Q}(\lambda)$.

The details of this method are unattractive, but the strategy is as follows. (For a full account, see [**11**].) Without loss of generality we can assume that the $c_i(\lambda)$ are in $\mathbf{Z}[\lambda]$. For any particular integral value λ_0 of λ, the bad places for the equation (9) are the bad places for the system together with the primes which divide one of the $c_i(\lambda_0) - c_j(\lambda_0)$. It was explained in Section 4 how to implement the 2-descent for (9). We shall eventually use Schinzel's Hypothesis to choose λ_0 so that the value of each irreducible factor of any $c_i(\lambda) - c_j(\lambda)$ at $\lambda = \lambda_0$ is the product of some bad primes for the system with one good prime. We call the latter a *Schinzel prime*; though it is a good prime for the system, it is a bad prime for the curve obtained by writing $\lambda = \lambda_0$ in (9). The effect of restricting λ_0 in this way is that we know those 2-coverings of (9) for $\lambda = \lambda_0$ which are locally soluble at all good primes for the curve; they form a finite group of 2-coverings $\mathscr{C}'_\lambda$ of J_λ which does not depend on the choice of λ_0 provided it satisfies the condition above.

This group certainly contains the original $\mathscr{C}_\lambda$ and the 2-coverings which correspond to the 2-division points. The next step, which involves a sophisticated analysis of the 2-descent process and also requires us to introduce additional well chosen bad primes for the system, is to find local conditions on λ_0 at the bad primes of the system which ensure that for $\lambda = \lambda_0$:

- the only elements of this group which are locally soluble at all the bad places of the system lie in the subgroup generated by the original $\mathscr{C}_\lambda$ and the 2-coverings generated by the 2-division points; and
- the original $\mathscr{C}_\lambda$ is locally soluble at all the bad places of the system.

This is not always possible; if it is impossible, that provides an obstruction to this method of attack on the problem though not necessarily to the solubility of the system. In general this obstruction is not much stronger than the Brauer–Manin obstruction, and in some cases they are provably the same; so this is not too serious a blemish on the method. If we achieve the two properties above then solubility at the Schinzel primes turns out to be automatic. (This is an example of what seems to be a rather general phenomenon, that if one thing goes wrong, so must at least one other which is related to it.) With our enlarged set of bad places for the system, we now choose λ_0 to satisfy the local conditions and the Schinzel condition in the previous paragraph. By Lemma 3, the curve (9) is soluble for this value of λ. But in contrast to what happens for pencils of conics, this kind of argument appears to provide no information about weak approximation.

Question 13. *Can one modify the method above so that it works without any assumption about the 2-division points of J_λ?*

Unfortunately it is not known (and probably is not even true) that an arbitrary Del Pezzo surface of degree 4 contains a pencil of curves of genus 1 of this particular type — and the situation for Del Pezzo surfaces of degree 3 is much worse, in that the natural curves to consider are 3-coverings rather than 2-coverings.

However, for Del Pezzo surfaces of degree 4 some progress has been made. Salberger and Skorobogatov [**35**] have shown that the Brauer–Manin obstruction is the only obstruction to weak approximation. (Recall that weak approximation presupposes the existence of at least one rational point.) As for the Hasse principle, the present situation is as follows. A Del Pezzo surface of degree 4 defined over the algebraic number field K has the form

$$V : F_1(X_0, \dots, X_5) = F_2(X_0, \dots, X_5) = 0,$$

where the coefficients of F_1 and F_2 are in K. By a linear transformation defined over an extension K_1 of odd degree over K, we can separate one of the variables; and over K_1 we can find a pencil of curves $\mathscr{C}_\lambda$ of genus 1 on V for which each J_λ has one 2-division point defined over $K_1(\lambda)$. Using the trick described in [**2**], and subject to an obstruction typical for the method, it can now be shown that V contains a point defined over K_1. A straightforward geometric argument, which does not rely on K being an algebraic number field, now shows that V contains a point defined over K. Unfortunately the overall arguments are so complicated (and so unnatural) that it is not clear whether the obstruction to the method is still simply the Brauer–Manin obstruction to the solubility of V over K; but even if it is stronger, it is not much stronger.

To use and then collapse a field extension in this way is a device which probably has a number of uses. For such a collapse step to be feasible, the degree of the field extension needs to be prime to the degree of the variety; and this leads one to phrase the same property somewhat differently.

Question 14. *Let V be a variety defined over a field K, not necessarily of a number-theoretic kind. For what families of V is it true that if V contains a 0-cycle of degree 1 defined over K then it contains a point defined over K?*

As stated above, this is true for Del Pezzo surfaces of degree 4. For pencils of conics it is in general false, even for algebraic number fields K. For Del Pezzo surfaces of degree 3 the question is open: I expect it to be true for algebraic number fields K but false for general fields.

A variant of the method above can be applied to diagonal cubic surfaces

$$V : a_0X_0^3 + a_1X_1^3 = a_2X_2^3 + a_3X_3^3, \tag{10}$$

subject to one very counterintuitive condition, which is that K, the field of definition of V, must not contain the primitive cube roots of unity. Write V in the form

$$a_0X_0^3 + a_1X_1^3 = \lambda Y^3, \quad a_2X_2^3 + a_3X_3^3 = \lambda Y^3 \tag{11}$$

where λ is at our disposal. We now have two pencils of curves of genus 1, each of which is a $\sqrt{-3}$-covering of its Jacobian; and we have to apply the method simultaneously to both curves. This introduces considerable additional complications, for which see **[45]**; and the obstruction to the method, though weak, is certainly strictly stronger than the Brauer–Manin obstruction. The reason for the condition on K is that otherwise the curves (11) would admit complex multiplication, and the latter acts on the $\sqrt{-3}$-Selmer groups; thus the order of the latter would always be an odd power of 3, whereas it has to be reduced to 9 for the method to work. (Here one factor 3 arises because of the 3-division points on the Jacobian defined over $\mathbf{Q}(\sqrt{-3})$.) On the other hand, in this argument we only need apply Schinzel's Hypothesis to the single polynomial X, so that it can be replaced by Dirichlet's theorem on primes in arithmetic progression.

All this relates to Question (A). For Question (B) the only known results are for quadrics, for which see the remarks after Theorem 1. It seems reasonable to ask whether there is an analogous result for other kinds of rational surfaces; this is another problem for which the first step should probably be to use numerical search to generate a plausible conjecture. For this purpose, one needs to examine a system with not too many parameters; and this leads to the following question:

Question 15. *For the surface V given by* (10) *with the a_i in $\mathbf{Z}$, is there a polynomial P in the $|a_i|$ such that if V is soluble in $\mathbf{Z}$ it then has such a solution for which each summand is absolutely bounded by P?*

The ideal answer to Question (C) would be to provide a birational map $V \to \mathbf{P}^2$ defined over $\mathbf{Q}$. However, it can be shown that such a map exists for nonsingular cubic surfaces V if and only if $V(\mathbf{Q})$ is not empty and V contains a divisor defined over $\mathbf{Q}$ which consists of 2, 3 or 6 skew lines. (For Del Pezzo surfaces of degree 4, the second condition must be replaced by the statement that V contains a divisor defined over $\mathbf{Q}$ which consists of one or more skew lines.) For Châtelet surfaces, which have the form

$$X_2(X_0^2 - cX_1^2) = f(X_2, X_3)$$

where c is a non-square and f is homogeneous cubic, it is known (see **[12]**) that there is a finite set of parametric solutions (each in 4 inhomogeneous variables) such that each point of $V(\mathbf{Q})$ is represented by one of them, though in an infinity of different ways. But in general more than one parametric solution is needed, and it was already shown in **[28]** that parametric solutions in only two variables cannot play a useful part in the process.

Question 16. *Is there a larger class of cubic surfaces (ideally, the class of all nonsingular cubic surfaces) for which analogous results hold?*

The question of parametric solutions is linked to the idea of *R-equivalence*. Let V be a variety defined over $\mathbf{Q}$; then R-equivalence is defined as the finest equivalence relation such that two points given by the same parametric solution are equivalent. Alternatively, it is the finest equivalence relation such that for any map $f : \mathbf{P}^1 \to V$ and points P_1, P_2, all defined over $\mathbf{Q}$, the points $f(P_1)$ and $f(P_2)$ are equivalent. A good deal is known about R-equivalence on cubic surfaces; in particular, it is shown in **[46]** that the closure of an R-equivalence class in $V(\mathbf{A})$ is computable, and that the closures of two R-equivalence classes are either the same or disjoint. There is an example in **[12]** of a Châtelet surface V containing two distinct R-equivalence classes, each of which has the whole of $V(\mathbf{A})$ as its closure. Work of Coray and Tsfasman shows that this V is birationally equivalent to a nonsingular cubic surface.

Question 17. *Is the set of R-equivalence classes on a nonsingular cubic surface finite?*

7. K3 surfaces and Kummer surfaces

K3 surfaces are the simplest kind of variety about whose number-theoretic properties very little is known; indeed they still present many unsolved problems even to the geometer. There are infinitely many families of K3 surfaces; the simplest of them, and the only one which will be considered in the present article, consists of all nonsingular quartic surfaces. An important special type of K3 surfaces consists of the *Kummer surfaces*, a phrase which can carry either of two related meanings:

- The quotient of an abelian surface A by the automorphism -1; this has 16 singular points corresponding to the 16 2-division points of A.
- The nonsingular surface obtained by blowing up the 16 singular points in the previous definition.

One advantage of Kummer surfaces in comparison with general K3 surfaces is that for the former it is easy to determine $\mathrm{Pic}(\bar{V})$.

Some K3 surfaces contain one or more pencils of curves of genus 1, and these pencils may even be of the kind discussed in the previous section; but one should not confine one's attention to K3 surfaces with this additional property. For the time being, there is merit in concentrating on diagonal quartics

$$V : a_0X_0^4 + a_1X_1^4 + a_2X_2^4 + a_3X_3^4 = 0, \tag{12}$$

because these contain few enough parameters to make systematic numerical experimentation possible. However, the number theory of such surfaces may be exceptional, because the geometry certainly is. Indeed $\mathrm{Pic}(\bar{V})$ has rank 20, which is the largest possible value for any K3 surface, and it is generated by the classes of the 48 lines on $\bar{V}$; moreover V is a Kummer surface up to isogeny, and indeed is the Kummer surface of $E \times E$ where E is a certain elliptic curve which admits complex multiplication. One consequence of this is that V is rigid in the sense of algebraic geometry.

There is an obvious map from V to the quadric surface

$$W : a_0 Y_0^2 + a_1 Y_1^2 + a_2 Y_2^2 + a_3 Y_3^2 = 0.$$

If $a_0 a_1 a_2 a_3$ is a square, and V is everywhere locally soluble, each of the two families of lines on W is defined over the ground field, and each such line lifts to a curve of genus 1 on V; moreover the Jacobians of these curves have the form (9), so that the methods of the previous section can be applied.

Martin Bright [**7**] has computed and tabulated $\mathrm{Br}_1(V)/\mathrm{Br}(\mathbf{Q})$ for all V of the form (12); it is necessary to do this by computer, because there are 546 distinct cases. Assuming Schinzel's Hypothesis and the finiteness of Ш, I had previously shown in [**44**] that over $\mathbf{Q}$ the Brauer–Manin obstruction is the only obstruction to the Hasse principle in the most general case in which $a_0 a_1 a_2 a_3$ is a square. (Most general in this context means that none of the $\pm a_i a_j$ is a square and $a_0 a_1 a_2 a_3$ is not a fourth power.) It seems reasonable to hope that the same property will still hold in all the cases for which $a_0 a_1 a_2 a_3$ is a square; but there are too many of them to examine individually. On the other hand, there is strong numerical evidence that when $a_0 a_1 a_2 a_3$ is not a square the obstruction coming from $\mathrm{Br}_1(V)$ is not in general the only obstruction to the Hasse principle.

Question 18. *What is the additional obstruction in this case?*

One particularly interesting example is the surface

$$X_0^4 + 2X_1^4 = X_2^4 + 4X_3^4; \tag{13}$$

this has two obvious rational points, but appears to have no others.

8. Density of rational points

So far I have ignored Question (D). It differs from the others in that it is not a birational question, but is associated with a particular embedding of the variety V in projective space. For simplicity we work over $\mathbf{Q}$. A point P in $\mathbf{P}^n$

defined over $\mathbf{Q}$ has a representation $(x_0, \ldots, x_n)$ where the x_i are integers with no common factor; and this representation is unique up to changing the signs of all the x_i. We define the *height* of P to be $\max |x_i|$; a linear transformation on the ambient space multiplies heights by numbers which lie between two positive constants depending on the linear transformation. Denote by $N(H, V)$ the number of points of $V(\mathbf{Q})$ whose height is less than H; then it is natural to ask how $N(H, V)$ behaves as $H \to \infty$. This is the core question for the Hardy–Littlewood method, which when it is applicable is the best (and often the only) way of proving that $V(\mathbf{Q})$ is not empty. In very general circumstances that method provides estimates of the form

$$N(H, V) = \text{ leading term } + \text{ error term.}$$

The leading term is usually the same as one would obtain by probabilistic arguments. But such results are only valuable when it can be shown that the error term is small compared to the leading term, and to achieve this the dimension of V needs to be large compared to its degree. The extreme case of this is the following theorem, due to Birch [**3**].

Theorem 5. *Let $r_1, \ldots, r_m$ be positive odd integers, not necessarily all different. Then there exists $N_0(r_1, \ldots, r_m)$ with the following property. For any $N \geqslant N_0$ let $F_i(X_0, \ldots, X_N)$ be homogeneous polynomials with coefficients in $\mathbf{Z}$ and $\deg F_i = r_i$ for $i = 1, \ldots, m$. Then the F_i have a common nontrivial zero in $\mathbf{Z}^N$.*

The proof falls into two parts. First, the Hardy–Littlewood method is used to prove the result in the special case when $m = 1$ and F_1 is diagonal — that is, to show that if r is odd and $N \geqslant N_1(r)$, then

$$c_0 X_0^r + \ldots + c_N X_N^r = 0$$

has a nontrivial integral solution. Then the general case is reduced to this special case by purely elementary methods. The requirement that all the r_i should be odd arises from difficulties connected with the real place; over a totally complex algebraic number field there is a similar theorem for which the r_i can be any positive integers.

Question 19. *In Theorem 5, can the condition that all the r_i are odd be replaced by the requirement on the F_i that the projective variety given by*

$$F_1 = \ldots = F_m = 0$$

has a nonsingular real point?

The Hardy–Littlewood method was designed for a single equation in which the variables are separated — for example, an equation of the form

$$f_1(X_1) + \ldots + f_N(X_N) = c$$

where the f_i are polynomials, the X_i are integers, and one wishes to prove solubility for all integers c, or all large enough c, or almost all c. But it has also been applied both to several simultaneous equations and to equations in which the variables are not separated. The following theorem of Hooley **[22, 23, 24]** is the most impressive result in this direction.

Theorem 6. *Homogeneous nonsingular nonary cubics over* $\mathbf{Q}$ *satisfy both the Hasse principle and weak approximation.*

It appears that the Hardy–Littlewood method can only work for families for which $N(H,V)$ is asymptotically equal to its probabilistic value; in particular it seems unlikely that it can be made to work for families for which weak approximation fails. Manin has put forward a conjecture about the asymptotic density of rational solutions for certain geometrically interesting families of varieties for which weak approximation is unlikely to hold: more precisely, for Fano varieties embedded in $\mathbf{P}^n$ by means of their anticanonical divisors. For simplicity, we describe his conjecture only for Del Pezzo surfaces V of degrees 3 and 4. To ask about $N(H,V)$ is now the wrong question, for V may contain lines L defined over $\mathbf{Q}$, and for any line $N(H,L) \sim AH^2$ for some nonzero constant A. This is much greater than the order-of-magnitude estimate for $N(H,V)$ given by a probabilistic argument. For the latter suggests an estimate $AH\prod(N(p)/(p+1))$, where the product is taken over all primes less than a certain bound which depends on H. In view of what is said in Section 3, this product ought to be replaced by something which depends on the behaviour of $L_2(s,V)$ near $s = 1$. More precisely, the way in which the leading term in the Hardy–Littlewood method is obtained suggests that here we should take $s-1$ to be comparable with $(\log H)^{-1}$. Remembering the Tate conjecture, this gives the right hand side of (14) as a conjectural estimate for $N(H,V)$. But if this argument were valid, L would contain more rational points than V, even though $V \supset L$. Manin's way to resolve this absurdity is to study not $N(H,V)$ but $N(H,U)$, where U is the open subset of V obtained by deleting the 27 or 16 lines on V. Manin conjectured that

(14) $\qquad N(H,U) \sim AH(\log H)^{r-1}$ where r is the rank of $\mathrm{Pic}(V)$;

and Peyre **[31]** has given a conjectural formula for A. (But note that there exist Fano varieties of dimension greater than 2 for which (14) is certainly false; and it is not clear how Manin's conjecture should be modified to cover such cases.)

Various people have proved this conjecture for the cone $X_0X_1X_2 = X_3^3$, and there are also results for the singular cubic surface

$$X_0X_1X_2 + X_0X_1X_3 + X_0X_2X_3 + X_1X_2X_3 = 0,$$

to which attention had been drawn by Birch. Heath-Brown **[21]** has proved that

$$A_1H(\log H)^6 < N(H,U) < A_2H(\log H)^6$$

for suitable constants A_1, A_2; but he doubts whether his method is capable of proving an asymptotic formula. Using quite different ideas, Rudge has sketched a proof of the asymptotic formula; but the details are not yet complete.

Question 20. *Are there nonsingular Del Pezzo surfaces of degree* 3 *or* 4 *for which the Manin conjecture can be proved?*

In the first instance, it would be wise to address this problem under rather restrictive hypotheses about $\mathrm{Pic}(V)$, not least because the Brauer–Manin obstruction to weak approximation occurs in the conjectural formula for A and therefore the problem is likely to be easier for families of V for which weak approximation holds. A one-sided estimate for one such family is given in **[41]**.

For Del Pezzo surfaces, the value of c for which $N(H,U) \sim AH(\log H)^c$ is defined by the geometry rather than by the number theory, though that is not true of A. For other varieties, the corresponding statement need no longer be true. We start with curves. For a curve of genus 0 and degree d, we have $N(H,V) \sim AH^{2/d}$; and for a curve of genus greater than 1 Faltings' theorem is equivalent to the statement that $N(H,V) = O(1)$. But if V is an elliptic curve then $N(H,V) \sim A(\log H)^{r/2}$ where r is the rank of the Mordell–Weil group. (For elliptic curves there is a more canonical definition of height, which is invariant under bilinear transformation; this is used to prove the result above.)

For pencils of conics, Manin's question is probably not the best one to ask, and it would be better to proceed as follows. A pencil of conics is a surface V together with a map $V \to \mathbf{P}^1$ whose fibres are conics. Let $N^*(H,V)$ be the number of points on $\mathbf{P}^1$ of height less than H for which the corresponding fibre contains rational points.

Question 21. *What is the conjectural estimate for $N^*(H,V)$ and under what conditions can one prove it?*

It may be worth asking the same questions for pencils of curves of genus 1.

For surfaces of general type, the Bombieri Lang conjecture implies that questions about $N(H,V)$ are really questions about certain curves on V; and for abelian surfaces (and indeed abelian varieties in any dimension) the obvious generalisation of the theorem for elliptic curves holds. But K3 surfaces pose new problems — and not ones on which any practicable amount of computation is likely to shed light. If V is a K3 surface, then we have to study not $N(H,V)$ but $N(H,U)$ where U is obtained from V by deleting the curves of genus 0 on V defined over $\mathbf{Q}$, of which there may be an infinite number. One can expect that $N(H,U) \sim A(\log H)^c$ for some constants A and c; and it seems reasonable to hope that c will be a half-integer. The surface (13) suggests that we can have $c = 0$, and it must be certain (though perhaps difficult to prove) that c can sometimes be strictly positive.

Question 22. *Can the value of c be obtained from the L-series $L_2(s,V)$?*

Question 23. *If V is a Kummer surface obtained from the abelian surface A, is c related to the rank of the Mordell–Weil group of A?*

I am indebted to Jean-Louis Colliot-Thélène for many valuable comments; but he bears no responsibility for the opinions expressed.

References

[1] A. Beauville – *Complex Algebraic Surfaces*, second ed., London Mathematical Society Student Texts, vol. 34, Cambridge University Press, Cambridge, 1996.

[2] A. O. Bender & P. Swinnerton-Dyer – Solubility of certain pencils of curves of genus 1, and of the intersection of two quadrics in $\mathbb{P}^4$, *Proc. London Math. Soc. (3)* **83** (2001), no. 2, 299–329.

[3] B. J. Birch – Homogeneous forms of odd degree in a large number of variables, *Mathematika* **4** (1957), 102–105.

[4] S. Bloch & K. Kato – L-functions and Tamagawa numbers of motives, The Grothendieck Festschrift, vol. I, Prog. Math., vol. 86, Birkhäuser Boston, Boston, MA, 1990, 333–400.

[5] S. J. Bloch – *Higher regulators, algebraic K-theory, and zeta functions of elliptic curves*, CRM Monograph Series, vol. 11, American Mathematical Society, Providence, RI, 2000.

[6] A. Borel – Cohomologie de SL_n et valeurs de fonctions zeta aux points entiers, *Ann. Scuola Norm. Sup. Pisa Cl. Sci. (4)* **4** (1977), no. 4, 613–636.

[7] M. Bright – Ph.D Thesis, Cambridge University, 2002.

[8] M. Bright & S. P. Swinnerton-Dyer – Computing the Brauer–Manin obstructions, to appear, 2002.

[9] F. E. Browder (ed.) – *Mathematical developments arising from Hilbert problems*, American Mathematical Society, Providence, R. I., 1976.

[10] J. W. S. Cassels – Second descents for elliptic curves, *J. Reine Angew. Math.* **494** (1998), 101–127.

[11] J.-L. Colliot-Thélène, A. N. Skorobogatov & P. Swinnerton-Dyer – Hasse principle for pencils of curves of genus one whose Jacobians have rational 2-division points, *Invent. Math.* **134** (1998), no. 3, 579–650.

[12] J.-L. Colliot-Thélène, J.-J. Sansuc & P. Swinnerton-Dyer – Intersections of two quadrics and Châtelet surfaces. II, *J. Reine Angew. Math.* **374** (1987), 72–168.

[13] J.-L. Colliot-Thélène & P. Swinnerton-Dyer – Hasse principle and weak approximation for pencils of Severi–Brauer and similar varieties, *J. Reine Angew. Math.* **453** (1994), 49–112.

[14] J. Denef, L. Lipshitz, T. Pheidas & J. Van Geel (eds.) – *Hilbert's tenth problem: relations with arithmetic and algebraic geometry*, Contemporary Mathematics, vol. 270, American Mathematical Society, Providence, RI, 2000, Papers from the workshop held at Ghent University, Ghent, November 2–5, 1999.

[15] G. Faltings, G. Wüstholz, F. Grunewald, N. Schappacher & U. Stuhler – *Rational points*, third ed., Aspects of Mathematics, E6, Friedr. Vieweg & Sohn, Braunschweig, 1992.

[16] J. Gebel & H. G. Zimmer – Computing the Mordell–Weil group of an elliptic curve over **Q**, Elliptic curves and related topics, CRM Proc. Lecture Notes, vol. 4, Amer. Math. Soc., Providence, RI, 1994, 61–83.

[17] B. Gross, W. Kohnen & D. Zagier – Heegner points and derivatives of L-series. II, *Math. Ann.* **278** (1987), no. 1-4, 497–562.

[18] B. H. Gross – Kolyvagin's work on modular elliptic curves, L-functions and arithmetic (Durham, 1989), London Math. Soc. Lecture Note Ser., vol. 153, Cambridge Univ. Press, Cambridge, 1991, 235–256.

[19] B. H. Gross & D. B. Zagier – Heegner points and derivatives of L-series, *Invent. Math.* **84** (1986), no. 2, 225–320.

[20] D. Harari – Obstructions de Manin transcendantes, Number theory (Paris, 1993–1994), London Math. Soc. Lecture Note Ser., vol. 235, Cambridge Univ. Press, Cambridge, 1996, 75–87.

[21] D. R. Heath-Brown – The density of rational points on Cayley's cubic surface, to appear.

[22] C. Hooley – On nonary cubic forms, *J. Reine Angew. Math.* **386** (1988), 32–98.

[23] ———, On nonary cubic forms. II, *J. Reine Angew. Math.* **415** (1991), 95–165.

[24] ———, On nonary cubic forms. III, *J. Reine Angew. Math.* **456** (1994), 53–63.

[25] W. W. J. Hulsbergen – *Conjectures in arithmetic algebraic geometry*, second ed., Aspects of Mathematics, E18, Friedr. Vieweg & Sohn, Braunschweig, 1994, A survey.

[26] S. L. Kleiman – Algebraic cycles and the Weil conjectures, Dix esposés sur la cohomologie des schémas, North-Holland, Amsterdam, 1968, 359–386.

[27] V. A. Kolyvagin – Finiteness of $E(\mathbf{Q})$ and $\mathrm{SH}(E, \mathbf{Q})$ for a subclass of Weil curves, *Izv. Akad. Nauk SSSR Ser. Mat.* **52** (1988), no. 3, 522–540, 670–671.

[28] Y. I. MANIN – *Cubic Forms*, second ed., North-Holland Mathematical Library, vol. 4, North-Holland Publishing Co., Amsterdam, 1986.

[29] B. MAZUR – Rational isogenies of prime degree (with an appendix by D. Goldfeld), *Invent. Math.* **44** (1978), no. 2, 129–162.

[30] L. MEREL – Bornes pour la torsion des courbes elliptiques sur les corps de nombres, *Invent. Math.* **124** (1996), no. 1-3, 437–449.

[31] E. PEYRE – Hauteurs et mesures de Tamagawa sur les variétés de Fano, *Duke Math. J.* **79** (1995), no. 1, 101–218.

[32] M. RAPOPORT, N. SCHAPPACHER & P. SCHNEIDER (eds.) – *Beilinson's conjectures on special values of L-functions*, Perspectives in Mathematics, vol. 4, Academic Press Inc., Boston, MA, 1988.

[33] K. RUBIN – Elliptic curves with complex multiplication and the conjecture of Birch and Swinnerton-Dyer, Arithmetic theory of elliptic curves (Cetraro, 1997), Lecture Notes in Math., vol. 1716, Springer, Berlin, 1999, 167–234.

[34] K. RUBIN & A. SILVERBERG – Ranks of elliptic curves, *Bull. Amer. Math. Soc. (N.S.)* **39** (2002), no. 4, 455–474 (electronic).

[35] P. SALBERGER & A. N. SKOROBOGATOV – Weak approximation for surfaces defined by two quadratic forms, *Duke Math. J.* **63** (1991), no. 2, 517–536.

[36] J.-P. SERRE – *Facteurs locaux des fonctions zêta des variétés algébriques (définitions et conjectures)*, Séminaire Delange-Pisot-Poitou, 1969/70, exp. 19.

[37] C. SIEGEL – Normen algebraischer Zahlen, Werke, Band IV, Vandenhoeck & Ruprecht, 1983, 250–268.

[38] A. SILVERBERG – Open questions in arithmetic algebraic geometry, Arithmetic algebraic geometry (Park City, UT, 1999), IAS/Park City Math. Ser., vol. 9, Amer. Math. Soc., Providence, RI, 2001, 83–142.

[39] A. SKOROBOGATOV – *Torsors and rational points*, Cambridge Tracts in Mathematics, vol. 144, Cambridge University Press, Cambridge, 2001.

[40] A. N. SKOROBOGATOV – Beyond the Manin obstruction, *Invent. Math.* **135** (1999), no. 2, 399–424.

[41] J. B. SLATER & P. SWINNERTON-DYER – Counting points on cubic surfaces. I, *Astérisque* (1998), no. 251, 1–12, Nombre et répartition de points de hauteur bornée (Paris, 1996).

[42] P. SWINNERTON-DYER – The conjectures of Birch and Swinnerton-Dyer, and of Tate, Proc. Conf. Local Fields (Driebergen, 1966), Springer, Berlin, 1967, 132–157.

[43] P. SWINNERTON-DYER – Rational points on pencils of conics and on pencils of quadrics, *J. London Math. Soc. (2)* **50** (1994), no. 2, 231–242.

[44] ———, Arithmetic of diagonal quartic surfaces. II, *Proc. London Math. Soc. (3)* **80** (2000), no. 3, 513–544.

[45] ———, The solubility of diagonal cubic surfaces, *Ann. Sci. École Norm. Sup. (4)* **34** (2001), no. 6, 891–912.

[46] ———, Weak approximation and R-equivalence on cubic surfaces, Rational points on algebraic varieties, Prog. Math., vol. 199, Birkhäuser, Basel, 2001, 357–404.

[47] J. TATE – On the conjectures of Birch and Swinnerton-Dyer and a geometric analog, Séminaire Bourbaki, Vol. 9, Soc. Math. France, Paris, 1995, Exp. No. 306, 415–440.

[48] R. C. VAUGHAN – *The Hardy–Littlewood Method*, second ed., Cambridge Tracts in Mathematics, vol. 125, Cambridge University Press, Cambridge, 1997.

Arithmetic of Higher-dimensional Algebraic Varieties
(B. Poonen, Yu. Tschinkel, eds.), p. 37–42
Progress in Mathematics, Vol. 226, © 2004 Birkhäuser Boston, Cambridge, MA

RATIONAL POINTS AND ANALYTIC NUMBER THEORY

Roger Heath-Brown
Mathematical Institute, 24-29 St Giles, Oxford, OX1 3LB, UK
E-mail : rhb@maths.ox.ac.uk

Abstract. We discuss connections between analytic number theory and geometry of higher-dimensional algebraic varieties.

1. Introduction

There are a number of distinct ways in which analytic number theory can be used to provide information about rational points on algebraic varieties. Conversely, there are also a number of ways in which hoped for results on the distribution of rational points could be used in classical problems from analytic number theory. Thus the analytic number theorist hopes not only to contribute to the theory of rational points, but also to get something back in return!

Key words and phrases. Rational points, circle method.

2. Contributions from analytic number theory

The best known application of analytic methods to the distribution of rational points is the Hardy–Littlewood circle method. We shall not go into this in detail here. Given a projective variety V defined over $\mathbb{Q}$, the goal is to estimate the number $N(B)$ of rational points on V which have height at most B. When the method succeeds it usually establishes an estimate of the form

$$N(B) \sim \sigma_\infty \prod_p \sigma_p \, B^m, \quad (B \to \infty) \tag{1}$$

for an appropriate integer exponent m. Here σ_p is the p-adic density of points on V, and σ_∞ is the corresponding real density. The method tends to work when the dimension of the variety is reasonably large compared with its degree, and much of the research on the circle method is designed to weaken this constraint. The form of the asymptotic formula (1) is closely related to the Hasse Principle. Indeed one can usually show that the factors σ_∞ and σ_p are all positive when the variety has points everywhere locally. Under these circumstances the asymptotic formula (1) shows in particular that there are infinitely many rational points. Moreover when the circle method works it can usually be adapted to prove a weak approximation result as well.

It is therefore natural to ask what happens for varieties which do not satisfy weak approximation. Should we still expect (1) to hold? Since we expect that whenever we can apply the circle method we can also establish weak approximation, it follows that we should not expect the circle method to succeed on such varieties. As a test case consider the variety defined by

$$\begin{cases} L_1(x_1,x_2)L_2(x_1,x_2) = x_3^2 + x_4^2, \\ L_3(x_1,x_2)L_4(x_1,x_2) = x_5^2 + x_6^2, \end{cases} \tag{2}$$

where the L_i are suitable linear forms. It is known that weak approximation may fail for such varieties, as indeed may the Hasse Principle. Nonetheless we can still establish an asymptotic formula for $N(B)$, which takes the shape

$$N(B) \sim \kappa\sigma_\infty \prod_p \sigma_p \, B^2. \tag{3}$$

The novelty here lies in the factor κ. This is a rational number in the range $\kappa \in [0,2]$. Moreover it is constructed out of p-adic densities for certain "descent varieties" related to (2). One can show that the Hasse Principle fails exactly when $\kappa = 0$. The asymptotic formula (3) is proved essentially by passing to the "descent varieties" and using a variant of the circle method on these. In fact the classical circle method does not quite work and a delicate alternative method has to be used. Full details are given in the author's work **[2]**.

A second area where discussion of rational points encounters issues in analytic number theory is in treatments of the Hasse Principle assuming Schinzel's Hypothesis. Schinzel's Hypothesis concerns the representation of primes by polynomials in one variable. The only instance of the hypothesis which is known to be true is Dirichlet's theorem on primes represented as a linear polynomial $aX + b$. The classical proof of the Hasse Principle for quadratic forms in 4 variables uses Dirichlet's theorem, and one can view more recent developments as an extension of this idea. Thus Colliot-Thélène and Sansuc [**1**] used Schinzel's Hypothesis to prove the following result. Let $a_1, \ldots, a_r$ be non-zero rationals, and let $P_1, \ldots, P_r$ be irreducible polynomials over $\mathbb{Q}$. Then the system of equations

$$0 \neq P_i(t) = x_i^2 - a_i y_i^2, \quad (1 \leq i \leq r),$$

satisfies the Hasse Principle, and weak approximation.

In some problems one can use versions of Schinzel's Hypothesis which refer to polynomials in 2 or more variables. Here there has been recent progress in prime number theory, which enables us to handle primes represented by binary cubic forms, for example. As an application one can show the following, due to Heath-Brown and Moroz [**3**].

Theorem 1. *Let a and b be coprime rational integers satisfying one of the following congruence conditions:*

$$a \text{ or } b = \pm 2 \text{ or } +3 \pmod 9,$$

or

$$a = \pm b \pmod 9.$$

Then there is a nontrivial rational point on the surface

$$x_0^3 + 2x_1^3 + ax_2^3 + bx_3^3 = 0.$$

This is one of the few completely unconditional results in the area. The proof depends on a result of Satgé [**4**, Proposition 3.3] which states that the curve

$$x_0^3 + 2x_1^3 = pZ^3$$

has a non-trivial rational point for any prime $p \equiv 2 \pmod 9$. Satgé's argument uses a Heegner point construction. For the result above it therefore suffices to show that the binary form $ax_2^3 + bx_3^3$ takes a prime value $p = 2 \pmod 9$. Progress in prime number theory is such that this has now been established. With the above congruence constraints on a and b there are always suitable rational values of x_2 and x_3, with denominators 1 or 3. Indeed we can find infinitely many primes of the required form, although the application requires only one such prime.

3. Potential applications to analytic number theory

When analytic number theorists attack problems on rational points they often run into other, related, questions. Consider the counting function

$$N(F;B)=\#\{\mathbf{x}\in\mathbb{Z}^n:\ F(\mathbf{x})=0\ \max_{1\le i\le n}|x_i|\le B\}.$$

For the diagonal cubic hypersurface corresponding to $F(\mathbf{x})=a_1x_1^3+\ldots+a_nx_n^3$ we can give an asymptotic formula for $N(F;B)$ as soon as $n\ge 8$, thanks to work on the circle method by Vaughan [**5**]. However we would like to handle smaller values of n. To deal with the case $n=7$ it would suffice to prove the following conjecture.

Conjecture 2. *We have*

$$N(F_0;B)\ll B^{\theta}$$

for some constant $\theta<7/2$, *where*

$$F_0(\mathbf{x})=x_1^3+x_2^3+x_3^3-x_4^3-x_5^3-x_6^3.$$

This estimate is known to hold for any $\theta>7/2$, by a classical result of Hua, and is believed to hold for any $\theta>3$. From a geometrical viewpoint there is no obvious reason why the form F_0 should be related to F. Nor is it immediately apparent from a geometric viewpoint how an upper estimate for $N(F_0;B)$ can lead to an asymptotic formula for $N(F;B)$. However these relationships are quite simple from the viewpoint of the circle method, which is what makes it such a distinctive and useful tool.

One may also ask what happens for the form

$$F_0(\mathbf{x})=x_1^d+x_2^d+x_3^d-x_4^d-x_5^d-x_6^d.$$

Again we conjecture that any $\theta>3$ is admissible. In fact one can take $\theta=\theta_d<7/2$ if d is large enough, but attempts to reduce the permissible size of d encounter some purely geometric questions. Typical of these is—what low degree curves are contained in the variety $F_0(\mathbf{x})=0$? For example when $d\ge 5$ the only lines are those that lie in trivial planes of the type $x_1=x_4, x_2=x_5, x_3=x_6$. It would be good, for example, to know that there were no curves of degree at most 4, other than those lying in such planes. We should remark that a resolution of the problems described by Salberger in his lecture would also result in significant progress in reducing the exponent θ.

As an example of a potential application to other areas of the subject, consider the variety $V(k,s) \subset \mathbb{P}^{2s-1}$ defined by the equations

$$x_1^j + \cdots + x_s^j = y_1^j + \cdots + y_s^j, \quad (1 \leqslant j \leqslant k).$$

This is a cone with vertex $(1,\ldots,1)$. The counting function $N(B)$ for this variety is the subject of Vinogradov's Mean Value Theorem. Upper bounds for $N(B)$ have various applications in analytic number theory, to the estimation of exponential sums in the first instance, and thence to bounds on the Riemann Zeta-function and the error term in the Prime Number Theorem. A great deal of effort has gone into improving the original upper bounds established by Vinogradov. It is not hard to show that

$$N(B) \gg \max\{B^s, B^{2s-k(k+1)/2}\}$$

for all $s,k \geq 1$. Moreover, if $s \leq k$ then all points have $x_1,\ldots,x_s$ a permutation of $y_1,\ldots,y_k$, so that $N(B) \sim c(k,s)B^s$ in this case. Moreover Vaughan and Wooley **[6]** have established the same asymptotic formula for $s = k+1$. On the other hand we know that

$$N(B) \sim c(k,s)B^{2s-k(k+1)/2}$$

for

$$s \geq s_0(k) = k^2(\log k + 2\log\log k + O(1))$$

(see Wooley **[7]**). It would be good to know how $N(B)$ behaves for values of s of intermediate size. It seems likely that a better understanding of the geometry of $V(k,s)$ would help. As an example, when $k=4$ and $s=6$ we ask the following. Let L be a linear space of projective dimension l, and C an irreducible component of $V(4,6) \cap L$. Assume that C is not contained in the 'diagonal' set where $(x_1,\ldots,x_6)$ is a permutation of $(y_1,\ldots,y_6)$. Then is $\dim C \leqslant (2l-1)/3$?

References

[1] J.-L. Colliot-Thélène & J.-J. Sansuc – Sur le principe de Hasse et l'approximation faible, et sur une hypothèse de Schinzel, *Acta Arith.* **41** (1982), no. 1, 33–53.

[2] D. R. Heath-Brown – Linear relations amongst sums of two squares, to appear.

[3] D. R. Heath-Brown & B. Z. Moroz – On the representation of primes by cubic polynomials in two variables, to appear in *Proc. London Math. Soc.*

[4] P. Satgé – Un analogue du calcul de Heegner, *Invent. Math.* **87** (1987), no. 2, 425–439.

[5] R. C. Vaughan – On Waring's problem for cubes, *J. Reine Angew. Math.* **365** (1986), 122–170.

[6] R. C. VAUGHAN & T. D. WOOLEY – A special case of Vinogradov's mean value theorem, *Acta Arith.* **79** (1997), no. 3, 193–204.

[7] T. D. WOOLEY – Some remarks on Vinogradov's mean value theorem and Tarry's problem, *Monatsh. Math.* **122** (1996), no. 3, 265–273.

Arithmetic of Higher-dimensional Algebraic Varieties
(B. Poonen, Yu. Tschinkel, eds.), p. 43–60
Progress in Mathematics, Vol. 226, © 2004 Birkhäuser Boston, Cambridge, MA

WEAK APPROXIMATION ON ALGEBRAIC VARIETIES

David Harari
D.M.A., E.N.S., 45 rue d'Ulm, 75005 Paris, France
E-mail : David.Harari@ens.fr

Abstract. We give an introduction to the study of weak approximation on algebraic varieties.

1. Introduction

This survey paper consists of two parts. The first is an introduction to the topic: definitions and first properties, classical examples and counterexamples. The second is about cohomological methods : Brauer–Manin obstruction, descent theory of Colliot-Thélène and Sansuc, nonabelian descent theory.

Throughout, we let k be a number field with algebraic closure $\bar{k}$ and absolute Galois group $\Gamma = \mathrm{Gal}(\bar{k}/k)$. We denote by Ω_k the set of all places of k (including the archimedean places) and by k_v the completion of k at the place v. The ring of integers of k (resp. k_v for v finite) is denoted by $\mathscr{O}_k$ (resp. $\mathscr{O}_v$).

Acknowledgements: This paper is an expanded version of two talks given at the workshop *Rational and integral points on higher-dimensional varieties*

Key words and phrases. Weak approximation, Brauer–Manin obstruction, torsors.

(Palo-Alto, December 2002). The author thanks the AIM and the organizers of this conference for their warm hospitality.

2. Classical results

2.1. Basic facts. The following result is a refinement of the Chinese remainder theorem.

***Theorem 2.1.1* (Weak Approximation).** *Let $\Sigma \subset \Omega_k$ be a finite set of places of k. Let $\alpha_v \in k_v$ for $v \in \Sigma$. Then there is an $\alpha \in k$ which is arbitrarily close to α_v for $v \in \Sigma$.*

For a complete proof, see [**27**], Theorem 1 p. 35. One reformulation of this theorem is as follows: the diagonal embedding $k \hookrightarrow \prod_{v\in\Omega_k} k_v$ is dense, the product being equipped with the product of the v-adic topologies.

We have a slight variant: $\mathbf{P}^1(k)$ is dense in $\prod_{v\in\Omega_k} \mathbf{P}^1(k_v)$ (here we have just replaced the affine line $\mathbf{A}^1_k$ by the projective line $\mathbf{P}^1_k$).

Definition 2.1.2. Let X/k be a geometrically integral algebraic variety. Then X satisfies *weak approximation* if given $\Sigma \subset \Omega_k$ a finite set of places and $M_v \in X(k_v)$ for $v \in \Sigma$, there exists a k-rational point $M \in X(k)$ which is arbitrarily close to M_v for $v \in \Sigma$.

Care must be taken if $\prod_{v\in\Omega_k} X(k_v)$ is empty; by convention, we will say that in this case X satisfies weak approximation even though $X(k)$ is empty. When $\prod_{v\in\Omega_k} X(k_v) \neq \emptyset$ but $X(k) = \emptyset$, one says that the *Hasse principle* fails[(1)].

We see that weak approximation is equivalent to the statement that $X(k)$ is dense in $\prod_{v\in\Omega_k} X(k_v)$ (equipped with the product of the v-adic topologies).

Remark 2.1.3. Let $\mathscr{X} \to \operatorname{Spec} \mathscr{O}_k$ be a flat model of X over $\operatorname{Spec} \mathscr{O}_k$; denote by $X(\mathbb{A}_k)$ the set of *adelic points* of X, that is, the restricted product of the sets $X(k_v)$ ($v \in \Omega_k$) with respect to the sets $\mathscr{X}(\mathscr{O}_v)$ (it is clearly independent of the choice of $\mathscr{X}$). If X is projective, then $X(\mathbb{A}_k) = \prod_{v\in\Omega_k} X(k_v)$ and weak approximation is equivalent to strong approximation, namely, $X(k)$ is dense in $X(\mathbb{A}_k)$ for the adelic topology.

Remark 2.1.4. Let X, X' be smooth. Assume that X is k-birational to X'. Then X satisfies weak approximation if and only if X' satisfies weak

(1) As Swinnerton-Dyer says, this corresponds to weak approximation failing dramatically.

approximation. This an easy consequence of the implicit function theorem for k_v (a reference for this well known result is [**38**], p. 85).

We can therefore speak about weak approximation for a function field $k(X)$: this means that weak approximation holds for any smooth (projective) model of X (such a model exists by Hironaka's Theorem on resolution of singularities).

Example 2.1.5. It follows immediately from Theorem 2.1.1 that the affine line, the projective line, and more generally the affine space $\mathbf{A}_k^n$ and the projective space $\mathbf{P}_k^n$ satisfy weak approximation, as does any k-rational variety (see Remark 2.1.4), e.g., a smooth quadric with a k-point.

2.2. More examples. We begin with the most classical example of "local-global principle":

Theorem 2.2.1. *Let $Q \subset \mathbf{P}_k^n$ be a smooth projective quadric. Then Q satisfies weak approximation.*

Here we do not assume that there is a k-rational point. This is the difficult part, proving the *Hasse principle*, that is: the existence of points everywhere locally implies the existence of a rational point. In the case of quadrics, this is the famous Hasse–Minkowski theorem (proven by Hasse around 1924). A detailed proof of this theorem for $k = \mathbf{Q}$ (the general case works the same way) can be found in Serre's book [**37**].

Here are some other results for complete intersections in $\mathbf{P}_k^n$:

Example 2.2.2. A smooth intersection of two quadrics $X \subset \mathbf{P}_k^n$ satisfies weak approximation if $n \geqslant 8$, or if $n \geqslant 4$ and there exists a pair of skew conjugate lines on X (Colliot-Thélène, Sansuc, Swinnerton-Dyer 1987 [**12**], Th. 10.1 and Prop. 5.2).

Example 2.2.3. Châtelet surfaces: let V be the affine surface $y^2 - az^2 = P(x)$, where $\deg P = 4$, $a \in k^* - k^{*2}$. If P is irreducible, then a smooth and projective model X of V satisfies weak approximation ([**12**], Th. 8.11).

Example 2.2.4. Let $X \subset \mathbf{P}_k^n$ a smooth cubic hypersurface, then weak approximation holds if $n \geqslant 16$ (Skinner 1997 [**40**]).

An interesting fact is that the proofs of the three previous results use different tools. The first statement is proved with the fibration method (see Subsection 2.3), the second one with descent theory (see Subsection 3.2). To deal with Example 2.2.4 one needs the Hardy–Littlewood circle method, which is especially efficient when the number of variables is substantially bigger than the degree. We shall not discuss further this analytic technique in these notes.

There are also results for linear algebraic groups.

Example 2.2.5. Let K/k be a cyclic field extension. Define the torus T by the equation

$$N_{K/k}(x_1\omega_1 + \cdots + x_r\omega_r) = 1$$

in the variables $x_1, \ldots, x_r$, where $(\omega_1, ..., \omega_r)$ is a basis of K/k. Then T satisfies weak approximation ([**47**], 11.5). The Hasse principle for equations $N_{K/k}(x_1\omega_1 + \cdots + x_r\omega_r) = a$, $a \in k^*$ goes back to Hasse (1924).

Example 2.2.6. If T is a k-torus, and $\dim T \leqslant 2$, then T satisfies weak approximation because T is k-rational (Voskresenskiĭ, [**47**], IV.9).

Example 2.2.7. If G is a semisimple, simply connected linear k-group, then G satisfies weak approximation. This is due to Kneser ([**25**], [**26**]), Harder ([**22**]), and Platonov ([**31**], [**32**]). The same result holds for semisimple adjoint groups (see [**33**], Theorem 7.8).

We conclude this subsection by two classical conjectures.

Conjecture 2.2.8. A smooth intersection of 2 quadrics in $\mathbf{P}^n$ for $n \geqslant 5$ satisfies weak approximation.

This is known when the variety has a rational point ([**12**], Th. 3.11). Thus the difficulty is now to prove the Hasse principle.

Conjecture 2.2.9. A smooth cubic hypersurface (of dimension at least 3) satisfies weak approximation.

Here the Hasse principle is known for *diagonal* hypersurfaces over $\mathbf{Q}$ (and more generally over any number field k which does not contain the primitive cube roots of 1) if we assume the finiteness of Tate–Shafarevich groups of elliptic curves (Swinnerton-Dyer [**46**]).

We shall see later (Subsection 2.4) that the similar conjectures for surfaces are false.

2.3. The fibration method. The general idea of this method is quite natural: consider a pencil of varieties satisfying weak approximation over a base which also satisfies weak approximation. Does this imply that weak approximation holds for the total space of the fibration? In general, the answer is no (even for examples for conic bundles over $\mathbf{P}^1_k$, see Example 2.4.3 below) but with additional assumptions the result becomes true. Here is a useful statement in this direction:

Theorem 2.3.1. *Let $p : X \to B$ be a projective, flat surjective morphism with X smooth. Assume that*

1. *B is projective and satisfies weak approximation.*
2. *Almost all k-fibres of p satisfy weak approximation.*
3. *All fibres of p are geometrically integral.*

Then X satisfies weak approximation.

(Here almost all means on a Zariski-dense open subset; the hypothesis X smooth is not essential, but it makes the statement simpler; one can also weaken the third assumption by replacing "geometrically integral" with "split" [(2)]).

There are refinements when B is the projective space: one can allow degenerate fibres on one hyperplane (using the strong approximation theorem for the affine space), see [**41**].

The idea of this method goes back to the proof of Hasse–Minkowski Theorem (more precisely, the step consisting of going from four variables to five). The first subtle application of Theorem 2.3.1 appeared in [**12**] for intersection of two quadrics in $\mathbf{P}^n$: when $n \geqslant 8$ (here the authors used a fibration in Châtelet surfaces), and also when $n \geqslant 5$ when the intersection of two quadrics contains a pair of skew conjugate lines (the point is to go from $n = 4$ to $n \geqslant 5$ by induction). Another example is provided by cubic hypersurfaces of dimension $\geqslant 4$ with 3 conjugate singular points (Colliot-Thélène, Salberger 1989 [**9**]).

Sketch of proof of Theorem 2.3.1. Start with a smooth k_v-point M_v for any $v \subset \Omega_k$ on X and fix a finite set of places Σ. Project M_v to $P_v := p(M_v) \in B(k_v)$. Using weak approximation on B, we can approximate P_v by $P \in B(k)$ for $v \in \Sigma$. Consider the fiber $X_P := p^{-1}(P) \subset X$; then X_P has a k_v-point M'_v close to M_v for $v \in \Sigma$ by the implicit function theorem. To apply weak approximation on X_P, it remains to check that $X_P(k_v) \neq \emptyset$ for each $v \notin \Sigma$; this is possible if Σ is sufficiently large by the Weil estimates: here we use that all k-fibers are geometrically irreducible, which implies that the same holds for the reduction mod v of X_P, for a sufficiently large v (independent of P). $\square$

2.4. Some counterexamples. It has been known for a long time that for example elliptic curves do not satisfy weak approximation (the defect of weak

(2) A k-variety is *split* if it contains a nonempty Zariski open subset which is geometrically integral. This notion was introduced by Skorobogatov in [**42**].

approximation is described by Cassels' dual exact sequence [**5**]; see also Theorem 2.4.5 below). It is more difficult to find counterexamples to weak approximation among rational varieties (that is: varieties X such that $\overline{X} := X \times_k \bar{k}$ is $\bar{k}$-birational to the projective space). Here are some examples of this situation:

Example 2.4.1. Some cubic surfaces do not satisfy the Hasse principle: the surface $5x^3+9y^3+10z^3+12w^3=0$ is a counterexample (Cassels and Guy 1966 [**6**]). The existence of a rational point does not imply weak approximation; a counterexample is given by the smooth locus of the surface defined in $\mathbf{P}^3_{\mathbf{Q}}$ by the equation

$$t(x^2+y^2) = (4z-7t)(z^2-2t^2)$$

(Swinnerton-Dyer 1962 [**45**]).

Example 2.4.2. In general, a smooth intersection X of two quadrics in $\mathbf{P}^4_k$ does not satisfy the Hasse principle, and weak approximation does not hold even if $X(k) \neq \emptyset$. For example, the variety defined in $\mathbf{P}^4_{\mathbf{Q}}$ by the equations

$$x_0x_1 - (x_2^2 - 5x_3^2) = 0$$
$$(x_0+x_1)(x_0+2x_1) - (x_2^2 - 5x_4^2) = 0$$

does not satisfy the Hasse principle (Birch and Swinnerton-Dyer [**2**]) and the variety X:

$$x_0x_1 - (x_2^2 + x_3^2) = 0$$
$$(4x_1 - 3x_0)(4x_0 - x_1) - (x_2^2 + x_4^2) = 0$$

is a counterexample to weak approximation with $X(\mathbf{Q}) \neq \emptyset$ ([**12**], 15.5).

Example 2.4.3. Let us explain how to construct counterexamples to weak approximation among Châtelet surfaces (which are special cases of conic bundles over $\mathbf{P}^1_k$). Consider the equation

$$X : y^2 + z^2 = f_1(x)f_2(x) \neq 0,$$

over the field $k = \mathbf{Q}$ of rational numbers where $\deg(f_1) = \deg(f_2) = 2$ and $\gcd(f_1, f_2) = 1$. Set $K = \mathbf{Q}(\sqrt{-1})$ and $K_v = K \otimes_{\mathbf{Q}} \mathbf{Q}_v$; then there exists a finite set $\Sigma_0 \subset \Omega_k$ such that if $v \notin \Sigma_0$ and $M_v \in X(\mathbf{Q}_v)$, then $f_1(M_v)$ is a norm of $K_v/\mathbf{Q}_v$ (use a computation with valuations). If one can find a $v_0 \in \Sigma_0$ with the properties:

(i) there exists an M_{v_0} such that $f_1(M_{v_0})$ is not a local norm,

(ii) for $v \neq v_0$ there exists an M_v such that $f_1(M_v)$ is a local norm, then there is no weak approximation, thanks to the global reciprocity law of class field theory, namely the exactness of the sequence

$$\mathbf{Q}^*/NK^* \to \bigoplus_{v\in\Omega_k} \mathbf{Q}_v^*/NK_v^* \to \mathbf{Z}/2.$$

An explicit example of this situation is given by the equation

$$y^2 + z^2 = ((x-2)^2 - 3)((x+2)^2 + 3).$$

Here there is an obvious rational point $P = (0,0,1)$ such that $f_1(P) = 1$ is a global norm, hence this gives for any v a local point P_v such that $f_1(P_v)$ is a local norm. For $v = 2$ it is easy to construct a local point M_v such that $f_1(M_v)$ is not a local norm (take $x = 2$ and use [**37**], p. 39).

It is even possible to obtain a counterexample to the Hasse principle, e.g., $y^2 + z^2 = (x^2 - 2)(3 - x^2)$ (Iskovskih [**24**]). In this example, $f_1(M_v)$ is always a norm of $K_v/\mathbf{Q}_v$, except for $v = 2$, where it cannot be a norm, hence by the reciprocity law there is no rational point.

Example 2.4.4. The results of Example 2.2.5 cannot be extended to arbitrary tori. Let K/k be a biquadratic extension, then there are counterexamples to weak approximation for T: $N_{K/k}(x_1w_1 + \cdots + x_4w_4) = 1$, where $w_1, \ldots, w_4$ is a basis of K/k; this holds e.g. for $k = \mathbf{Q}$, $K = \mathbf{Q}(\sqrt{-1}, \sqrt{2})$, see [**47**], 11.6., Example 3. Neither does weak approximation hold for arbitrary semi-simple connected linear groups; the first counterexample was given by Serre: it consists of the group $\mathbf{R}_{K/\mathbf{Q}}\mathrm{SL}_8/(\mathbf{Z}/8)$, where K is the extension of $\mathbf{Q}$ obtained by adjoining a primitive 8th root of unity, and $\mathbf{R}_{K/\mathbf{Q}}$ denotes Weil's restriction of scalars from K to $\mathbf{Q}$.

All the previous counterexamples are related to reciprocity laws in global class field theory. In Subsection 3.2 we will describe a general framework for these, namely the Brauer–Manin obstruction.

We conclude this section with the following negative result ([**30**])

Theorem 2.4.5 **(Minchev).** *Let X be a projective and smooth k-variety and put $\overline{X} = X \otimes \bar{k}$. Assume that the geometric étale fundamental group $\pi_1(\overline{X})$ is not trivial and that $X(k) \neq \emptyset$. Then X does not satisfy weak approximation.*

Sketch of proof. Enlarge the situation over $\operatorname{Spec} \mathscr{O}_{k,\Sigma_0}$ where Σ_0 is a finite set of places. By assumption, there is a nontrivial geometrically connected covering $Y \to X$, which for models gives $\mathscr{Y} \to \mathscr{X}$. Take an arbitrary $M \subset X(k)$, then the fiber Y_M can be written as $Y = \operatorname{Spec} L$ where L is an étale algebra $L = k_1 \times \cdots \times k_r$; each k_i is unramified outside Σ_0, hence only finitely many k_i are possible (by Hermite's Theorem, cf.[**27**], Theorem 5 p. 121). Find $v \notin \Sigma_0$

with v totally split for each k_i (such a v does exist by the Cebotarev Density Theorem, [**27**] Theorem 10 p. 169); find M_v such that the fiber of Y at M_v is not totally split (this is possible because Y is geometrically connected, via a "geometric" Cebotarev-like Theorem as in [**14**], Lemma 1.2). Then M_v cannot be approximated by a rational point M (use Krasner's Lemma, [**27**], Proposition 3 p. 43). □

Here the obstruction to weak approximation cannot always be related to a reciprocity law as above. See Subsection 3.5 and [**16**].

3. Cohomological methods

Let X be a smooth and geometrically integral variety over k. From now on suppose that X is projective. We denote by $\overline{X(k)}$ the closure of $X(k)$ in $\prod_{v\in\Omega_k} X(k_v) = X(\mathbb{A}_k)$. Here our aim is to:

(i) explain the counterexamples to weak approximation;
(ii) find "intermediate" sets E between $\overline{X(k)}$ and $X(\mathbb{A}_k)$;
(iii) in some cases, prove that $E = \overline{X(k)}$.

3.1. General setting.

Let G/k be an algebraic group (usually linear, but not necessarily connected, e.g., G finite). If G is commutative, define the étale cohomology groups $H^i(X,G)$ ($i = 1,2$; the cohomological dimension of a nonarchimedean local field is two, making the higher cohomology groups uninteresting). In general, we have only the pointed set $H^1(X,G)$ (defined by Cech cocycles for the étale topology). If $X = \operatorname{Spec} k$, $H^1(X,G) = H^1(\Gamma, G(\bar{k}))$. If G is linear, then $H^1(X,G)$ corresponds to G-torsors (G-principal homogeneous spaces) over X up to isomorphism (cf. [**29**], III.4 and [**44**], Chapter 2).

Take $f \in H^i(X,G)$, and define

$$X(\mathbb{A}_k)^f = \{(M_v) \in X(\mathbb{A}_k) : (f(M_v)) \in \operatorname{Im}[H^i(k,G) \to \prod_{v\in\Omega_k} H^i(k_v,G)]\}.$$

Obviously $X(k) \subset X(\mathbb{A}_k)^f$. We will see that in many cases $\overline{X(k)} \subset X(\mathbb{A}_k)^f$.

Example 3.1.1. Let $\operatorname{Br} X = H^2(X, \mathbf{G}_m)$ be the (cohomological) Brauer group of X; define the *Brauer–Manin* set of X by

$$X(\mathbb{A}_k)^{\operatorname{Br}} = \bigcap_{f\in \operatorname{Br} X} X(\mathbb{A}_k)^f.$$

Then $\overline{X(k)} \subset X(\mathbb{A}_k)^{\mathrm{Br}}$. Indeed X is projective and $\mathrm{Br}\,\mathscr{O}_v = 0$ for each finite place v ([**29**], IV.2.13), so for each $\alpha \in \mathrm{Br}\,X$ there exists a finite set of places Σ_0 (the places of bad reduction for X or α) such that for any $v \notin \Sigma$ and any $M_v \in X(k_v)$, we have $\alpha(M_v) = 0$. Let $(P_v) \in X(\mathbb{A}_k)$; if $P \in X(k)$ is sufficiently close to P_v for $v \in \Sigma_0$ then $\alpha(P) = \alpha(P_v)$ for any $v \in \Omega_k$. Thus $\sum_{v\in\Omega_k} j_v(\alpha(P_v)) = 0$ because P is rational.

Manin ([**28**]) showed in 1970 that for a genus one curve with finite Tate–Shafarevich group, the condition $X(\mathbb{A}_k)^{\mathrm{Br}} \neq \emptyset$ implies the existence of a rational point. A similar statement for abelian varieties is true and there is also an analog about weak approximation ([**48**]).

Remark 3.1.2. One does not get any refinement of the Brauer–Manin conditions by enlarging the ground field. Indeed let L/k be a finite field extension, and suppose that an adelic point $(M_v)_{v\in\Omega_k}$ belongs to $X(\mathbb{A}_k)^{\mathrm{Br}}$. Let $(M_w)_{w\in\Omega_L}$ be the image of $(M_v)_{v\in\Omega_k}$ in $X_L(\mathbb{A}_L)$ (via the natural map $A_k \hookrightarrow A_L$), where $X_L := X \times_k L$. Then $(M_w)_{w\in\Omega_L}$ belongs to $X_L(\mathbb{A}_L)^{\mathrm{Br}}$: to see this, consider the corestriction $\alpha \in \mathrm{Br}\,X$ of an element $\alpha_L \in \mathrm{Br}\,X_L$, and note that by local class field theory, the corestriction map $\mathrm{Br}\,L_w \to \mathrm{Br}\,k_v$ induces a commutative diagram

$$\begin{array}{ccc} \mathrm{Br}\,L_w & \xrightarrow{j_w} & \mathbf{Q}/\mathbf{Z} \\ \downarrow & & \downarrow{\scriptstyle \mathrm{Id}} \\ \mathrm{Br}\,k_v & \xrightarrow{j_v} & \mathbf{Q}/\mathbf{Z} \end{array}$$

for any place w of L dividing $v \in \Omega_k$.

Example 3.1.3. Let $f : Y \to X$ be a Galois, geometrically connected, nontrivial étale covering with group G. We can view f as an element of $H^1(X, G)$, where G is considered as a constant group scheme. Essentially the proof of Minchev's result (Theorem 2.4.5) consists in showing $\overline{X(k)} \subset X(\mathbb{A}_k)^f$ (this is the step which uses Hermite's Theorem), and then finding an $(M_v) \notin X(\mathbb{A}_k)^f$, by a geometric Cebotarev Theorem.

Remark 3.1.4. Since $H^3(k, \mathbf{G}_m) = 0$ for any number field k, the Hochschild–Serre spectral sequence $H^p(k, H^q(\overline{X}, \mathbf{G}_m)) \Rightarrow H^{p+q}(X, \mathbf{G}_m)$ yields an exact sequence

$$\mathrm{Br}\,k \to \mathrm{Ker}\,[\mathrm{Br}\,X \to \mathrm{Br}\,\overline{X}] \to H^1(k, \mathrm{Pic}\,\overline{X}) \to 0$$

where $\overline{X} = X \times_k \bar{k}$. Denote by $\mathrm{Br}\,X/\mathrm{Br}\,k$ the quotient of $\mathrm{Br}\,X$ by the image of the canonical map $\mathrm{Br}\,k \to \mathrm{Br}\,X$ (even though this map is not necessarily injective if $X(k) = \emptyset$). If X is rational, then $\mathrm{Br}\,X/\mathrm{Br}\,k = H^1(k, \mathrm{Pic}\,\overline{X})$ is finite. Since for a constant element f of $\mathrm{Br}\,X$ (i.e., an element coming from $\mathrm{Br}\,k$) we obviously have $X(\mathbb{A}_k) = X(\mathbb{A}_k)^f$, we obtain that in this case $X(\mathbb{A}_k)^{\mathrm{Br}}$ is (at least in theory) "computable".

***Theorem 3.1.5* (Harari, Skorobogatov).** *Let X be a projective, smooth and geometrically integral k-variety, G a linear k-group and $f \in H^1(X,G)$. Then $\overline{X(k)} \subset X(\mathbb{A}_k)^f$ (and $X(\mathbb{A}_k)^f$ is "computable").*

The idea of the proof is to apply Borel–Serre Finiteness Theorem ([**39**], III.4.6) instead of Hermite's Theorem. See [**18**] (Th. 4.7) or [**44**] (5.3) for the details.

3.2. Abelian descent theory. This was developed by Colliot-Thélène and Sansuc [**11**], and recently completed by Skorobogatov [**43**]. Recall that a *group of multiplicative type* S over k is a commutative linear k-group which is an extension of a finite group by a torus. The *module of characters* of S is the abelian group $\widehat{S} = \mathrm{Hom}(\overline{S}, \mathbf{G}_m)$, equipped with the action of the Galois group Γ, where $\overline{S} = S \times_k \bar{k}$. One of the main results of the theory consists of the following:

Theorem 3.2.1. *Let X be a projective, smooth, and geometrically integral k-variety. Define*

$$X(\mathbb{A}_k)^{\mathrm{Br}_1} = \bigcap_{f \in \mathrm{Br}_1 X} X(\mathbb{A}_k)^f$$

where $\mathrm{Br}_1 X = \mathrm{Ker}(\mathrm{Br}\, X \to \mathrm{Br}\, \overline{X})$. *Assume further that $X(\mathbb{A}_k)^{\mathrm{Br}_1} \neq \emptyset$. Then:*

1. *We have*

$$X(\mathbb{A}_k)^{\mathrm{Br}_1} = \bigcap_{\substack{f \in H^1(X,S) \\ S \text{ of multiplicative type}}} X(\mathbb{A}_k)^f.$$

2. *Assume further that* $\mathrm{Pic}\, \overline{X}$ *is of finite type, let S_0 be the group of multiplicative type with module of characters* $\mathrm{Pic}\, \overline{X}$*; then there exists a torsor $f_0 : Y \to X$ under S_0 (a* universal torsor*), such that*

$$X(\mathbb{A}_k)^{\mathrm{Br}_1} = X(\mathbb{A}_k)^{f_0}.$$

Intuitively, universal means "as nontrivial as possible"; in particular if there exists a universal torsor $f_0 : Y \to X$, then for any torsor $f : Z \to X$ under S_0 there exists a unique morphism of X-torsors $\varphi : Y \to Z$ such that $f_0 = f \circ \varphi$. See [**44**], 2.3.3. for more details about the definition of universal torsors (this notion is due to Colliot-Thélène and Sansuc [**11**]).

Theorem 3.2.1 is difficult: see Skorobogatov's book [**44**] for a complete account of the subject. One of the ideas is to recover the Brauer group of X (mod $\operatorname{Br} k$) by making cup-products $[Y] \cup a$, where $a \in H^1(k, \widehat{S}_0)$ and $[Y]$ is the class of Y in $H^1(X, S_0)$. Another step (which takes much work to carry out) is to show that the condition $X(\mathbb{A}_k)^{\operatorname{Br}_1} \neq \emptyset$ implies the existence of a universal torsor.

Now assume that X is a rational variety, so $X(\mathbb{A}_k)^{\operatorname{Br}} = X(\mathbb{A}_k)^{\operatorname{Br}_1}$ (since $\operatorname{Br} \overline{X} = 0$). Assume $X(\mathbb{A}_k)^{\operatorname{Br}} \neq \emptyset$. Consider a universal torsor $f : Y \to X$. If $\sigma \in H^1(k, S_0)$, one can define the *twisted torsor* $f^\sigma : Y^\sigma \to X$ where

$$[Y^\sigma] = [Y] - \sigma \in H^1(X, S_0).$$

Then

$$X(\mathbb{A}_k)^f = \bigcup_{\sigma \in H^1(k, S_0)} f^\sigma(Y^\sigma(\mathbb{A}_k)).$$

The universal torsors are precisely the torsors $Y^\sigma, \sigma \in H^1(k, S_0)$. If one can prove that they satisfy weak approximation, then $\overline{X(k)} = X(\mathbb{A}_k)^f = X(\mathbb{A}_k)^{\operatorname{Br}}$, which means that the Brauer–Manin obstruction to weak approximation is the only one for X. In practice it is important to obtain *explicit* equations for the universal torsors (this is done in [**11**] Th. 2.3.1, see also [**44**], 4.3.1). Once the universal torsors are described by these equations, one can hope to prove (e.g., using fibration methods) that weak approximation holds for them because their Brauer group is trivial (that is: consists of constant elements), hence the Brauer–Manin obstruction vanishes for them. Here are some examples where this approach works completely:

Example 3.2.2. Consider a Châtelet surface: $y^2 - az^2 = P(x)$, $a \in k^* - k^{*2}$, $\deg P = 4$. Colliot-Thélène, Sansuc, Swinnerton-Dyer showed in [**12**] (Th. 8.11) that for a projective and smooth model X, the equality $\overline{X(k)} = X(\mathbb{A}_k)^{\operatorname{Br}}$ holds. Here weak approximation on universal torsors follows from the similar statement for intersections of two quadrics in $\mathbf{P}^n_k$ ($n \geqslant 4$) containing a pair of skew conjugate lines (cf. Example 2.2.2).

If P is irreducible, then $\operatorname{Br} X / \operatorname{Br} k = 0$ ([**44**], Prop. 7.1.1), so X satisfies weak approximation. It is worth noting that it seems impossible to deal with this special case without using descent, even though the Brauer–Manin obstruction already vanishes on X.

If P is reducible, we can have a counterexample to weak approximation, cf. Example 2.4.3. Here the obstruction is given by the Hilbert symbol $f = (a, f_1)$. This reinterprets the reciprocity obstruction explained in Example 2.4.3 as a special case of the Brauer–Manin obstruction.

Example 3.2.3. Let X be a conic bundle surface over $\mathbf{P}^1$ with at most 5 degenerate fibers. Then $\overline{X(k)} = X(\mathbb{A}_k)^{\mathrm{Br}}$. Works by Salberger ([**34**]) and Colliot-Thélène ([**8**]) covered at most 4 degenerate fibers via the descent method. Salberger and Skorobogatov ([**35**]) treated the case of 5 bad fibers, using descent and K-theory. It is widely believed that the Brauer–Manin obstruction to weak approximation is the only one for a conic bundle over $\mathbf{P}^1$ with an arbitrary number of bad fibers. This was proved by Serre (unpublished) under Schinzel's hypothesis[(3)] in 1992 (Serre's proof holds more generally for families of Severi-Brauer varieties over $\mathbf{P}^1_k$). Another proof and several extensions of his result (in particular an unconditional zero-cycle version) can be found in [**13**]. The first application of Schinzel's hypothesis to rational points on algebraic varieties was given by Colliot-Thélène and Sansuc ([**10**]) in the case of surfaces $y^2 - az^2 = P(x)$ over $\mathbf{Q}$.

We conclude this subsection with the following general result about algebraic groups ([**36**]):

***Theorem 3.2.4* (Sansuc).** *Let G be a connected linear algebraic k-group and X a smooth compactification of G. Then the Brauer–Manin obstruction to weak approximation on X is the only one:*

$$\overline{X(k)} = X(\mathbb{A}_k)^{\mathrm{Br}}.$$

This result was extended by Borovoi ([**4**], [**3**]) to homogeneous spaces of connected linear groups with connected stabilizers (resp. of simply connected semisimple groups with abelian stabilizers). The case of flag varieties G/P goes back to Harder (1968).

3.3. Open descent. In the previous subsection, we considered descent over projective varieties. But the general results of the theory still hold over a geometrically integral variety U as soon as the only invertible functions on $\overline{U}$ are constant; this is often useful for obtaining torsors described by nice equations. Descent over an open subset U of a projective variety X was introduced in 2000 by Colliot-Thélène and Skorobogatov. In particular they showed:

***Proposition 3.3.1* ([7], Prop. 1.1).** *Let X be a smooth, proper and geometrically integral k-variety. Let U be a nonempty Zariski open subset of X. Assume that $\mathrm{Br}\,U/\mathrm{Br}\,k$ is of finite index in $\mathrm{Br}\,X/\mathrm{Br}\,k$. Then $U(\mathbb{A}_k)^{\mathrm{Br}}$ is dense in $X(\mathbb{A}_k)^{\mathrm{Br}}$ in the adelic topology.*

(3) Schinzel's hypothesis is a (rather wild) generalization of Dirichlet's Theorem on primes in an arithmetic progression, see [**13**].

Note that elements of $\operatorname{Br} U$ do not necessarily belong to $\operatorname{Br} X$. This proposition is a consequence of the "formal lemma" ([**19**], 2.6.1; see also the next subsection). With the help of Proposition 3.3.1, it is sometimes possible to prove that $\overline{X(k)} = X(\mathbb{A}_k)^{\mathrm{Br}}$ with a descent over a well chosen U instead of the whole X; this works for example for certain varieties fibred over the projective line ([**7**], Th. A and B).

Another application of the open descent is the following recent result ([**23**]); a new tool is to use the circle method to prove that universal torsors over U satisfy weak approximation.

Theorem 3.3.2 **(Heath-Brown, Skorobogatov).** *Let $K/\mathbf{Q}$ be a finite field extension. Consider the affine variety V, defined by a norm-type equation*

$$t^{a_0}(1-t)^{a_1} = N_{K/k}(x_1\omega_1 + \cdots + x_r\omega_r)$$

where $(\omega_1, \ldots, \omega_r)$ is a basis of $K/\mathbf{Q}$, a_0, a_1 are two coprime integers, and $t, x_1, \ldots, x_r$ are variables. Then the Brauer–Manin obstruction to weak approximation is the only one for a smooth and projective model X of V.

3.4. Back to fibration methods. If $p : X \to B$ is a fibration, we saw that if the base and the fibres satisfy weak approximation, then, under certain circumstances, X satisfies weak approximation.

Here we consider a projective, surjective morphism $p : X \to \mathbf{P}^1$ with smooth generic fibre X_η. Assume also that X_η has a $k(\eta)$-point (this technical condition is satisfied in most applications, e.g., if X_η is geometrically rationally connected, by a recent result of Graber, Harris and Starr [**15**]). A natural question is the following: If $\overline{X_P(k)} = X_P(\mathbb{A}_k)^{\mathrm{Br}}$ for almost all fibres X_P, $P \in \mathbf{P}^1(k)$, can one prove that $\overline{X(k)} = X(\mathbb{A}_k)^{\mathrm{Br}}$? The following result ([**19**], [**20**]) gives a partial answer to this question.

Theorem 3.4.1. *With the notations and assumptions as above, we have*

$$\overline{X(k)} = X(\mathbb{A}_k)^{\mathrm{Br}},$$

provided that:

1. $\operatorname{Pic} \overline{X_\eta}$ *is torsion-free, where* $\overline{X_\eta} = X_\eta \times_K \overline{K}$, $K = k(\eta)$; *e.g.* X_η *rational, or a smooth complete intersection of dimension at least three.*
2. $\operatorname{Br} \overline{X_\eta}$ *is finite.*
3. *Either all fibres, but one, are geometrically integral, or* X_η *has a* $k(\eta)$*-point.*

Here again it is possible to replace "geometrically integral" by "split" in the third condition [(4)].

If we compare the proof of Theorem 3.4.1 with the proof of Theorem 2.3.1, there are two additional ingredients:

1. Show that the specialization map $\operatorname{Br} X_\eta/\operatorname{Br} K \to \operatorname{Br} X_P/\operatorname{Br} k$ is an isomorphism for many k-fibres X_P ('many' in the sense of Hilbert's irreducibility theorem). This is a consequence of assumptions 1. and 2. ([**19**], 3.5.1. and [**20**], 2.3.1.).
2. If $\alpha_1, \dots, \alpha_r$ are elements of $\operatorname{Br} X_\eta$ which generate $\operatorname{Br} X_\eta/\operatorname{Br} k(\eta)$, choose an open subset $U \subset X$ such that $\alpha_i \in \operatorname{Br} U$. Then apply the following "formal lemma" ([**19**], 2.6.1.): Let $(M_v) \in X(\mathbb{A}_k)^{\operatorname{Br}}$, $M_v \in U$, and Σ_0 a finite set of places; then there exists $(P_v) \in X(\mathbb{A}_k)$, $P_v \in U$, and $\Sigma \supset \Sigma_0$ finite such that:
   (a) $P_v = M_v$ for $v \in \Sigma_0$;
   (b) $\sum_{v \in \Sigma} j_v(\alpha_i(P_v)) = 0$ for $1 \leqslant i \leqslant r$, where $j_v : \operatorname{Br} k_v \to \mathbf{Q}/\mathbf{Z}$ is the local invariant.

The formal lemma is a consequence of

Theorem 3.4.2 **([19], 2.1.1.).** *Let $\alpha \in \operatorname{Br} U$, suppose $\alpha \notin \operatorname{Br} X$. Then there exist infinitely many places v of k such that the image of the evaluation map $[U(k_v) \to \operatorname{Br} k_v, M_v \mapsto \alpha(M_v)]$ is not zero.*

Theorem 3.4.1 has several applications:

Example 3.4.3. We can recover Sansuc's result just knowing the case of a torus (which essentially goes back to [**47**]). Here we apply Theorem 3.4.1 in a situation when X_η has a $k(\eta)$-point, [**19**], 5.3.1.

Example 3.4.4. If $\overline{X(k)} = X(\mathbb{A}_k)^{\operatorname{Br}}$ for any smooth projective cubic surface (this is a widely believed conjecture), then by induction the same holds for cubic hypersurfaces ([**19**], 5.2.2.); therefore if $\dim X \geqslant 3$, then X satisfies weak approximation (the Brauer group of smooth hypersurfaces of dimension at least 3 is trivial: see the first appendix to the Poonen–Voloch paper in this volume).

It is also possible to combine open descent with the fibration method to obtain generalizations of Theorem 3.4.1 when at most 2 (or 3 in very special cases) fibres are degenerate (see [**21**]).

(4) This refinement is especially useful if we have to deal with a nonprojective morphism because the "split" condition remains valid after compactification of the morphism. See [**20**], Proof of Prop. 3.1.1.

3.5. Nonabelian descent. In the last few years it has become apparent that the Brauer–Manin obstruction can be refined if we consider nonabelian cohomology. In particular if G/k is a finite but not commutative k-group, it may happen that for $f \in H^1(X,G)$, we have $X(\mathbb{A}_k)^f \not\supset X(\mathbb{A}_k)^{\mathrm{Br}}$. The following result was the first unconditional counterexample to the Hasse principle not accounted for by the Brauer–Manin obstruction ([**43**]).

***Theorem 3.5.1* (Skorobogatov).** *There exists a bielliptic surface X over* $\mathbf{Q}$ *such that $X(\mathbf{Q}) = \emptyset$, $X(\mathbb{A}_{\mathbf{Q}})^{\mathrm{Br}} \neq \emptyset$.*

Recall that a *bielliptic surface* X is a surface such that $\overline{X}$ is the quotient of the product of two elliptic curves by the free action of a finite group. (In Skorobogatov's example this finite group is $\mathbf{Z}/2$. Similar examples with bigger groups were constructed later by Basile and Skorobogatov [**1**]).

Actually one can show ([**18**], 5.1) that the surface in the theorem satisfies $X(\mathbb{A}_{\mathbf{Q}})^f = \emptyset$ for some $f \in H^1(X,G)$, where G is a finite k-group satisfying $G(\overline{\mathbf{Q}}) = (\mathbf{Z}/4\mathbf{Z})^2 \rtimes \mathbf{Z}/2\mathbf{Z}$.

There are similar statements for weak approximation ([**16**])), e.g. take X/k any bielliptic surface, $X(k) \neq \emptyset$, then $\overline{X(k)} \subsetneq X(\mathbb{A}_k)^{\mathrm{Br}}$.

Nevertheless the Brauer–Manin condition is quite strong, as we can see from the following result ([**17**]; compare Th. 3.2.1 above):

Theorem 3.5.2. *Let X be a projective, smooth, and geometrically integral k-variety. Then:*

1. *If G/k is a linear connected k-group, $f \in H^1(X,G)$, then*
$$X(\mathbb{A}_k)^{\mathrm{Br}} \subset X(\mathbb{A}_k)^f.$$
2. *If G is any commutative k-group, $f \in H^2(X,G)$, then*
$$X(\mathbb{A}_k)^{\mathrm{Br}} \subset X(\mathbb{A}_k)^f.$$

Let us conclude with an open question: is the first part of this theorem still true for a (noncommutative) G which is an extension of a finite *abelian* group by a connected linear group (e.g. a torus)? My guess is "no".

References

[1] C. Basile & A. Skorobogatov – On the Hasse principle for bielliptic surfaces, to appear.

[2] B. J. Birch & P. Swinnerton-Dyer – The Hasse problem for rational surfaces, *J. Reine Angew. Math.* **274/275** (1975), 164–174.

[3] M. Borovoi – The Brauer–Manin obstructions for homogeneous spaces with connected or abelian stabilizer, *J. Reine Angew. Math.* **473** (1996), 181–194.

[4] M. V. Borovoi – Abelianization of the second nonabelian Galois cohomology, *Duke Math. J.* **72** (1993), no. 1, 217–239.

[5] J. W. S. Cassels – Arithmetic on curves of genus 1. VII. The dual exact sequence, *J. Reine Angew. Math.* **216** (1964), 150–158.

[6] J. W. S. Cassels & M. J. T. Guy – On the Hasse principle for cubic surfaces, *Mathematika* **13** (1966), 111–120.

[7] J.-L. Colliot-Thélène & A. Skorobogatov – Descent on fibrations over $\mathbf{P}^1_k$ revisited, *Math. Proc. Cambridge Philos. Soc.* **128** (2000), no. 3, 383–393.

[8] J.-L. Colliot-Thélène – Surfaces rationnelles fibrées en coniques de degré 4, Séminaire de Théorie des Nombres, Paris 1988–1989, Prog. Math., vol. 91, Birkhäuser Boston, Boston, MA, 1990, 43–55.

[9] J.-L. Colliot-Thélène & P. Salberger – Arithmetic on some singular cubic hypersurfaces, *Proc. London Math. Soc. (3)* **58** (1989), no. 3, 519–549.

[10] J.-L. Colliot-Thélène & J.-J. Sansuc – Sur le principe de Hasse et l'approximation faible, et sur une hypothèse de Schinzel, *Acta Arith.* **41** (1982), no. 1, 33–53.

[11] ———, La descente sur les variétés rationnelles. II, *Duke Math. J.* **54** (1987), no. 2, 375–492.

[12] J.-L. Colliot-Thélène, J.-J. Sansuc & P. Swinnerton-Dyer – Intersections of two quadrics and Châtelet surfaces. II, *J. Reine Angew. Math.* **374** (1987), 72–168.

[13] J.-L. Colliot-Thélène & P. Swinnerton-Dyer – Hasse principle and weak approximation for pencils of Severi–Brauer and similar varieties, *J. Reine Angew. Math.* **453** (1994), 49–112.

[14] T. Ekedahl – An effective version of Hilbert's irreducibility theorem, Séminaire de Théorie des Nombres, Paris 1988–1989, Prog. Math., vol. 91, Birkhäuser Boston, Boston, MA, 1990, 241–249.

[15] T. Graber, J. Harris & J. Starr – Families of rationally connected varieties, *J. Amer. Math. Soc.* **16** (2003), no. 1, 57–67 (electronic).

[16] D. Harari – Weak approximation and non-abelian fundamental groups, *Ann. Sci. École Norm. Sup. (4)* **33** (2000), no. 4, 467–484.

[17] ———, Groupes algébriques et points rationnels, *Math. Ann.* **322** (2002), no. 4, 811–826.

[18] D. Harari & A. Skorobogatov – Non-abelian cohomology and rational points, *Compositio Math.* **130** (2002), no. 3, 241–273.

[19] D. Harari – Méthode des fibrations et obstruction de Manin, *Duke Math. J.* **75** (1994), no. 1, 221–260.

[20] ———, Flèches de spécialisations en cohomologie étale et applications arithmétiques, *Bull. Soc. Math. France* **125** (1997), no. 2, 143–166.

[21] D. Harari & A. Skorobogatov – The Brauer group of torsors and its arithmetic applications, Ann. Inst. Fourier, to appear.

[22] G. Harder – Eine Bemerkung zum schwachen Approximationssatz, *Arch. Math. (Basel)* **19** (1968), 465–471.

[23] R. Heath-Brown & A. Skorobogatov – Rational solutions of certain equations involving norms, *Acta Math.* **189** (2002), no. 2, 161–177.

[24] V. A. Iskovskih – A counterexample to the Hasse principle for systems of two quadratic forms in five variables, *Mat. Zametki* **10** (1971), 253–257.

[25] M. Kneser – Starke Approximation in algebraischen Gruppen. I, *J. Reine Angew. Math.* **218** (1965), 190–203.

[26] ———, Strong approximation, Algebraic Groups and Discontinuous Subgroups (Proc. Sympos. Pure Math., Boulder, Colo., 1965), Amer. Math. Soc., Providence, R.I., 1966, 187–196.

[27] S. Lang – *Algebraic Number Theory*, second ed., Graduate Texts in Mathematics, vol. 110, Springer-Verlag, New York, 1994.

[28] Y. I. Manin – Le groupe de Brauer–Grothendieck en géométrie diophantienne, Actes du Congrès International des Mathématiciens (Nice, 1970), Tome 1, Gauthier-Villars, Paris, 1971, 401–411.

[29] J. S. Milne – *Étale Cohomology*, Princeton Mathematical Series, vol. 33, Princeton University Press, Princeton, N.J., 1980.

[30] K. P. Minchev – Strong approximation for varieties over an algebraic number field, *Dokl. Akad. Nauk BSSR* **33** (1989), no. 1, 5–8, 92.

[31] V. P. Platonov – The problem of strong approximation and the Kneser–Tits hypothesis for algebraic groups, *Izv. Akad. Nauk SSSR Ser. Mat.* **33** (1969), 1211–1219.

[32] ———, A supplement to the paper "The problem of strong approximation and the Kneser–Tits hypothesis for algebraic groups", *Izv. Akad. Nauk SSSR Ser. Mat.* **34** (1970), 775–777.

[33] V. Platonov & A. Rapinchuk – *Algebraic groups and number theory*, Pure and Applied Mathematics, vol. 139, Academic Press Inc., Boston, MA, 1994.

[34] P. Salberger – Some new Hasse principles for conic bundle surfaces, Séminaire de Théorie des Nombres, Paris 1987–88, Prog. Math., vol. 81, Birkhäuser Boston, Boston, MA, 1990, 283–305.

[35] P. Salberger & A. Skorobogatov – Weak approximation for surfaces defined by two quadratic forms, *Duke Math. J.* **63** (1991), no. 2, 517–536.

[36] J.-J. Sansuc – Groupe de Brauer et arithmétique des groupes algébriques linéaires sur un corps de nombres, *J. Reine Angew. Math.* **327** (1981), 12–80.

[37] J.-P. Serre – *Cours d'Arithmétique*, Collection SUP: "Le Mathématicien", vol. 2, Presses Universitaires de France, Paris, 1970.

[38] ———, *Lie Algebras and Lie Groups*, second ed., Lecture Notes in Mathematics, vol. 1500, Springer-Verlag, Berlin, 1992, 1964 lectures given at Harvard University.

[39] ———, *Cohomologie Galoisienne*, fifth ed., Lecture Notes in Mathematics, vol. 5, Springer-Verlag, Berlin, 1994.

[40] C. M. Skinner – Forms over number fields and weak approximation, *Compositio Math.* **106** (1997), no. 1, 11–29.

[41] A. Skorobogatov – On the fibration method for proving the Hasse principle and weak approximation, Séminaire de Théorie des Nombres, Paris 1988–1989, Prog. Math., vol. 91, Birkhäuser Boston, Boston, MA, 1990, 205–219.

[42] A. Skorobogatov – Descent on fibrations over the projective line, *Amer. J. Math.* **118** (1996), no. 5, 905–923.

[43] ———, Beyond the Manin obstruction, *Invent. Math.* **135** (1999), no. 2, 399–424.

[44] ———, *Torsors and Rational Points*, Cambridge Tracts in Mathematics, vol. 144, Cambridge University Press, Cambridge, 2001.

[45] P. Swinnerton-Dyer – Two special cubic surfaces, *Mathematika* **9** (1962), 54–56.

[46] P. Swinnerton-Dyer – The solubility of diagonal cubic surfaces, *Ann. Sci. École Norm. Sup. (4)* **34** (2001), no. 6, 891–912.

[47] V. E. Voskresenskiĭ – *Algebraic Groups and their Birational Invariants*, Translations of Mathematical Monographs, vol. 179, American Mathematical Society, Providence, RI, 1998.

[48] L. Wang – Brauer–Manin obstruction to weak approximation on abelian varieties, *Israel J. Math.* **94** (1996), 189–200.

Arithmetic of Higher-dimensional Algebraic Varieties
(B. Poonen, Yu. Tschinkel, eds.), p. 61–81
Progress in Mathematics, Vol. 226, © 2004 Birkhäuser Boston, Cambridge, MA

COUNTING POINTS ON VARIETIES USING UNIVERSAL TORSORS

Emmanuel Peyre

Institut Fourier, UMR 5582 du CNRS, Université de Grenoble I, BP 74, 38402 Saint-Martin d'Hères, France • *E-mail :* Emmanuel.Peyre@ujf-grenoble.fr

Abstract. Around 1989, Manin initiated a program to understand the asymptotic behaviour of rational points of bounded height on Fano varieties. This program led to the search of new methods to estimate the number of points of bounded height on various classes of varieties. Methods based on harmonic analysis were successful for compactifications of homogeneous spaces. However, they do not apply to other types of varieties. Universal torsors, introduced by Colliot-Thélène and Sansuc in connection with the Hasse principle and weak approximation, turned out to be a useful tool in the treatment of other varieties. The aim of this short survey is to describe the use of torsors in various representative examples.

1. Introduction

If the rational points of a variety V over a number field k are Zariski dense, it is natural to equip V with a height H and to study the asymptotic distribution of the set of points of bounded height on V. In **[FMT89]** and **[BM90]**, Batyrev, Franke, Manin and Tschinkel gave strong evidence supporting conjectures relating the asymptotic behaviour of the number of points of bounded height on open subsets of V to geometrical invariants of V. This work motivated the

Key words and phrases. Rational points, heights, universal torsors.

development of several methods to estimate the number of points of bounded height for new classes of varieties. One of the most successful methods was the use of harmonic analysis on adelic groups. For example, it was used by Batyrev and Tschinkel in [**BT95**], [**BT96a**], and [**BT98**] to handle the case of projective toric varieties, by Strauch and Tschinkel in [**ST97**] and [**ST99**] for toric bundles over flag varieties, and by Chambert-Loir and Tschinkel in [**CLT00a**], [**CLT00b**], and [**CLT02**] for equivariant compactifications of vector spaces. However, these kind of methods apply only to equivariant compactifications of homogeneous spaces. One may say that almost all other methods have one step in common, namely the lifting to universal torsors. Universal torsors have been introduced by Colliot-Thélène and Sansuc in [**CTS76**], [**CTS77**], [**CTS80**], and [**CTS87**] to study the Hasse principle and the weak approximation. The interest of universal torsors is that, from an arithmetic point of view, these torsors should be much simpler than the variety itself. As an example, universal torsors over smooth projective toric varities are open subsets of an affine space. When the Fano variety V is a smooth complete intersection of dimension bigger than three in the projective space, the universal torsor may be described as the cone over the variety. In that case, if the dimension of the variety is big enough, the conjectural formula of Manin may be deduced from the formula given by the classical circle method. This reduction, which is described in [**FMT89**], may be seen as a particular case of the lifting to the universal torsor. Salberger in [**Sal98**] was the first to use explicitly universal torsors in relation with points of bounded height. In particular, he was able to give a new proof of the theorem of Batyrev and Tschinkel for smooth projective split toric varieties over $\mathbf{Q}$. This lifting to the universal torsor was then used by de la Bretèche in [**dlB01**] to give a better estimate for the number of points of bounded height on toric varieties. The lifting to universal torsors was later used by Salberger and de la Bretèche (see [**dlB02**]) to prove the asymptotic formula for the plane blown up in four points over $\mathbf{Q}$. In a more general setting, the author described in [**Pey98**] and [**Pey01**] how the conjectural asymptotic formula lifts naturally to universal torsors.

The aim of this short survey is to explain in a self-contained way the usefulness of universal torsors for counting points of bounded height. In Section 2, we describe the heights used throughout the paper, and in Section 3 we recall the empirical formula for the number of points on Fano varieties. In Section 4 we give a short list of cases for which this formula holds, in Section 5 we describe the counterexample of Batyrev and Tschinkel. In Section 6 we describe briefly both methods: harmonic analysis and universal torsors. Section 7 is devoted to the case of a hypersurface in $\mathbf{P}^n$. The next section contains the definition of universal torsors in general. In Section 9 we describe Cox's construction of universal torsors for toric varieties and explain how Salberger used it and in Section 10 we turn to the case of the plane blown up in four points, in which case the universal torsor was described by Salberger and Skorobogatov. The last section contains a short description of the generalization of these lifting arguments to a larger class of varieties.

Acknowledgements: I am very grateful to John Voight who typed the notes of my talks at the American Institute of Mathematics on which this survey is based.

2. Heights on projective varieties

Definition 2.1. The classical exponential height on the projective space over $\mathbf{Q}$ is defined as follows:

$$\begin{aligned} H_N\colon \mathbf{P}^N(\mathbf{Q}) &\to \mathbf{R}_{>0} \\ (x_0 : \ldots : x_N) &\mapsto \sup_{0 \leqslant i \leqslant N} |x_i|, \quad \text{if } \begin{cases} x_i \in \mathbf{Z}, \text{ and} \\ \gcd(x_i) = 1. \end{cases} \end{aligned}$$

If K is a number field, one generalizes this construction in the following way:

$$\begin{aligned} H_N : \mathbf{P}^N(K) &\to \mathbf{R}_{>0} \\ (x_0 : \ldots : x_N) &\mapsto \prod_{v \in \Omega_K} \sup_{0 \leqslant i \leqslant N} |x_i|_v \end{aligned}$$

where Ω_K is the set of places of K and for any $x \in K_v$,

$$|x|_v = |N_{K_v/\mathbf{Q}_p}(x)|_p \quad \text{if} \quad v \mid p.$$

Any morphism of varieties $\phi : V \to \mathbf{P}^N_K$ induces a height

$$\begin{aligned} H\colon V(K) &\to \mathbf{R}_{>0} \\ x &\mapsto H_N(\phi(x)). \end{aligned}$$

If $U \subset V$ is an open subset, then we would like to describe and understand the asymptotic behavior of the counting function

$$N_{U,H}(B) = \#\{x \in U(K) \mid H(x) \leqslant B\}.$$

Let us first give a few examples:

Example 2.2. If $V = \mathbf{P}^N(\mathbf{Q})$ and $\phi = \mathrm{id}$, then an easy Möbius inversion formula gives that

$$N_{V,H}(B) \sim \frac{2^N}{\zeta_{\mathbf{Q}}(N+1)} B^{N+1}$$

as $B \to \infty$ (see Figure 2.2).

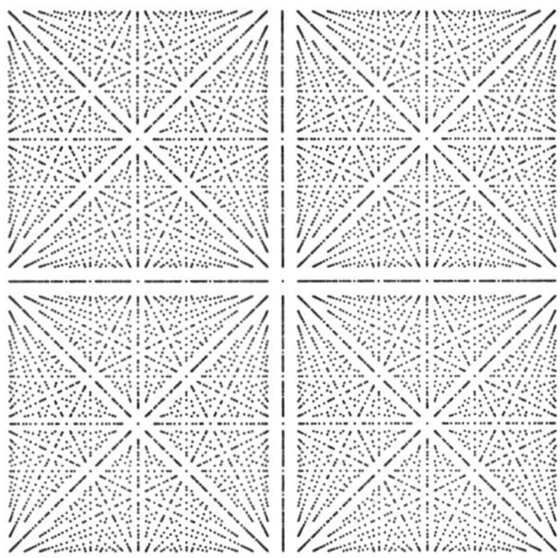

FIGURE 1. Projective space

This result was later generalized by Schanuel in **[Sch79]** to the projective space over any number field (see Figure 2.2).

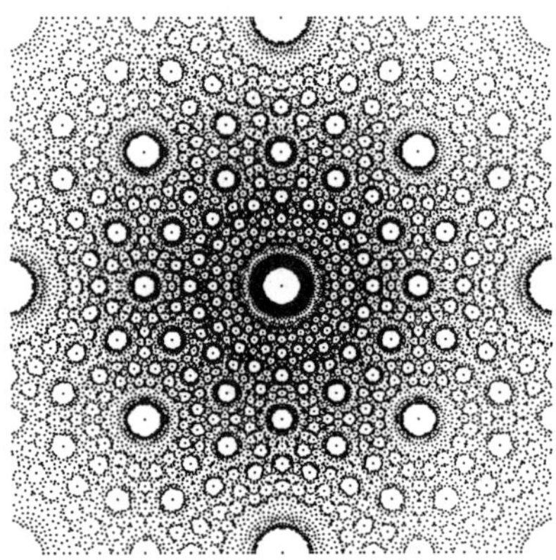

FIGURE 2. Projective line over $\mathbf{Q}(i)$

Example 2.3. If $V = V_1 \times V_2$, and $H_i\colon V_i(K) \to \mathbf{R}_{>0}$ are heights as above, and $U_i \subset V_i$ is an open subset for each i, then the height $H : V \to \mathbf{R}_{>0}$ defined by $H(x_1, x_2) = H_1(x_1)H_2(x_2)$ corresponds to the Segre embedding of $V_1 \times V_2$. Assume that

$$N_{U_i,H_i}(B) = C_i B(\log B)^{t_i-1} + O\left(B(\log B)^{t_i-2}\right).$$

Then, by [**FMT89**, Proposition 2]

$$N_{U_1\times U_2,H}(B) \sim \frac{(t_1-1)!(t_2-1)!}{(t_1+t_2-1)!} C_1 C_2 B(\log B)^{t_1+t_2-1}$$

as $B \to \infty$. For $\mathbf{P}^1 \times \mathbf{P}^1$, we get

$$N_{\mathbf{P}^1\times\mathbf{P}^1,H}(B) \sim CB^2 \log B$$

(see Figure 2.3).

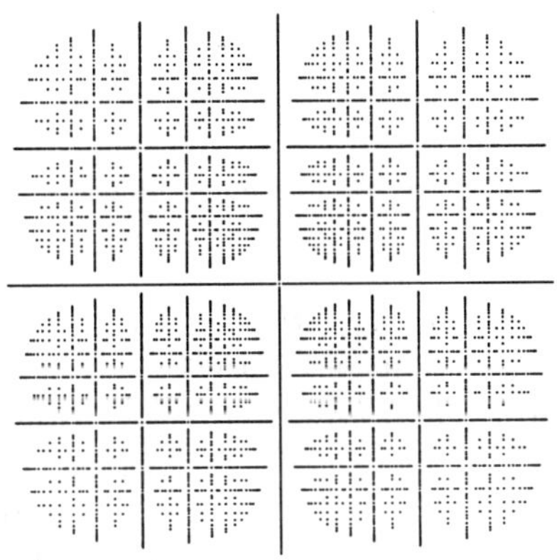

FIGURE 3. Product of two projective lines

Example 2.4. Let $V \to \mathbf{P}^2(\mathbf{Q})$ be the blow up of $\mathbf{P}^2$ at $P_1 = (1:0:0)$, $P_2 = (0:1:0)$, and $P_3 = (0:0:1)$. Then V may be seen as the hypersurface in $\mathbf{P}^1 \times \mathbf{P}^1 \times \mathbf{P}^1$ given by the equation $x_1x_2x_3 = y_1y_2y_3$. We put

$$H(P_1, P_2, P_3) = H_1(P_1)H_1(P_2)H_1(P_3)$$

which defines a height $H : V(\mathbf{Q}) \to \mathbf{R}_{>0}$.

On V, there are 6 exceptional lines $E_{i,j}\colon x_i = 0, y_j = 0$ for $i \neq j$. Let $U = V - \bigcup_{i\neq j} E_{ij}$. We have

$$N_{E_{i,j},H}(B) \sim CB^2$$

and

$$N_{U,H}(B) \sim \frac{1}{6}\left(\prod_p \left(1-\frac{1}{p}\right)^4\left(1+\frac{4}{p}+\frac{1}{p^2}\right)\right) B(\log B)^3$$

(see Figure 2.4).

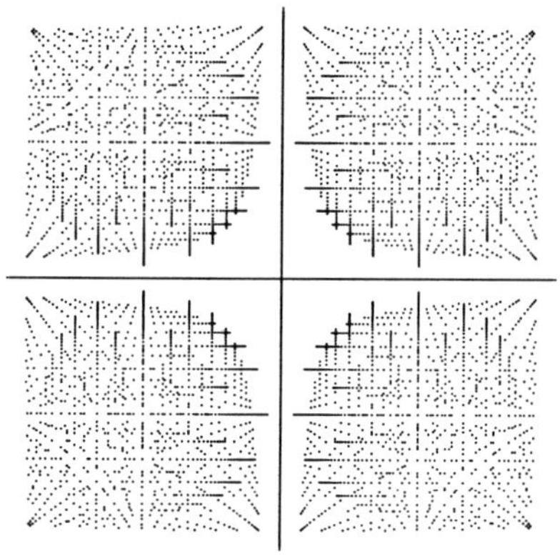

FIGURE 4. The plane blown up

We see that $N_{U,H}(B) = o(N_{E_{i,j},H}(B))$. Thus, in this case, the dominant term of the asymptotic behaviour of $N_{V,H}(B)$ is given by the number of points on the six lines. Therefore it cannot reflect the geometry of the whole of V. One of the basic ideas in the interpretation of the asymptotic behaviour of the number of points of bounded height is that one has to consider open subsets to be able to get a meaningful geometric interpretation.

In all examples known to the author for which it is possible to give a precise estimate of the number of points of bounded height, the asymptotic behaviour is of the form

$$N_{U,H}(B) \sim CB^a(\log B)^{b-1}$$

with $C \geqslant 0$, $a \geqslant 0$ and $b \in \frac{1}{2}\mathbf{Z}$, $b \geqslant 1$. Thus one wishes to give a geometric interpretation of a, b and C.

3. Manin's principle

We assume that V is a smooth, geometrically integral projective variety of dimension n over the number field K. We also assume that $\omega_V^{-1} = \Lambda^n T_V$ is very ample (in particular, V is a Fano variety). We look only at the height relative to this anticanonical divisor $\phi^*(\mathscr{O}_{\mathbf{P}^N}(1)) = \omega_V^{-1}$, and we assume that $V(K)$ is Zariski dense. The following question is a variant of the conjecture C′ in [**BM90**]:

Question 3.1. *Does there exist a dense open subset $U \subset V$ and a constant $C > 0$ such that*

$$N_{U,H}(B) \sim CB(\log B)^{t-1}$$

as $B \to \infty$, where t is the rank of the Picard group of V. (Since V is Fano, $\operatorname{Pic} V$ *is a free* $\mathbf{Z}$*-module of finite rank.)*

In fact, it is even possible to give a conjectural interpretation of C, but to describe this conjectural constant, we first need to express the height in terms of metrics.

Notation 3.2. Let V be a geometrically integral smooth projective variety and let H be the height corresponding to an embedding $\phi\colon V \to \mathbf{P}^N_K$. Let L be $\phi^*(\mathscr{O}_{\mathbf{P}^N}(1))$. We denote by $s_0, \dots, s_N$ the pull-backs in $\Gamma(V, L)$ of the sections $X_0, \dots, X_N$ of $\mathscr{O}_{\mathbf{P}^N}(1)$. We view L as a line bundle over V and define for any place v of K a v-adically continuous metric $\|\cdot\|_v : L(K_v) \to \mathbf{R}$ by the condition

$$\forall x \in V(K_v), \quad \forall s \in \Gamma(V, L), \quad \|s(x)\|_v = \inf_{\substack{0 \leqslant i \leqslant N \\ s_i(x) \neq 0}} \left| \frac{s(x)}{s_i(x)} \right|_v .$$

Then the height H may be characterized by

$$\forall x \in V(K), \quad \forall s \in \Gamma(V, L), \quad s(x) \neq 0 \Rightarrow H(x) = \prod_{v \in \Omega_K} \|s(x)\|_v^{-1}.$$

From now on we assume that the above line bundle L is the anticanonical line bundle ω_V^{-1}. We now define a measure on the adelic space $V(\mathbf{A}_K)$ which coincides with the product $\prod_{v \in \Omega_K} V(K_v)$, since V is projective.

Definition 3.3. For any place v of K, we normalize the Haar measure $\mathrm{d}x_v$ on K_v by the conditions:

- $\int_{\mathscr{O}_v} \mathrm{d}x_v = 1$ if v is finite,
- $\mathrm{d}x_v([0,1]) = 1$ if K_v is isomorphic to $\mathbf{R}$,
- $\mathrm{d}x_v = i\,\mathrm{d}z\,\mathrm{d}\overline{z} = 2\,\mathrm{d}x\,\mathrm{d}y$ if K_v is isomorphic to $\mathbf{C}$.

The measure $\boldsymbol{\omega}_v$ on $V(F_v)$ is defined locally by the formula

$$\boldsymbol{\omega}_v = \left\| \frac{\partial}{\partial x_1} \wedge \dots \wedge \frac{\partial}{\partial x_n} \right\|_v \mathrm{d}x_{1,v} \dots \mathrm{d}x_{n,v}$$

if $(x_1, \dots, x_n)$ is a local system of coordinates on $V(K_v)$ in the v-adic topology and where $\frac{\partial}{\partial x_1} \wedge \dots \wedge \frac{\partial}{\partial x_n}$ is viewed as a section of ω_V^{-1}. The fact that these expressions glue together follows from the chosen normalization of the absolute value. Indeed the formula for a change of variables is given by

$$\mathrm{d}y_{1,v}\cdots\mathrm{d}y_{n,v}=\left|\det\left(\frac{\partial y_i}{\partial x_j}\right)_{\substack{1\leqslant i\leqslant n\\1\leqslant j\leqslant n}}\right|_v\mathrm{d}x_{1,v}\cdots\mathrm{d}x_{n,v}$$

(see [**Wei82**, §2.2.1]).

Remark 3.4. At any real place this construction amounts to the classical recipe for producing a measure on a differential variety from a continuous section of its canonical line bundle. At almost all finite places, using ideas of Tamagawa and Weil, one may prove the following proposition

Proposition 3.5. *For almost all finite* $\mathfrak{p}$ *in* Ω_K,

$$\omega_{\mathfrak{p}}(V(K_{\mathfrak{p}}))=\frac{\#V(\mathbf{F}_{\mathfrak{p}})}{(\#\mathbf{F}_{\mathfrak{p}})^{\dim V}}$$

where $\mathbf{F}_{\mathfrak{p}}$ *is the residue field at* $\mathfrak{p}$.

In particular, this implies that the product $\prod_{\mathfrak{p}}\omega_{\mathfrak{p}}(V(K_{\mathfrak{p}}))$ diverges. Therefore we have to introduce convergence factors. These factors are suggested by the Grothendieck–Lefschetz formula.

Definition 3.6. We fix a finite set S of bad places containing all archimedean places and all places of bad reduction. Let $\overline{K}$ be an algebraic closure of K and put $\overline{V}=V\times_K\overline{K}$. Then for $\mathfrak{p}\in\Omega_K-S$ one defines

$$L_{\mathfrak{p}}(s,\operatorname{Pic}(\overline{V}))=\frac{1}{\det(1-q^{-s}\operatorname{Frob}_q\mid\operatorname{Pic}(V_{\overline{\mathbf{F}}_{\mathfrak{p}}})\otimes\mathbf{Q})}$$

where $q=\#\mathbf{F}_{\mathfrak{p}}$ and Frob_q is the q-power Frobenius automorphism of the field $\overline{\mathbf{F}}_{\mathfrak{p}}$, which induces a linear endomorphism of the $\mathbf{Q}$-vector space $\operatorname{Pic}(V_{\overline{\mathbf{F}}_{\mathfrak{p}}})\otimes\mathbf{Q}$. The global L-function is given by the Euler product

$$L_S(s,\operatorname{Pic}(\overline{V}))=\prod_{\mathfrak{p}\in\Omega_K-S}L_{\mathfrak{p}}(s,\operatorname{Pic}(\overline{V}))$$

which converges for $\operatorname{Re}(s)>1$ and admits a meromorphic continuation to $\mathbf{C}$. We define the convergence factors by

$$\lambda_v=\begin{cases}L_v(1,\operatorname{Pic}(\overline{V})), & \text{if } v\in\Omega_K-S,\\ 1, & \text{otherwise.}\end{cases}$$

The adelic measure on $V(\mathbf{A}_K)$ is then defined by the formula

$$\omega_H = \lim_{s\to 1}(s-1)^t L_S(s,\operatorname{Pic}(\overline{V}))\ \frac{1}{\sqrt{d_K}^{\dim V}} \prod_{v\in\Omega_K} \lambda_v^{-1}\omega_v,$$

where d_K is the absolute value of the discriminant of K.

Remarks 3.7.

(i) The convergence of the product $\prod_{v\in\Omega_K}\lambda_v^{-1}\omega_v$ follows from the Lefschetz trace formula and Weil's conjecture about the absolute value of the eigenvalues of the Frobenius operator which was proven by Deligne [**Del74**].

(ii) By definition, the measure ω_H does not depend on S.

(iii) Note that $\sqrt{d_K}$ is the volume of $\mathbf{A}_K/K$ for the measure $\prod_{v\in\Omega_K} dx_v$.

To define the conjectural constant it remains to multiply by two rational factors which are the object of the next definition.

Definition 3.8. Let $C^1_{\text{eff}}(V)$ be the cone in $\operatorname{Pic}(V)\otimes_{\mathbf{Z}}\mathbf{C}$ generated by the classes of effective divisors and $C^1_{\text{eff}}(V)^\vee$ the dual cone defined by

$$C^1_{\text{eff}}(V)^\vee = \{\, y\in \operatorname{Pic}(V)\otimes_{\mathbf{Z}}\mathbf{R}^\vee \mid \forall x\in C^1_{\text{eff}}(V),\ \langle x,y\rangle \geqslant 0\,\}.$$

Then

$$\alpha(V) = \frac{1}{(t-1)!}\int_{C^1_{\text{eff}}(V)^\vee} e^{\langle \omega_V^{-1},y\rangle}\,dy$$

where the measure on $\operatorname{Pic}(V)\otimes_{\mathbf{Z}}\mathbf{R}^\vee$ is normalized so that the covolume of the dual lattice $\operatorname{Pic}(V)^\vee$ is one. We also consider the integer

$$\beta(V) = \#H^1(K,\operatorname{Pic}(\overline{V})).$$

Remarks 3.9.

(i) The constant $\alpha(V)$ may also be defined as the volume of the domain

$$\{\, y\in C^1_{\text{eff}}(V)^\vee \mid \langle y,\omega_V^{-1}\rangle = 1\,\}$$

for a suitable measure on the affine hyperplane $\langle y,\omega_V^{-1}\rangle = 1$ (see [**Pey95**, §2.2.5]). Therefore, if there exists a finite family $(D_i)_{1\leqslant i\leqslant r}$ of effective divisors on V such that

$$C^1_{\text{eff}}(V) = \sum_{i=1}^{r}\mathbf{R}_{\geqslant 0}[D_i],$$

then the constant $\alpha(V)$ is rational.

(ii) The constant $\beta(V)$ was introduced by Batyrev and Tschinkel in [**BT95**].

The conjectural constant is then defined as follows.

Definition 3.10. We define

$$\theta_H(V) = \alpha(V)\beta(V)\omega_H(\overline{V(K)}),$$

where $\overline{V(K)}$ denotes the closure of the rational points in the adelic space $V(\mathbf{A}_K)$.

We can now give a refined version of Question 3.1:

Empirical formula 3.11. *With notation as in Question 3.1, there often exists a dense open subset $U_0 \subset V$ such that for any nonempty subset U of U_0, one has*

$$N_{U,H}(B) \sim \theta_H(V)B(\log B)^{t-1} \qquad \text{(F)}$$

4. Results

The formula (F) is true in the following cases:

- $V = G/P$, where G is a reductive algebraic group over K and P is a parabolic subgroup of G defined over K. It follows from the work of Langlands on Eisenstein series [**Lan76**] (see Franke, Manin, Tschinkel [**FMT89**] and [**Pey95**, §6]). We may take $U_0 = V$. In particular, it is true for any quadric.
- V is a smooth projective toric variety, that is an equivariant compactification of an algebraic torus (see [**Pey95**, §8–11] for particular cases, Batyrev and Tschinkel [**BT95**], [**BT96a**], and [**BT98**], Salberger [**Sal98**], and de la Bretèche [**dlB01**]). One may take the open orbit as U_0. This case includes the plane blown up in 1, 2, or 3 points, and Hirzebruch surfaces.
- V is an equivariant compactification of an affine space for the action of the corresponding vector space (see Chambert-Loir, Tschinkel [**CLT00a**], [**CLT00b**], and [**CLT02**]).
- $V = \mathbf{P}^2_{\mathbf{Q}}$ blown up at $(1:0:0), (0:1:0), (0:0:1), (1:1:1)$ (see Salberger for an upper bound, and de la Bretèche [**dlB02**]).

The formula (F) is compatible with:

- the circle method. In particular, it is true if $V \subset \mathbf{P}^n(\mathbf{Q})$ is a smooth hypersurface of degree d, if $n > 2^d(d-1)$ (see Birch [**Bir62**]);
- products of varieties (see [**FMT89**], [**Pey95**, §4]);
- numerical tests for some diagonal cubic surfaces [**PT01a**], [**PT01b**];
- lower bounds for some cubic surfaces (see Slater and Swinnerton-Dyer [**SSD98**]). The problem of finding an optimal upper bound for cubic surfaces is still open.

All these examples support the empirical formula. However, there are also counterexamples, which will be discussed in the next section.

5. The counterexample of Batyrev and Tschinkel

Take $V \subset \mathbf{P}^3 \times \mathbf{P}^3$ defined by

$$x_0 y_0^3 + x_1 y_1^3 + x_2 y_2^3 + x_3 y_3^3 = 0.$$

We have

$$\mathrm{Pic}(V) \simeq \mathrm{Pic}(\mathbf{P}^3 \times \mathbf{P}^3) = \mathbf{Z} \times \mathbf{Z}$$

and $\omega_V^{-1} = \mathscr{O}_V(3,1)$. In particular, V is a Fano variety. We may use the height

$$H\colon V(\mathbf{Q}) \to \mathbf{R}_{>0}$$

$$((x_0 : \ldots : x_3), (y_0 : \ldots : y_3)) \mapsto H_3(x)^3 H_3(y).$$

If (F) is true for V, then there is an open subset U and a constant C such that

$$N_{U,H}(B) \sim CB \log B$$

as $B \to \infty$. There is a projection onto the first coordinate $\pi_1 : V \to \mathbf{P}^3$. If $(x_0 : \ldots : x_3) \in \mathbf{P}^3$ is such that $\prod_{i=0}^3 x_i \neq 0$, then $\pi_1^{-1}(x)$ is a smooth cubic surface; if $x_1/x_0, x_2/x_0, x_3/x_0$ are cubes, then $\mathrm{rk}\,\mathrm{Pic}(\pi_1^{-1}(x)) = 4$. If (F) is true for the fiber, then

$$N_{\pi^{-1}(x),H}(B) \sim C_x B (\log B)^3$$

as $B \to \infty$, but these fibers are Zariski dense, so the answer to Question 3.1 can not be positive for both V and the fibers. In fact, Batyrev and Tschinkel prove the following more precise result:

Theorem 5.1 **(Batyrev and Tschinkel [BT96b]).** *If K contains a nontrivial cube root of unity, then for all nonempty $U \subset V$, (F) does not hold for U.*

6. Methods of counting

We now return to the methods used to prove the results given in Section 4.

Harmonic analysis: Assume that there exists a dense open subset U of V which is of the form G/H, where G is a reductive algebraic group, look at the height zeta function

$$\zeta_{U,H}(s) = \sum_{x \in U(K)} H(x)^{-s}$$

which converges when $\mathrm{Re}(s) \gg 0$.

The asymptotic behavior of $N_{U,H}(B)$ is given by the meromorphic properties of $\zeta_{U,H}(s)$. If $U = G$, one may use a Poisson formula. If $V = G/P$, $\zeta_{U,H}(s)$ is an Eisenstein series and we may apply the work of Langlands. In both cases the problem may be handled using harmonic analysis.

These methods do not apply when the variety is not an equivariant compactification of a homogeneous space. All other cases appearing in the list of section 4 have one preliminary step in common: they all use a lift to universal torsors.

Universal torsors: implicit in the case of a hypersurface in $\mathbf{P}^n(\mathbf{Q})$, it was made explicit by Salberger in [**Sal98**] in his alternative treatment of split toric varieties over $\mathbf{Q}$; it was then used by Salberger and de la Bretèche in the case of the plane blown up in 4 points. The end of this survey is devoted to the description of this preliminary step in those cases.

7. A basic example

In the case of hypersurfaces of large dimension and small degree, the principle of Manin follows from the following deep theorem which is based upon the Hardy–Littlewood circle method.

Theorem 7.1 **(Birch [Bir62]).** *Let $f \in \mathbf{Z}[x_0, \dots, x_N]$ be homogeneous of degree d, and let $W \subset \mathbf{A}^{N+1} - \{0\}$ be the cone defined by $f = 0$. Assume that:*

(i) *W is smooth,*
(ii) *$W(\mathbf{R}) \neq \emptyset$, and for all primes p, $W(\mathbf{Q}_p) \neq \emptyset$,*
(iii) *$N > 2^d(d-1)$.*

Let

$$M_W(B) = \#\Big\{x \in \mathbf{Z}^{N+1} - \{0\} \;\Big|\; f(x) = 0 \text{ and } \sup_{0 \leqslant i \leqslant N} |x_i| \leqslant B\Big\}.$$

Then there exist explicit $C > 0$ and $\delta > 0$ such that

$$M_W(B) = CB^{N+1-d} + O(B^{N+1-d-\delta}).$$

Let $\pi : \mathbf{A}^{N+1} - \{0\} \to \mathbf{P}^N$, and let $V = \pi(W)$ be the corresponding projective hypersurface. Then $\omega_V^{-1} = \mathscr{O}_V(N+1-d)$, so we may take the height

$$H(x) = H_N(x)^{N+1-d}$$

where H_N was defined in Section 2. Then

$$N_{V,H}(B) = \frac{1}{2}\#\left\{x \in \mathbf{Z}^{N+1} - \{0\} \,\middle|\, \begin{cases} f(x) = 0, \\ \sup |x_i|^{N+1-d} \leqslant B, \\ \gcd(x_i) = 1. \end{cases} \right\}.$$

Using Möbius inversion, we get

$$N_{V,H}(B) = \frac{1}{2}\sum_k \mu(k)\#\left\{x \in (k\mathbf{Z})^{N+1} - \{0\} \,\middle|\, \begin{cases} f(x) = 0, \\ \sup_i |x_i|^{N+1-d} \leqslant B \end{cases} \right\},$$

where $\mu : \mathbf{Z}_{>0} \to \{-1, 0, 1\}$ is the Möbius function. Then

$$\begin{aligned} N_{V,H}(B) &= \frac{1}{2}\sum_k \mu(k) M_W\Big(\frac{B^{1/(N+1-d)}}{k}\Big) \sim \frac{C}{2}\sum_k \frac{\mu(k)}{k^{N+1-d}} B \\ &= \frac{C}{2\zeta(N+1-d)} B. \quad \square \end{aligned}$$

The motivation behind the introduction of universal torsors was to generalize this simple descent argument to other varieties.

8. Universal torsors

Let V be a smooth, geometrically integral projective variety over K, where $\operatorname{char} K = 0$. Assume (for simplicity) that V is Fano, which means that ω_V^{-1} is ample. Thus, if $\overline{K}$ is an algebraic closure of K, and $\overline{V} = V \times_K \overline{K}$, then $\operatorname{Pic}(\overline{V})$ is a free abelian group of finite rank.

Assume $K = \overline{K}$ first. Let $L_1, \ldots, L_t$ be line bundles on V such that $[L_1], \ldots, [L_t]$ form a basis of $\operatorname{Pic}(V) = \operatorname{Pic}(\overline{V})$. Let $L_i^\times = L_i -$ zero section. Consider

$$\pi : L_1^\times \times_V L_2^\times \times_V \cdots \times_V L_t^\times \to V.$$

On the left we have an action of $\mathbf{G}_m^t$, this is "the" *universal torsor* of V.

Proposition 8.1. *If $K = \overline{K}$, the universal torsor constructed above does not depend, up to isomorphism, on the chosen basis of the Picard group.*

Proof. Let $L'_1, \ldots, L'_t$ be line bundles on V whose classes $[L'_1], \ldots, [L'_t]$ form another basis of the Picard group of V. Let $M = (m_{i,j})$ in $\mathrm{GL}_n(\mathbf{Z})$ be the matrix such that

$$[L'_i] = \sum_{i=1}^{t} m_{j,i}[L_j].$$

In other words, for each i in $\{1,\dots,t\}$ we may fix an isomorphism

$$\psi_i : \bigotimes_{j=1}^{t} L_j^{\otimes m_{j,i}} \xrightarrow{\sim} L'_i.$$

But if $E_1,\dots,E_m$ are one-dimensional vector spaces and $k_1,\dots,k_m$ integers there is a canonical map

$$\begin{array}{rcl} \times_{i=1}^{m}(E_i-\{0\}) & \to & \bigotimes_{i=1}^{m} E_i^{\otimes k_i} \\ (y_1,\dots,y_m) & \mapsto & \bigotimes_{i=1}^{m} y_i^{\otimes k_i} \end{array}$$

where for any vector space E of dimension one, and any nonzero y in E, $y^{\otimes -1}$ is the unique element of the dual $E^\vee$ of E such that $y^{\otimes -1}(y)=1$. In that way, composing with ψ_i, we get maps

$$\rho_i : \mathop{\times}_{j=1}^{t} L_j^\times \to {L'}_i^\times.$$

This map is equivariant for the action of $\mathbf{G}_m^t$ in the following sense:

$$\forall (z_1,\dots,z_t)\in \mathbf{G}_m^t(K),\ \forall y\in \mathop{\times}_{j=1}^{t} L_j^\times(K),\ \rho_i((z_1,\dots,z_t)\cdot y)=\prod_{j=1}^{t} z_j^{m_{j,i}}\cdot \rho_i(y).$$

Note that if ρ'_i is another map from $\times_{j=1}^{t} L_j^\times$ to ${L'}_i^\times$ with the same equivariance property, then there is a section $s\in\Gamma(V,\mathbf{G}_m)$ such that

$$\forall y\in \mathop{\times}_{j=1}^{t} L_j^\times(K),\quad \rho'_i(y)=s(\pi(y))\cdot\rho_i(y).$$

But, since V is projective, $\Gamma(V,\mathbf{G}_m)=K^*$ and ρ_i is unique up to multiplication by a constant. The maps ρ_i yield a map

$$\rho : \mathop{\times}_{i=1}^{t} L_i^\times \to \mathop{\times}_{i=1}^{t} {L'}_i^\times.$$

The matrix M defines a morphism of algebraic groups

$$\begin{array}{rcl} \widetilde{M} : \mathbf{G}_m^t & \to & \mathbf{G}_m^t \\ (z_1,\dots,z_t) & \mapsto & (\prod_{j=1}^{t} z_j^{m_{j,i}})_{1\leqslant i\leqslant t} \end{array}$$

and the map ρ is equivariant with respect to $\widetilde{M}$:

$$\forall z\in \mathbf{G}_m^t(K),\quad \forall y\in \mathop{\times}_{i=1}^{t} L_i^\times(K),\quad \rho(z\cdot y)=\widetilde{M}(z)\cdot\rho(y).$$

Moreover, if ρ' is another map with the same equivariance property, then there is $z\in\mathbf{G}_m^t(K)$ such that $\rho'=z.\rho$. Similarly we may define a map

$$\tau : \mathop{\times}_{i=1}^{t} {L'}_i^\times \to \mathop{\times}_{i=1}^{t} L_i^\times$$

which is equivariant with respect to $\widetilde{M}^{-1}$. Thus the composite map

$$\tau \circ \rho : \mathop{\times}_{i=1}^{t} L_i^\times \longrightarrow \mathop{\times}_{i=1}^{t} L_i^\times$$

is equivariant with respect to the identity map and therefore coincides with the action of an element of $\mathbf{G}_m^t(K)$. Thus $\tau \circ \rho$ and $\rho \circ \tau$ are isomorphisms. □

For arbitrary fields, a universal torsor may be described as a K-structure on the above torsor. Let us define this notion more precisely:

Recall that there is a contravariant equivalence of categories between the category of algebraic tori, that is, algebraic groups T such that $\overline{T}$ is isomorphic to $\mathbf{G}_{m,\overline{K}}^{\dim T}$, and the category of $\mathrm{Gal}(\overline{K}/K)$-lattices, that is, $\mathrm{Gal}(\overline{K}/K)$-modules which are free abelian groups of finite rank. The functors giving the equivalence are

$$T \mapsto X^*(T) = \mathrm{Hom}_{\text{alg.gp.}}(\overline{T}, \mathbf{G}_m) \quad \text{and} \quad \mathrm{Spec}(\overline{K}[M])^{\mathrm{Gal}(\overline{K}/K)} \leftarrow M.$$

Definition 8.2. Let the Néron–Severi torus, T_{NS}, be the torus corresponding to the $\mathrm{Gal}(\overline{K}/K)$-lattice $\mathrm{Pic}\,\overline{V}$.

If G is an algebraic group over K, then a *G-torsor* is a faithfully flat map $\pi : \mathcal{T} \to V$ with an action of G on $\mathcal{T}$ such that locally for the faithfully flat topology, $\mathcal{T} \times_V U \simeq G \times U$, where the isomorphism is compatible with the action of G. (In another language, these are principal homogeneous spaces.)

A T_{NS}-torsor $\mathcal{T} \to V$ is said to be *universal* if $\overline{\mathcal{T}} \to \overline{V}$ is isomorphic as a torsor to $L_1^* \times_V \cdots \times_V L_t^* \to V$.

Why are these universal torsors interesting? The following facts are due to Colliot-Thélène and Sansuc, who introduced this notion.

Proposition 8.3 **(Colliot-Thélène, Sansuc).** *With notation as above,*

- *For all $x \in V(K)$, there exists a unique (up to isomorphism) universal torsor $\pi : \mathcal{T} \to V$ such that $x \in \pi(\mathcal{T}(K))$.*
- *If K is a number field, there exist up to isomorphism only finitely many universal torsors $\pi : \mathcal{T} \to V$ such that $\mathcal{T}(K) \neq \emptyset$.*

This proposition gives us a nice decomposition of the set of rational points

$$V(K) = \bigsqcup_{1 \leqslant i \leqslant m} \pi_i(\mathcal{T}_i(K)).$$

Heuristic 8.4. *From the arithmetical point of view, universal torsors should be much simpler than the variety V.*

This heuristic can be justified by the following statement:

Proposition 8.5 **(Colliot-Thélène, Sansuc).** *If $\mathcal{T}^c$ is a smooth projective compactification of a universal torsor $\mathcal{T} \to V$, then $\mathcal{T}^c(\boldsymbol{A}_K)^{\mathrm{Br}} = \mathcal{T}^c(\boldsymbol{A}_K)$. In other words, there is no Brauer–Manin obstruction to the Hasse principle or weak approximation.*

Example 8.6. Let $V \subset \mathbf{P}^N$ be a hypersurface over $\mathbf{Q}$ with $\dim V \geqslant 3$, $\deg V = d$, and $N+1-d > 0$. Then the cone $W \subset \mathbf{A}^{N+1} - \{0\}$ above V is, up to isomorphism, the unique universal torsor over V.

9. Toric varieties

The following construction is due to Cox. Let T be an algebraic torus and V a smooth projective equivariant compactification of T. This means that there is an action of T on V, an open subset $U \subset V$, and an isomorphism from U to T compatible with the actions of T.

Denote by $\Sigma(1)$ the set of orbits of codimension 1 in $\overline{V}$. Then there is an exact sequence of $\mathrm{Gal}(\overline{K}/K)$-modules

$$0 \to X^*(T) \xrightarrow{\mathrm{div}} \mathbf{Z}^{\Sigma(1)} \xrightarrow{\rho} \mathrm{Pic}(\overline{V}) \to 0$$
$$e_\sigma \mapsto [D_\sigma]$$

where D_σ is the closure of the orbit σ in $\overline{V}$, which is an irreducible divisor of $\overline{V}$. Moreover, we have

$$\omega_V^{-1} = \sum_{\sigma \in \Sigma(1)} [D_\sigma].$$

By duality, we get an exact sequence of tori

$$1 \to T_{\mathrm{NS}} \to T_{\Sigma(1)} \xrightarrow{\pi} T \to 1.$$

But $T \subset V$ and we want to extend the map π to get a torsor over V. We do this in the following way: We consider the affine space

$$\mathbf{A}_{\Sigma(1)} = \mathrm{Spec}\big((\overline{K}[X_\sigma]_{\sigma\in\Sigma(1)})^{\mathrm{Gal}(\overline{K}/K)}\big)$$

and a closed subset $F \subset \mathbf{A}_{\Sigma(1)}$, defined over $\overline{K}$ as a union of affine subspaces,

$$\overline{F} = \bigcup_{\substack{I \subset \Sigma(1) \\ \bigcap_{\sigma\in I} D_\sigma = \emptyset}} \Big(\bigcap_{\sigma\in I} (X_\sigma = 0)\Big).$$

Note that $\overline{F}$ is stable under the action of the Galois group, so it is defined over K. We take $\mathcal{T} = \mathbf{A}_{\Sigma(1)} - F$.

Claim 9.1. *For all $x \in T(K)$, the map*

$$\begin{aligned} T_{\Sigma(1)} &\to T \\ t &\mapsto \pi(t) \cdot x \end{aligned}$$

extends to a map $\mathcal{T} \to V$ sending $1 \in T_{\Sigma(1)}$ to x.

Theorem 9.2. **(Colliot-Thélène, Sansuc, Salberger, Madore).** *The above construction gives a bijection between $T(K)/T_{\Sigma(1)}(K)$ and isomorphism classes of universal torsors over V.*

We return now to the problem of counting points. We assume that $K = \mathbf{Q}$, that the action of $\mathrm{Gal}(\overline{\mathbf{Q}}/\mathbf{Q})$ on $X^*(T)$ and $\Sigma(1)$ are trivial, and that ω_V^{-1} is generated by global sections.

Then we consider

$$\mathscr{M} = \{m \in \mathbf{Z}^{\Sigma(1)} \mid \forall \sigma \in \Sigma(1),\ m_\sigma \geqslant 0 \text{ and } \rho(m) = \omega_V^{-1} \in \operatorname{Pic} V\}.$$

For all $m \in \mathscr{M}$, let $X^m \in \mathbf{Q}[X_\sigma]_{\sigma\in\Sigma(1)}$ be the corresponding monomial. We lift the height to the universal torsor by

$$\widetilde{H}((y_\sigma)_{\sigma\in\Sigma(1)}) = \sup_{m\in\mathscr{M}} |X^m((y_\sigma)_{\sigma\in\Sigma(1)})|.$$

Theorem 9.3 **(Salberger [Sal98]).** *There exists a height H relative to ω_V^{-1} such that $N_{U,H}(B) = \widetilde{N}(B)/2^{\dim T_{\mathrm{NS}}}$, where $\widetilde{N}(B)$ is the number of $(y_\sigma)_{\sigma\in\Sigma(1)}$ in $\mathbf{Z}^{\Sigma(1)}$ such that*

$$\begin{cases} \widetilde{H}(y) \leqslant B, \\ \forall I \subset \Sigma(1),\ \bigcap_{\upsilon\in I} D_\sigma = \emptyset \Rightarrow \gcd_{\upsilon\in I}(y_\sigma) = 1. \end{cases}$$

To prove Manin's conjecture in that case one may then proceed as follows: By use of a Möbius inversion formula, reduce to give an estimate

$$\#\Big\{(y_\sigma)_{\sigma\in\Sigma(1)} \in (\mathbf{Z} - \{0\})^{\Sigma(1)} \;\Big|\; \widetilde{H}(y) \leqslant B\Big\}$$

and prove that when B goes to $+\infty$ this is asymptotic to

$$\operatorname{vol}\Big(\Big\{(y_\sigma)_{\sigma\in\Sigma(1)} \in \mathbf{R}^{\Sigma(1)} \;\Big|\; \widetilde{H}(y) \leqslant B\Big\}\Big) \sim CB(\log B)^{\operatorname{rk}\operatorname{Pic} V - 1},$$

which proves the conjecture in this case.

10. The plane blown up in 4 points

The construction is due to Salberger and Skorobogatov. We consider in this section the blowup $\pi : V \to \mathbf{P}^2$ of $P_1 = (1:0:0)$, $P_2 = (0:1:0)$, $P_3 = (0:0:1)$, and $P_4 = (1:1:1)$. The exceptional divisors on V are $E_{i,5} = \pi^{-1}(P_i)$ and $E_{i,j}$, the strict pullback of the line through P_k and P_l if $\{i,j,k,l\} = \{1,2,3,4\}$. Then $E_{i,j} \cap E_{k,l} = \emptyset$ if and only if $\{i,j\} \cap \{k,l\} \neq \emptyset$. Then we consider the Grassmannian variety $\mathrm{Gr}(2,5)$ of the planes in $\mathbf{Q}^5$; we may embed it into $\mathbf{P}(\Lambda^2\mathbf{Q}^5)$. The cone above it, $W \subset \Lambda^2\mathbf{Q}^5$ is given by the Plücker relations:

$$\begin{cases} X_{1,2}X_{3,4} - X_{1,3}X_{2,4} + X_{1,4}X_{2,3} &= 0, \\ X_{1,2}X_{3,5} - X_{1,3}X_{2,5} + X_{1,5}X_{2,3} &= 0, \\ X_{1,2}X_{4,5} - X_{1,4}X_{2,5} + X_{1,5}X_{2,4} &= 0, \\ X_{1,3}X_{4,5} - X_{1,4}X_{3,5} + X_{1,5}X_{3,4} &= 0, \\ X_{2,3}X_{4,5} - X_{2,4}X_{3,5} + X_{2,5}X_{3,4} &= 0. \end{cases}$$

Indeed, the vector space $\Lambda^2\mathbf{Q}^5$ has dimension 10, and we take $X_{i,j}$ as coordinates corresponding to the basis elements $e_i \wedge e_j$ for $i \neq j$. We consider the closed subset $F \subset W$ given by

$$F = \bigcup_{\{i,j\}\cap\{k,\ell\}\neq\emptyset} ((X_{i,j} = 0) \cap (X_{k,\ell} = 0)),$$

and define $\mathcal{T} = W - F$. There is an action $\mathbf{G}_m^5 \subset GL_5(\mathbf{Q})$ on $\mathcal{T}$, and $\mathcal{T}/\mathbf{G}_m^5$ is isomorphic to V and $\mathcal{T} \to V$ is up to isomorphism the only universal torsor.

We put

$$\mathscr{M} = \left\{ (m_{i,j})_{i<j} \in \mathbf{Z}^{10} \;\middle|\; \sum m_{i,j} E_{i,j} = \omega_V^{-1} \right\}$$

and for $m \in \mathscr{M}$, $X^m \in \mathbf{Q}[X_{i,j}, i<j]$. Then we may lift the height by using

$$\widetilde{H}((y_{i,j})) = \sup_{m \in \mathscr{M}} |X^m(y)|.$$

Proposition 10.1 **(Salberger).** *There is a height H relative to ω_V^{-1} such that $N_{U,H}(B) = \widetilde{N}(B)/2^{\dim T_{\mathrm{NS}}}$ where $\widetilde{N}(B)$ is the number of $(y_{i,j})$ in $W(\mathbf{Z})$ such that*

$$\begin{cases} \widetilde{H}(y_{i,j}) \leqslant B, \\ \{i,j\} \cap \{k,l\} \neq \emptyset \Rightarrow \gcd(y_{i,j}, y_{k,l}) = 1. \end{cases}$$

This description, which was made by Salberger in order to get an upper bound for the number of points of bounded height, was also the first step in the proof of the empirical formula (F), given by de la Bretèche in [**dlB02**].

11. Generalization

The papers **[Pey98]** and **[Pey01]**, showed how to generalize the lifting described in Sections 7, 9, and 10. The first fact enabling this generalization is the existence of a gauge form on each universal torsor:

Proposition 11.1. *If V is a Fano variety over K, and $\mathcal{T}$ a universal torsor over V, then*

- *the canonical bundle $\omega_{\mathcal{T}}$ is trivial,*
- $\Gamma(\mathcal{T}, \mathbf{G}_m) = K^*$.

As usual, a gauge form yields a measure:

Definition 11.2. Up to a constant, there exists a unique non-vanishing section of the canonical line bundle. Let $\breve{\omega}_{\mathcal{T}}$ be such a section. This section defines for any place v of K a measure $\boldsymbol{\omega}_{\mathcal{T},v}$ on $\mathcal{T}(K_v)$ which is locally given by

$$\boldsymbol{\omega}_{\mathcal{T},v} = \left|\left\langle \frac{\partial}{\partial x_1} \wedge \cdots \wedge \frac{\partial}{\partial x_N}, \breve{\omega}_{\mathcal{T}} \right\rangle\right|_v \mathrm{d}x_{1,v} \cdots \mathrm{d}x_{N,v},$$

where $(x_1, \ldots, x_N)$ is an analytic local system of coordinates on $\mathcal{T}(K_v)$ for v-adic topology. We then define a canonical measure on $\prod_{v \in \Omega_K} \mathcal{T}(K_v)$ by

$$\boldsymbol{\omega}_{\mathcal{T}} = \frac{1}{\sqrt{d_K}^{\dim \mathcal{T}}} \prod_{v \in \Omega_K} \boldsymbol{\omega}_{\mathcal{T},v}.$$

Remarks 11.3.

(i) If V is a hypersurface in $\mathbf{P}^n$, the measure $\boldsymbol{\omega}_{\mathcal{T},v}$ coincides with the classical Leray measure on $\mathcal{T}(K_v)$.

(ii) The measure $\boldsymbol{\omega}_{\mathcal{T}}$ does not depend on the choice of the section $\breve{\omega}_{\mathcal{T}}$.

(iii) The volume $\boldsymbol{\omega}_{\mathcal{T}}(\prod_{v \in \Omega_K} \mathcal{T}(K_v))$ is infinite, but if $S \subset \Omega_K$ is a finite set of places containing the archimedean ones and $\mathscr{T}$ a model of $\mathcal{T}$ over the ring of S-integers $\mathscr{O}_S$, then the product $\prod_{v \in \Omega_K - S} \boldsymbol{\omega}_{\mathcal{T},v}(\mathscr{T}(\mathscr{O}_v))$ converges.

Let $\mathcal{T}_1, \ldots, \mathcal{T}_r$ be torsors representing all isomorphism classes of universal torsors over V having a rational point. It is then possible to construct families of integrable functions $\Psi_{i,j,B} : \prod_{v \in \Omega_K} \mathcal{T}(K_v) \to \mathbf{R}_{>0}$ such that upper bounds for the difference

$$\left| \sum_{y \in \mathcal{T}_i(K)} \Psi_{i,j,B}(y) - \int_{\prod_{v \in \Omega_K} \mathcal{T}(K_v)} \Psi_{i,j,B}(y) \boldsymbol{\omega}_{\mathcal{T}} \right|$$

yield an upper bound for the difference

$$|N_{U,H}(B) - \theta_H(V) B (\log B)^{t-1}|.$$

The liftings described in previous sections may be seen as particular cases of this descent argument.

References

[Bir62] B. J. Birch – Forms in many variables, *Proc. Roy. Soc. Ser. A* **265** (1961/1962), 245–263.

[BM90] V. V. Batyrev & Y. I. Manin – Sur le nombre des points rationnels de hauteur bornée des variétés algébriques, *Math. Ann.* **286** (1990), no. 1-3, 27–43.

[BT95] V. V. Batyrev & Y. Tschinkel – Rational points of bounded height on compactifications of anisotropic tori, *Internat. Math. Res. Notices* (1995), no. 12, 591–635.

[BT96a] V. Batyrev & Y. Tschinkel – Height zeta functions of toric varieties, *J. Math. Sci.* **82** (1996), no. 1, 3220–3239.

[BT96b] V. V. Batyrev & Y. Tschinkel – Rational points on some Fano cubic bundles, *C. R. Acad. Sci. Paris Sér. I Math.* **323** (1996), no. 1, 41–46.

[BT98] ———, Manin's conjecture for toric varieties, *J. Algebraic Geom.* **7** (1998), no. 1, 15–53.

[CLT00a] A. Chambert-Loir & Y. Tschinkel – Points of bounded height on equivariant compactifications of vector groups. I, *Compositio Math.* **124** (2000), no. 1, 65–93.

[CLT00b] ———, Points of bounded height on equivariant compactifications of vector groups. II, *J. Number Theory* **85** (2000), no. 2, 172–188.

[CLT02] ———, On the distribution of points of bounded height on equivariant compactifications of vector groups, *Invent. Math.* **148** (2002), no. 2, 421–452.

[CTS76] J.-L. Colliot-Thélène & J.-J. Sansuc – Torseurs sous des groupes de type multiplicatif; applications à l'étude des points rationnels de certaines variétés algébriques, *C. R. Acad. Sci. Paris Sér. A-B* **282** (1976), no. 18, Aii, A1113–A1116.

[CTS77] ———, La descente sur une variété rationnelle définie sur un corps de nombres, *C. R. Acad. Sci. Paris Sér. A-B* **284** (1977), no. 19, A1215–A1218.

[CTS80] J.-L. Colliot-Thélène & J.-J. Sansuc – La descente sur les variétés rationnelles, Journées de Géométrie Algébrique d'Angers, Juillet 1979/Algebraic Geometry, Angers, 1979, Sijthoff & Noordhoff, 1980, 223–237.

[CTS87] J.-L. Colliot-Thélène & J.-J. Sansuc – La descente sur les variétés rationnelles. II, *Duke Math. J.* **54** (1987), no. 2, 375–492.

[Del74] P. Deligne – La conjecture de Weil. I, *Inst. Hautes Études Sci. Publ. Math.* (1974), no. 43, 273–307.

[dlB01] R. de la Bretèche – Compter des points d'une variété torique, *J. Number Theory* **87** (2001), no. 2, 315–331.

[dlB02] ———, Nombre de points de hauteur bornée sur les surfaces de del Pezzo de degré 5, *Duke Math. J.* **113** (2002), no. 3, 421–464.

[FMT89] J. Franke, Y. I. Manin & Y. Tschinkel – Rational points of bounded height on Fano varieties, *Invent. Math.* **95** (1989), no. 2, 421–435.

[Lan76] R. P. Langlands – *On the functional equations satisfied by Eisenstein series*, Springer-Verlag, Berlin, 1976, Lecture Notes in Mathematics, Vol. 544.

[Pey95] E. Peyre – Hauteurs et mesures de Tamagawa sur les variétés de Fano, *Duke Math. J.* **79** (1995), no. 1, 101–218.

[Pey98] ———, Terme principal de la fonction zêta des hauteurs et torseurs universels, *Astérisque* (1998), no. 251, 259–298.

[Pey01] ———, Torseurs universels et méthode du cercle, Rational points on algebraic varieties, Prog. Math., vol. 199, Birkhäuser, Basel, 2001, 221–274.

[PT01a] E. Peyre & Y. Tschinkel – Tamagawa numbers of diagonal cubic surfaces, numerical evidence, *Math. Comp.* **70** (2001), no. 233, 367–387.

[PT01b] ———, Tamagawa numbers of diagonal cubic surfaces of higher rank, Rational points on algebraic varieties, Prog. Math., vol. 199, Birkhäuser, Basel, 2001, 275–305.

[Sal98] P. Salberger – Tamagawa measures on universal torsors and points of bounded height on Fano varieties, *Astérisque* (1998), no. 251, 91–258.

[Sch79] S. H. Schanuel – Heights in number fields, *Bull. Soc. Math. France* **107** (1979), no. 4, 433–449

[SSD98] J. B. Slater & P. Swinnerton-Dyer – Counting points on cubic surfaces. I, *Astérisque* (1998), no. 251, 1–12.

[ST97] M. Strauch & Y. Tschinkel – Height zeta functions of twisted products, *Math. Res. Lett.* **4** (1997), no. 2-3, 273–282.

[ST99] ———, Height zeta functions of toric bundles over flag varieties, *Selecta Math. (N.S.)* **5** (1999), no. 3, 325–396.

[Wei82] A. Weil – *Adeles and Algebraic Groups*, Prog. Math., vol. 23, Birkhäuser Boston, Mass., 1982.

Part II

Research Articles

Arithmetic of Higher-dimensional Algebraic Varieties
(B. Poonen, Yu. Tschinkel, eds.), p. 85–103
Progress in Mathematics, Vol. 226, © 2004 Birkhäuser Boston, Cambridge, MA

THE COX RING OF A DEL PEZZO SURFACE

Victor V. Batyrev

Mathematisches Institut Universität Tübingen, Auf der Morgenstelle 10, Tübingen D-72076, Germany • *E-mail :* victor.batyrev@uni-tuebingen.de

Oleg N. Popov

Department of Algebra, Faculty of Mathematics, Moscow State University, Moscow 117234, Russia • *E-mail :* popov@mccme.ru

Abstract. Let X_r be a smooth Del Pezzo surface obtained from $\mathbb{P}^2$ by blowing up $r \leqslant 8$ points in general position. It is well known that for $r \in \{3, 4, 5, 6, 7, 8\}$ the Picard group $\mathrm{Pic}(X_r)$ contains a canonical root system $R_r \in \{A_2 \times A_1, A_4, D_5, E_6, E_7, E_8\}$. We prove some general properties of the Cox ring of X_r ($r \geqslant 4$) and show its similarity to the homogeneous coordinate ring of the orbit of the highest weight vector in some irreducible representation of the algebraic group G associated with the root system R_r.

1. Introduction

Let X be a projective algebraic variety over a field $\Bbbk$. Assume that the Picard group $\mathrm{Pic}(X)$ is a finitely generated abelian group. Consider the vector space

$$\Gamma(X) := \bigoplus_{[D] \in \mathrm{Pic}(X)} H^0(X, \mathscr{O}(D)).$$

Key words and phrases. Del Pezzo surfaces, torsors, homogeneous spaces, algebraic groups.

We want to make it a $\Bbbk$-algebra graded by the monoid of effective classes in $\mathrm{Pic}(X)$ in such a way that the algebra structure is compatible with the natural bilinear map

$$b_{D_1,D_2} \,:\, H^0(X,\mathcal{O}(D_1)) \times H^0(X,\mathcal{O}(D_2)) \to H^0(X,\mathcal{O}(D_1+D_2)).$$

However, there exist some problems with the realization of this idea. First, there is no natural isomorphism between $H^0(X,\mathcal{O}(D))$ and $H^0(X,\mathcal{O}(D'))$ if $[D]=[D']$. There exists only a canonical bijection between the linear systems $|D| \cong |D'|$ (where $|D|$ is the projectivization of the $\Bbbk$-vector space $H^0(X,\mathcal{O}(D))$). As a consequence, the bilinear map b_{D_1,D_2} depends not only on the classes $[D_1],[D_2],[D_1+D_2] \in \mathrm{Pic}(X)$, but also on their particular representatives. One can easily see that only the morphism

$$s_{[D_1],[D_2]} \,:\, |D_1| \times |D_2| \to |D_1+D_2|$$

of the product of two projective spaces $|D_1| \times |D_2|$ to another projective space $|D_1+D_2|$ is well defined. For this reason, it is much more natural to consider the graded set of projective spaces

$$|\Gamma(X)| := \bigsqcup_{[D]\in \mathrm{Pic}(X)} |D|$$

together with all possible morphisms $s_{[D_1],[D_2]}$ for any two effective classes $[D_1],[D_2] \in \mathrm{Pic}(X)$.

Inspired by the paper of Cox on the homogeneous ring of a toric variety **[Cox95]**, Hu and Keel **[HK00]** suggested a definition of a *Cox ring*

$$\mathrm{Cox}(X) = R(X,L_1,\ldots,L_r) := \bigoplus_{(m_1,\ldots,m_r)\in\mathbb{Z}^r} H^0(X,\mathcal{O}(m_1L_1+\cdots+m_rL_r))$$

which uses a choice of some $\mathbb{Z}$-basis $L_1,\ldots,L_r$ in $\mathrm{Pic}(X)$ (e.g., if $\mathrm{Pic}(X)\cong\mathbb{Z}^r$ is a free abelian group). Using such a $\mathbb{Z}$-basis, one obtains a particular representative for each class in $\mathrm{Pic}(X)$ together with a well-defined multiplication, so $R(X,L_1,\ldots,L_r)$ becomes a well-defined $\Bbbk$-algebra. If $L_1',\ldots,L_r'$ is another $\mathbb{Z}$-basis of $\mathrm{Pic}(X)$, then the corresponding Cox algebra $R(X,L_1',\ldots,L_r')$ is isomorphic to $R(X,L_1,\ldots,L_r)$. Unfortunately, we cannot expect to choose a $\mathbb{Z}$-basis of $\mathrm{Pic}(X)$ in a natural canonical way. More often one can choose in a natural way some effective divisors $D_1,\ldots,D_n$ on X such that $\mathrm{Pic}(X)$ is generated by $[D_1],\ldots,[D_n]$. If we set

$$U := X \setminus (D_1 \cup \cdots \cup D_n)$$

and assume that X is smooth, then $\mathrm{Pic}(U) = 0$ and we obtain the exact sequence

$$1 \to \Bbbk^* \to \Bbbk[U]^* \to \bigoplus_{i=1}^{n} \mathbb{Z}[D_i] \to \mathrm{Pic}(X) \to 0.$$

Choosing a $\Bbbk$-rational point p in U, we can split the monomorphism $\Bbbk^* \to \Bbbk[U]^*$, so that one has an isomorphism

$$\Bbbk[U]^* \cong \Bbbk^* \oplus G,$$

where $G \subset \Bbbk[U]^*$ is a free abelian group of rank $n-r$. The choice of a $\Bbbk$-rational point $p \in U$ gives us another way to construct the graded space $\Gamma(X)$ and the Cox algebra:

Definition 1.1. Let $X, U, p, D_1, \ldots, D_n$ be as above. We consider the graded $\Bbbk$-algebra

$$\Gamma(X, U, p) := \bigoplus_{(m_1,\ldots,m_n)\in\mathbb{Z}^n} H^0(X, \mathscr{O}(m_1D_1 + \cdots + m_nD_n))$$

and define

$$\mathrm{Cox}(X, U, p) := \Gamma(X, U, p)_G$$

as the quotient of $\Gamma(X, U, p)$ modulo the ideal generated by

$$\{x - gx \mid x \in \Gamma(X, U, p), g \in G\}.$$

Since $\mathrm{Pic}(X) \cong \mathbb{Z}^n/G$, we obtain a natural $\mathrm{Pic}(X)$-grading on $\mathrm{Cox}(X, U, p)$.

We expect that the algebra $\mathrm{Cox}(X, U, p)$ can be applied to some arithmetic questions about $\Bbbk$-rational points in $U \subset X$.

Remark 1.2. The definition of the ring $\mathrm{Cox}(X, U, p)$ depends on the choice of an open subset $U \subset X$ and a $\Bbbk$-rational point $p \in U$. A similar idea was used by Colliot-Thélène and Sansuc in **[CTS87]** for constructing universal torsors and deriving explicit equations for them. It is the absence of a canonical choice in the above construction what makes descending the universal torsor an interesting problem. Some applications of universal torsors over Del Pezzo surfaces of degree 5 were considered by Skorobogatov in **[Sko93]** (see also **[Sko01]**). Recently, Hassett and Tschinkel have investigated the Cox rings and the universal torsors for some interesting singular cubic surfaces **[HT03]**.

Remark 1.3. If X is a smooth projective toric variety and $U \subset X$ is the open dense torus orbit, then the choice of a point $p \in U$ defines an isomorphism of U with the algebraic torus T, so that the subgroup $G \subset \Bbbk[U]^*$ can be identified with the character group of T. In this way, one can show that $\mathrm{Cox}(X, U, p)$

is isomorphic to a polynomial ring in n variables, where n is the number of irreducible components of $X \setminus U$, cf. [**Cox95**].

Remark 1.4. The field of fractions of the ring $\mathrm{Cox}(X, U, p)$ is a purely transcendental extension of degree r of the field of rational functions on X. Therefore, $\dim \operatorname{Spec} \Gamma(X, U, p) = \dim X + r$, if $\Gamma(X, U, p)$ is a finitely generated $\Bbbk$-algebra.

Let X_r be a smooth Del Pezzo surface obtained from $\mathbb{P}^2$ by blow-up of $r \leqslant 8$ points in general position. It is well known that for $r \in \{3, 4, 5, 6, 7, 8\}$ the Picard group $\mathrm{Pic}(X_r)$ contains a canonical root system

$$R_r \in \{A_2 \times A_1, A_4, D_5, E_6, E_7, E_8\}.$$

Moreover, the natural embedding $\mathrm{Pic}(X_{r-1}) \hookrightarrow \mathrm{Pic}(X_r)$ induces the inclusion of root systems $R_{r-1} \hookrightarrow R_r$. If $G(R_r)$ is a connected algebraic group corresponding to the root system R_r, then the embedding $R_{r-1} \hookrightarrow R_r$ defines a maximal parabolic subgroup $P(R_{r-1}) \subset G(R_r)$ [**Hum75**]. We expect that for $r \geqslant 4$ there should be some relation between a Del Pezzo surface X_r and the GIT-quotient of the homogeneous space $G(R_r)/P(R_{r-1})$ modulo the action of a maximal torus T_r of $G(R_r)$.

Our starting observation is the well-known isomorphism $X_4 \cong G(3,5)//T_4$ which follows from an isomorphism between the homogeneous coordinate ring of the Grassmannian $G(3,5) = G(A_4)/P(A_2 \times A_1) \subset \mathbb{P}^9$ and the Cox ring of X_4 (see 4.1). Another proof of this fact uses the identification of X_4 with the moduli space $\overline{M_{0,5}}$ of stable rational curves with 5 marked points [**Kap93**].

In this paper, we investigate the Cox ring of Del Pezzo surfaces X_r for $r \geqslant 4$. It is natural to choose the set of classes of all exceptional curves $E_1, \ldots, E_{N_r} \subset X_r$ as a generating set for the Picard group $\mathrm{Pic}(X_r)$. There is a natural $\mathbb{Z}_{\geqslant 0}$-grading on $\mathrm{Pic}(X_r)$ defined by the intersection with the anticanonical divisor $-K$.

We prove several general properties of Cox rings of Del Pezzo surfaces X_r for $r \geqslant 4$ and show their similarity to homogeneous coordinate rings of $G(R_r)/P(R_{r-1})$. We remark that the homogeneous space $G(R_r)/P(R_{r-1})$ can be interpreted as the orbit of the highest weight vector in some natural irreducible representation of $G(R_r)$.

Remark 1.5. Some other connections between Del Pezzo surfaces and the corresponding algebraic groups were considered recently by Friedman and Morgan in [**FM02**] (see also Leung in [**Leu00**]).

In this paper, we show that the Cox ring of a Del Pezzo surface X_r is generated by elements of degree 1. This implies that the homogeneous coordinate ring of $G(R_r)/P(R_{r-1})$ is naturally graded by the monoid of effective divisor classes on the surface X_r (the same monoid defines the multigrading of the Cox ring of X_r). Moreover, we obtain some results of the quadratic relations between the generators of the Cox ring of X_r.

The authors would like to thank Yu. Tschinkel, A. Skorobogatov, E. S. Golod, S. M. Lvovski and E. B. Vinberg for useful discussions and encouraging remarks.

2. Del Pezzo surfaces

Let us summarize briefly some well-known classical results on Del Pezzo surfaces which can be found in [**Man86, DP80, Nag60, Nag61**].

For $r \leqslant 8$, one says that r points $p_1, \dots, p_r$ in $\mathbb{P}^2$ are in *general position* if no 3 of them lie a line, no 6 of them lie on a conic ($r \geqslant 6$), and any cubic containing 7 of the points having one of the 7 points as a double point does not pass through the 8th point ($r = 8$).

Denote by X_r ($r \geqslant 3$) the Del Pezzo surfaces obtained from $\mathbb{P}^2$ by blowing up of r points $p_1, \dots, p_r$ in general position. If $\pi : X_r \to \mathbb{P}^2$ is the corresponding projective morphism, then the Picard group $\mathrm{Pic}(X_r) \cong \mathbb{Z}^{r+1}$ contains a $\mathbb{Z}$-basis $l_i, (0 \leqslant i \leqslant r)$, where $l_0 = [\pi^* \mathcal{O}(1)]$ and $l_i := [\pi^{-1}(p_i)]$, $i = 1, \dots, r$. With respect to this basis, the intersection form $(*, *)$ on $\mathrm{Pic}(X_r)$ is given by a diagonal matrix: $(l_0, l_0) = 1$, $(l_i, l_i) = -1$ for $i \geqslant 1$, and $(l_i, l_j) = 0$ for $i \neq j$. The anticanonical class of X_r equals $-K = 3l_0 - l_1 - \cdots - l_r$. The number $d := (K, K) = 9 - r$ is called the *degree* of X_r. The anticanonical system $|-K|$ of a Del Pezzo surface X_r is very ample if $r \leqslant 6$, it determines a two-fold covering of $\mathbb{P}^2$ if $r = 7$, and it has one base point, determining a rational map to $\mathbb{P}^1$ if $r = 8$. Smooth rational curves $E \subset X_r$ such that $(E, E) = -1$ and $(E, -K) = 1$ are called *exceptional curves*.

Theorem 2.1. [**Man86**] *The exceptional curves on X_r are the following:*

1. *blown-up points $p_1, \dots, p_r$;*
2. *lines through pairs of points p_i, p_j;*
3. *conics through 5 points from $\{p_1, \dots, p_r\}$ ($r \geqslant 5$);*
4. *cubics containing 7 points and 1 of them double ($r \geqslant 7$);*
5. *quartics containing 8 points and 3 of them double ($r = 8$);*
6. *quintics containing 8 points and 6 of them double ($r = 8$);*
7. *sextics containing 8 of those points, 7 of them double and 1 triple ($r = 8$).*

(In items 2 through 7, we mean the strict transforms of the given curves in $\mathbb{P}^2$.) The number N_r of exceptional curves on X_r is given by the following table:

r	3	4	5	6	7	8
N_r	6	10	16	27	56	240

The root system $R_r \subset \mathrm{Pic}(X_r)$ is defined as

$$R_r := \{\, \alpha \in \mathrm{Pic}(X_r) \;:\; (\alpha, \alpha) = -2,\ (\alpha, -K) = 0 \,\}.$$

It is easy to show that R_r is exactly the set of all classes $\alpha = [E_i] - [E_j]$ where E_i and E_j are two exceptional curves on X_r such that $E_i \cap E_j = \emptyset$.

The corresponding Weyl group W_r is generated by the reflections $\sigma : x \mapsto x + (x, \alpha)\alpha$ for $\alpha \in R_r$. There are so called *simple roots* $\alpha_1, \ldots, \alpha_r$ such that the corresponding reflexions $\sigma_1, \ldots, \sigma_r$ form a minimal generating subset of W_r. The set of simple roots can be chosen as

$$\alpha_1 = l_1 - l_2, \alpha_2 = l_2 - l_3, \alpha_3 = l_0 - l_1 - l_2 - l_3,$$

$$\alpha_i = l_{i-1} - l_i, \quad i \geqslant 4.$$

The blow-up morphism $X_r \to X_{r-1}$ determines an isometric embedding of Picard lattices $\mathrm{Pic}(X_{r-1}) \hookrightarrow \mathrm{Pic}(X_r)$. This induces embeddings for root systems, simple roots and Weyl groups W_r. For $r \geqslant 3$, the Dynkin diagram of R_r can be considered as the subgraph on the vertices α_i $(i \leqslant r)$ of the following graph:

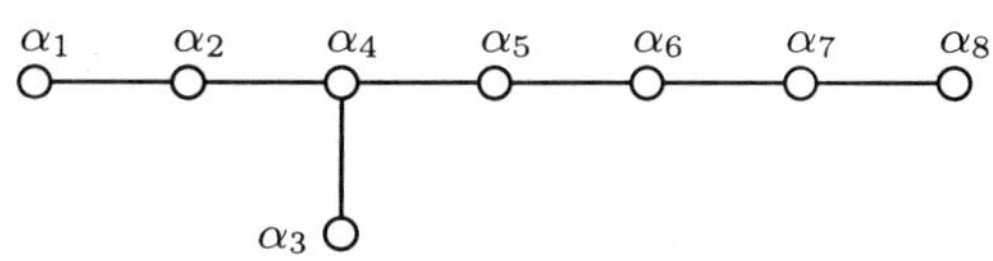

In particular, we obtain $R_3 = A_2 \times A_1$, $R_4 = A_4$, $R_5 = D_5$, $R_6 = E_6$, $R_7 = E_7$, $R_8 = E_8$.

Denote by $\varpi_1, \ldots, \varpi_r$ the dual basis to the $\mathbb{Z}$-basis $-\alpha_1, \ldots, -\alpha_r$. Each ϖ_i is the highest weight of an irreducible representation of $G(R_r)$ called a *fundamental representation.* We shall denote by $V(\varpi)$ the representation space of $G(R_r)$ with the highest weight ϖ.

Definition 2.2. A dominant weight ϖ is called *minuscule* if all weights of $V(\varpi)$ are nonzero and the W_r-orbit of the highest weight vector is a $\Bbbk$-basis of $V(\varpi)$ [**Ses78**]). A dominant weight ϖ is called *quasiminuscule* [**LMS79**], if all nonzero weights of $V(\varpi)$ have multiplicity 1 and form an W_r-orbit of ϖ (the zero weight of $V(\varpi)$ may have some positive multiplicity).

One can see from the explicit description of the root systems R_r that ϖ_r is minuscule for $3 \leqslant r \leqslant 7$, and ϖ_8 is quasiminuscule.

The dimension d_r of of the irreducible representation $V(\varpi_r)$ of $G(R_r)$ is given by the following table:

r	4	5	6	7	8
d_r	10	16	27	56	248

We will need the following statement:

Proposition 2.3. *Let D be a divisor on a Del Pezzo surface X_r $(2 \leqslant r \leqslant 8)$ such that $(D, E) \geqslant 0$ for every exceptional curve $E \subset X_r$. Then the following statements hold:*
(i) the linear system $|D|$ has no base points on any exceptional curve $E \subset X_r$;
(ii) if $r \leqslant 7$, then the linear system $|D|$ has no base points on X_r.

Proof. Induction on r. If $r = 2$, then there exist exactly 3 exceptional curves E_0, E_1, E_2, whose classes in the standard basis are $l_0 - l_1 - l_2, l_1, l_2$. Moreover $[E_0]$, $[E_1]$ and $[E_2]$ form a basis of the Picard lattice $\mathrm{Pic}(X_2)$. The dual basis with respect to the intersection form is l_0, $l_0 - l_1$, $l_0 - l_2$. Therefore the above conditions on D imply that

$$[D] = n_0 l_0 + n_1(l_0 - l_1) + n_2(l_0 - l_2), \quad n_0, n_1, n_2 \in \mathbb{Z}_{\geqslant 0}.$$

So it suffices to check that the linear systems with the classes $l_0, l_0 - l_1, l_0 - l_2$ have no base points. This holds since the first linear system defines the birational morphism $X_2 \to \mathbb{P}^2$ contracting E_1 and E_2, and the second and third linear systems define conic bundle fibrations over $\mathbb{P}^1$.

For $r > 2$, we use an inner induction on $\deg D = (D, -K)$.

If there is an exceptional curve $E \subset X_r$ with $(D, E) = 0$, then the invertible sheaf $\mathscr{O}(D)$ is the inverse image of an invertible sheaf $\mathscr{O}(D')$ on the Del Pezzo surface X_{r-1} obtained by contracting E. Since the pullback of any exceptional curve on X_{r-1} under the birational morphism $\pi_E : X_r \to X_{r-1}$ is again an exceptional curve on X_r, we obtain that D' satisfies all conditions of the proposition on X_{r-1}. By the induction assumption $(r - 1 \leqslant 7)$, the linear system $|D'|$ has no base points on X_{r-1}. Therefore, $|D| = |\pi_E^* D'|$ has no base points on X_r.

If there is no exceptional curve $E \subset X_r$ with $(D, E) = 0$, then we denote by m the minimal intersection number (D, E) where E runs over all exceptional curves. Since we have $(E, -K) = 1$ for all exceptional curves, the divisor $D' := D + mK$ has nonnegative intersections with all exceptional curves, and there exists an exceptional curve $E \subset X_r$ with $(D', E) = 0$. Since $\deg D' = (D', -K) = (D, -K) - m(K, K) < (D, -K) = \deg D$, by the induction assumption, we obtain that $|D'|$ is base point free. If $r \leqslant 7$, then the anticanonical linear system $|-K|$ has no base points. Therefore, $|D| = |D' + m(-K)|$ is also base point free. In the case $r = 8$, $|-K|$ does have a base point $p \in X_8$. However, p cannot lie on an exceptional curve E, because the short exact sequence

$$0 \to H^0(X_8, \mathcal{O}(-K - E)) \to H^0(X_8, \mathcal{O}(-K)) \to H^0(E, \mathcal{O}(-K)|_E) \to 0$$

induces an isomorphism $H^0(X_8, \mathcal{O}(-K)) \cong H^0(E, \mathcal{O}(-K)|_E)$, since $\deg(-K - E) = 0$ and $H^0(X_8, \mathcal{O}(-K - E)) = 0$. □

3. Generators of $\mathrm{Cox}(X_r)$

Let $\{E_1, \ldots, E_{N_r}\}$ be the set of all exceptional curves on a Del Pezzo surface X_r. We choose a $\Bbbk$-rational point $p \in U := X_r \setminus (\bigcup_{i=1}^{N_r} E_i)$ and denote the ring $\mathrm{Cox}(X_r, U, p)$ (see 1.1) simply by $\mathrm{Cox}(X_r)$.

The ring

$$\mathrm{Cox}(X_r) = \bigoplus_{[D] \in M_{\mathrm{eff}}(X_r)} \mathrm{Cox}(X_r)^{[D]}$$

is graded by the semigroup $M_{\mathrm{eff}}(X_r) \subset \mathrm{Pic}(X_r)$ of classes $[D]$ of effective divisors D on X_r. There is a coarser grading on $\mathrm{Cox}(X_r)$ given by

$$\mathrm{Cox}(X_r)^d := \bigoplus_{\deg [D] = d} \mathrm{Cox}(X_r)^{[D]},$$

where $\deg [D] := (D, -K)$.

Proposition 3.1. *The graded ring* $\mathrm{Cox}(X_3)$ *is isomorphic to a polynomial ring in 6 variables* $\Bbbk[x_1, \ldots, x_6]$, *where* x_i *are sections defining all* 6 *exceptional curves on* X_3.

Proof. The Del Pezzo surface X_3 is a toric variety which can be described as the blow-up of $\mathbb{P}^2$ at 3 torus-invariant points $(1 : 0 : 0)$, $(0 : 1 : 0)$, and $(0 : 0 : 1)$. So we can apply a general result of Cox on toric varieties **[Cox95]** (see also 1.3). □

Theorem 3.2. *For* $3 \leqslant r \leqslant 8$, *the ring* $\mathrm{Cox}(X_d)$ *is generated by elements of degree 1. If* $r \leqslant 7$, *then the generators of* $\mathrm{Cox}(X_d)$ *are global sections of invertible sheaves defining the exceptional curves. If* $r = 8$, *then we should add to the above set of generators two linearly independent global sections of the anticanonical sheaf on* X_8.

Proof. Induction on r. The case $r = 3$ is settled by the previous proposition.

For $r > 3$ we choose an effective divisor D on X_r. We call a section $s \in H^0(X_r, \mathcal{O}(D))$ a *distinguished global section* if its support is contained in the union of exceptional curves of X_r $(r \leqslant 7)$, or if its support is contained in the union of exceptional curves of X_8 and some anticanonical curves on X_8. Our purpose is to show that the vector space $H^0(X_r, \mathcal{O}(D))$ is spanned by distinguished global sections.

This will be proved by induction on $\deg D := (D, -K) > 0$.

We consider several cases:

- If there exists an exceptional curve E such that $(D, E) < 0$, then
$$H^0(X_r, \mathcal{O}(D)|_E) = 0.$$
The exact sequence
$$H^1(X_r, \mathcal{O}(D)|_E) \to H^0(X_r, \mathcal{O}(D - E)) \to H^0(X_r, \mathcal{O}(D)) \to 0$$
implies that the multiplication by a nonzero distinguished global section of $\mathcal{O}(E)$ induces a surjection $H^0(X_r, \mathcal{O}(D-E)) \to H^0(X_r, \mathcal{O}(D))$. Since $\deg(D-E) = \deg D - 1$, the inductive hypothesis for $D' = D - E$ implies the required statement for D.

- If there exists an exceptional curve E such that $(D, E) = 0$, then $\mathcal{O}(D)$ is the inverse image of a sheaf $\mathcal{O}(D')$ on the Del Pezzo surface X_{r-1} obtained by the contraction of E. Therefore we have an isomorphism $H^0(X_r, \mathcal{O}(D)) \cong H^0(X_{r-1}, \mathcal{O}(D'))$. By the inductive hypothesis for $r-1$, we obtain the required statement for D, because distinguished global sections of $\mathcal{O}(D')$ lift to distinguished global sections of $\mathcal{O}(D)$.

- If $D = -K$ (or equivalently, if $(D, E) = 1$ for every exceptional curve E), then $\mathcal{O}(D)|_E$ is isomorphic to $\mathcal{O}_E(1)$. Then $H^1(X_r, \mathcal{O}(D)|_E) = 0$ so we obtain an exact sequence
$$0 \to H^0(X_r, \mathcal{O}(D - E)) \to H^0(X_r, \mathcal{O}(D)) \to H^0(X_r, \mathcal{O}(D)|_E) \to 0,$$
in which $H^0(X_r, \mathcal{O}(D)|_E)$ is 2-dimensional. Since
$$\deg(D - E) = \deg D - 1 < \deg D,$$
we can apply the inductive hypothesis for $D' = D - E$. It remains to show that there exist two linearly independent distinguished global sections of $\mathcal{O}(D)$ whose restrictions to E are linearly independent global sections of $\mathcal{O}(D)|_E$. We describe these two distinguisched sections explicitly for each value of $r \in \{4, 5, 6, 7, 8\}$. Without loss of generality we can assume that $[E] = l_1$.

If $r = 4$, then we write the anticanonical class $-K = 3l_0 - l_1 - \cdots - l_4$ in the following two ways:

$$\begin{aligned} -K &= (l_0 - l_1 - l_2) + (l_0 - l_3 - l_4) + (l_0 - l_2 - l_3) + l_2 + l_3 \\ &= (l_0 - l_1 - l_3) + (l_0 - l_2 - l_4) + (l_0 - l_2 - l_3) + l_2 + l_3. \end{aligned}$$

These two decompositions of $-K$ determine two distinguished global sections of $\mathcal{O}(-K)$ with support on 5 exceptional curves. The projections of these sections under the morphism $X_4 \to \mathbb{P}^2$ are shown below in Figure 1.

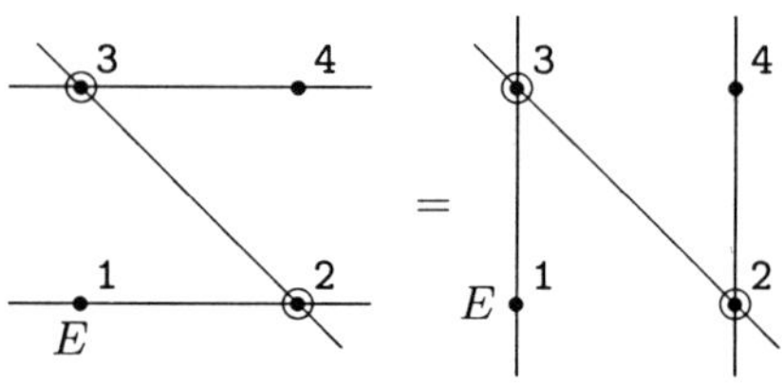

FIGURE 1. Two distinguished anticanonical classes for $r = 4$.

The restriction of the first section to E vanishes at the point q_1 where E meets the exceptional curve having class $l_0 - l_1 - l_2$. The restriction of the second section to E vanishes at the point q_2 where E meets the exceptional curve having class $l_0 - l_1 - l_3$. Since $q_1 \neq q_2$, the distinguished anticanonical sections are linearly independent.

If $r = 5$, then we write the anticanonical class as

$$\begin{aligned} -K &= 3l_0 - l_1 - \cdots - l_5 \\ &= (l_0 - l_1 - l_2) + (l_0 - l_3 - l_4) + (l_0 - l_4 - l_5) + l_4 \\ &= (l_0 - l_1 - l_5) + (l_0 - l_2 - l_3) + (l_0 - l_3 - l_4) + l_3. \end{aligned}$$

The corresponding distinguished anticanonical sections vanish at two different intersection points of E with the exceptional curves belonging to the classes $l_0 - l_1 - l_2$ and $l_0 - l_1 - l_5$.

If $r = 6$, then we write the anticanonical class as

$$\begin{aligned} -K &= 3l_0 - l_1 - \cdots - l_6 \\ &= (l_0 - l_1 - l_2) + (l_0 - l_3 - l_4) + (l_0 - l_5 - l_6) \\ &= (l_0 - l_1 - l_6) + (l_0 - l_5 - l_4) + (l_0 - l_3 - l_2). \end{aligned}$$

The corresponding distinguished anticanonical sections vanish at two different intersection points of E with the exceptional curves belonging to the classes $l_0 - l_1 - l_2$ and $l_0 - l_1 - l_6$.

If $r = 7$, then we write the anticanonical class as

$$\begin{aligned} -K &= 3l_0 - l_1 - \cdots - l_7 \\ &= (2l_0 - l_1 - l_2 - l_3 - l_4 - l_5) + (l_0 - l_6 - l_7) \\ &= (2l_0 - l_7 - l_6 - l_5 - l_4 - l_3) + (l_0 - l_2 - l_1). \end{aligned}$$

The corresponding distinguished anticanonical sections vanish at two different intersection points of E with the exceptional curves belonging to the classes $2l_0 - l_1 - l_2 - l_3 - l_4 - l_5$ and $l_0 - l_2 - l_1$.

If $r = 8$, then $\deg D - E = 0$. Therefore, $H^0(X_8, \mathcal{O}(D - E)) = 0$ (see the proof of 2.3) and we have an isomorphism

$$H^0(X_8, \mathcal{O}(D)) \cong H^0(X_8, \mathcal{O}(D)|_E).$$

So $H^0(X_8, \mathcal{O}(D)|_E)$ is generated by the restrictions of the anticanonical sections and we are done.

– If $(D, E) \geqslant 1$ for all exceptional curves E and $D \neq -K$, then we denote by m the minimum of the numbers (D, E) for all exceptional curves. Let E_0 be an exceptional curve such that $(D, E_0) = m \geqslant 1$. We define $D' = D - E_0$ and $D'' := D + mK$. By 2.3, $|D'|$ and $|D''|$ have no base points (if $r \leqslant 7$). In particular, D'' is represented by an effective divisor. Since $\deg D'' = \deg D - m(K, K) < \deg D$, the divisor D'' can be seen as the zero locus of a distinguished global section $s \in H^0(X_r, \mathcal{O}(D + mK))$ whose support does not contain the exceptional curve E_0 (if $r \leqslant 8$). We have the short exact sequence

$$0 \to H^0(X_r, \mathcal{O}(D')) \to H^0(X_r, \mathcal{O}(D)) \to H^0(X_r, \mathcal{O}(D)|_{E_0}) \to 0.$$

By the inductive hypothesis, the space $H^0(X_r, \mathcal{O}(D'))$ is generated by distinguished global sections. It remains to show that there exist distinguished global sections of $\mathcal{O}(D)$ whose restrictions to E_0 generate the space $H^0(X_r, \mathcal{O}(D)|_{E_0})$. Since $(-mK, E_0) = (D, E_0) = m$, the space $H^0(X_r, \mathcal{O}(D)|_{E_0})$ is isomorphic to $H^0(X_r, \mathcal{O}(-mK)|_{E_0})$. Since $(D'', E_0) = 0$ the distinguished global section $s \in H^0(X_r, \mathcal{O}(D + mK))$ is nonzero at any point of E_0. Therefore multiplication by the distinguished global section s defines a homomorphism

$$H^0(X_r, \mathcal{O}(-mK)) \to H^0(X_r, \mathcal{O}(D))$$

whose restriction to E_0 is an isomorphism

$$H^0(X_r, \mathcal{O}(-mK)|_{E_0}) \cong H^0(X_r, \mathcal{O}(D)|_{E_0}.$$

Therefore it is enough to show that the restrictions of the distinguished global sections of $\mathcal{O}(-mK)$ to E_0 generate the space $H^0(X_r, \mathcal{O}(-mK)|_{E_0})$.

Our previous considerations have shown this for $m = 1$. The general case $m \geqslant 1$ follows, since the homomorphism $H^0(X_r, \mathcal{O}(-K)) \to H^0(E_0, \mathcal{O}_{E_0}(1))$ is surjective and the space $H^0(E_0, \mathcal{O}_{E_0}(m))$ is spanned by tensor products of m elements from $H^0(E_0, \mathcal{O}_{E_0}(1))$.

□

Corollary 3.3. *The semigroup $M_{\text{eff}}(X_r) \subset \operatorname{Pic}(X_r)$ of classes of effective divisors on a Del Pezzo surfaces X_r ($2 \leqslant r \leqslant X_r$) is generated by elements of degree 1. These elements are exactly the classes of exceptional curves if $r \leqslant 7$ and the classes of exceptional curves together with the anticanoncal class for $r = 8$.*

Proposition 3.4. *If D is an effective divisor of degree $\geqslant 2$ on X_8, then the vector space $H^0(X_8, \mathcal{O}(D))$ is spanned by distinguished global sections of $\mathcal{O}(D)$ with supports only on exceptional curves.*

Proof. By 3.2 and 3.3, it is sufficient to check the statement for $D = -2K$ and for $D = -K+E$ for any exceptional curve. The latter case immediately follows from 3.2, because $D = -K + E$ is the pullback of the anticanonical sheaf on the variety X_7 obtained by contracting E. In the case $D = -2K$, we obtain 120 distinguished global sections of $\mathcal{O}(D)$ from 120 pairs of exceptional curves E_i, E_j such that $(E_i, E_j) = 3$:

$$-2K = 6l_0 - 2l_1 - \ldots - 2l_8 = l_1 + (6l_0 - 3l_1 - 2l_2 \ldots - 2l_8).$$

It is wellknown (see e.g. [**DP80**]) that X_8 can be realized as a hypersurface of degree 6 in the weighted projective space $\mathbb{P}(3,2,1,1)$. In particular, the linear system $|-2K|$ defines a double covering of X_8 over a singular quadratic cone $\mathcal{Q} \cong \mathbb{P}(2,1,1) \subset \mathbb{P}^3$. The single singular point $p \in Q$ is the image of the base point $b \in X_8$ of $|-K|$ on X_8. Let $C \subset \mathcal{Q}$ be the ramification locus (C is a curve of degree 6 in $\mathbb{P}(2,1,1)$). Then the 120 pairs of exceptional curves E_i, E_j on X_8 such that $[E_i] + [E_j] = 2[-K]$ are in bijection with the conics in $\mathbb{P}(2,1,1)$ that are 3-tangent to the ramification curve C. Every such conic in $\mathcal{Q}$ is $\mathcal{Q} \cap H$ for a unique plane $H \subset \mathbb{P}^3$. Therefore the distinguished sections in $H^0(X_8, \mathcal{O}(-2K))$ can be identified (up to a scalar multiple) with the above planes $H \subset \mathbb{P}^3$. It remains to show that for a generic choice of the sextic $C \subset \mathbb{P}(2,1,1)$, these 120 planes H do not all pass through a single point $x \in \mathbb{P}^3$. This can be checked by standard dimension arguments. □

Remark 3.5. Since $H^0(X_r, \mathcal{O}(E))$ is 1-dimensional for each exceptional curve $E \subset X_r$, we can choose a nonzero section $x_E \in H^0(X_r, \mathcal{O}(E))$ which is determined up to multiplication by a nonzero scalar. Therefore the affine algebraic variety $\mathbb{A}(X_r) := \operatorname{Spec}\operatorname{Cox}(X_r)$ is embedded into the affine space $\mathbb{A}^{N_r}$ on which the maximal torus $T_r \subset G(R_r)$ acts in a canonical way, compatible with an identification of the space $\mathbb{A}^{N_r}$ with the representation space $V(\varpi_r)$ of the algebraic group $G(R_r)$ (if $r \leqslant 7$). In the case $r = 8$, the 240 exceptional curves on X_8 can be similarly identified with the nonzero weights of the adjoint representation of $G(E_8)$ in $V(\varpi_8)$. The space $V(\varpi_8)$ contains the weight-0 subspace of dimension 8, but the ring $\mathrm{Cox}(X_r)$ has only a 2-dimensional space of anticanonical sections. Thus we cannot identify the degree-1 homogeneous component of $\mathrm{Cox}(X_8)$ with the representation space $V(\varpi_8)$ of $G(E_8)$.

Since the kernel of the surjective homomorphism

$$\deg : \mathrm{Pic}(X_r) \to \mathbb{Z},$$

can be identified with the character group $\mathfrak{X}^*(T_r)$ of a maximal torus $T_r \subset G(R_r)$ and the torus T_r acts on the homogeneous space $G(R_r)/P(R_{r-1})$ embedded into the projective space $\mathbb{P}(V(\varpi_r))$, we obtain a natural $\mathrm{Pic}(X_r)$-grading of the homogeneous coordinate ring of the projective variety $G(R_r)/P(R_{r-1})$.

Theorem 3.6. *Let λ be an element in* $\mathrm{Pic}(X_r)$. *The weight-λ subspace in the homogeneous coordinate ring of the projective variety $G(R_r)/P(R_{r-1})$ is nonzero if and only if λ is represented by an effective divisor on X_r (i.e., $\lambda \in M_{\mathrm{eff}}(X_r)$).*

Proof. It is known that the projective variety $G(R_r)/P(R_{r-1})$ is arithmetically normal and Cohen–Macaulay [**Ram85, LS86**]. In particular, the homogeneous coordinate ring of $G(R_r)/P(R_{r-1})$ is generated by elements of degree 1. Therefore, the weight-λ subspace in the cooordinate ring is nonzero if and only if λ is a nonnegative integral linear combination of $\mathrm{Pic}(X_r)$-weights having positive multiplicity in $V(\varpi_r)$. By 3.3 and 3.5, the latter is equivalent to $\lambda \in M_{\mathrm{eff}}(X_r)$. □

4. Quadratic relations in $\mathrm{Cox}(X_r)$

Let us denote $\mathbb{P}(X_r) := \mathrm{Proj}\ \mathrm{Cox}(X_r)$. If $4 \leqslant r \leqslant 7$, then the projective variety $\mathbb{P}(X_r)$ is canonically embedded into the projective space $\mathbb{P}^{N_r-1}$ (N_r is the number of exceptional curves on X_r). The affine variety $\mathbb{A}(X_r) \subset \mathbb{A}^{N_r}$ is the affine cone over $\mathbb{P}(X_r)$.

Proposition 4.1. *The ring* $\mathrm{Cox}(X_4)$ *is isomorphic to the subring of all* 3×3*-minors of a generic* 3×5 *matrix. In particular, the projective variety* $\mathbb{P}(X_4) \subset \mathbb{P}^9$ *is isomorphic to the Plücker embedding of the Grassmannian* $Gr(3,5)$.

Proof. In order to describe the multiplication in $\mathrm{Cox}(X_4)$, one needs to choose a basis in $\mathrm{Pic}(X_4)$.

Let $x : y : z$ be the homogeneous coordinates on $\mathbb{P}^2$. We choose the basis $l_0, \ldots, l_4$, as in Section 2; i.e., l_0 is the preimage of the line $z = 0$ at infinity, and l_1, l_2, l_3, l_4 are classes of the exceptional fibers over 4 points $p_1, \ldots, p_4 \in \mathbb{P}^2$. We identify the representatives of each class in $\mathrm{Pic}(X_4)$ with the subsheaves $\mathscr{O}(\sum_{i=0}^4 m_i l_i)$ of the constant sheaf $\mathscr{K}(X_4)$ of rational functions on X_4. Then the ring multiplication in $\mathrm{Cox}(X_4)$ is just the multiplication of the corresponding rational functions in $\mathscr{K}(X_4)$.

Let $(x_i : y_i : z_i)$ be the coordinates of the blown-up point $p_i \in \mathbb{P}^2$ ($i = 1, \ldots, 4$). Consider the 3×5 matrix

$$M = \begin{pmatrix} x_1 & x_2 & x_3 & x_4 & x/z \\ y_1 & y_2 & y_3 & y_4 & y/z \\ z_1 & z_2 & z_3 & z_4 & 1 \end{pmatrix}.$$

For any 3-element subset $I = \{i, j, k\} \subset \{1, \ldots, 5\}$, we denote by M_I the maximal minor of M consisting of the columns with numbers in I taken in the natural order.

We choose the rational functions in $\mathscr{K}(X_4)$ representing the generators x_E of $\mathrm{Cox}(X_4)$ as follows:

$$x_{l_1} := M_{\{2,3,4\}},\ x_{l_1} := M_{\{1,3,4\}},\ x_{l_3} := M_{\{1,2,4\}},\ x_{l_1} := M_{\{1,2,3\}},$$
$$x_{l_0 - l_i - l_j} := M_{\{i,j,5\}},\ \ 1 \leqslant i < j \leqslant 4.$$

All these functions are nonzero because the points $p_1, \ldots, p_4$ are in general position.

It is known that the generators of the homogeneous coordinate ring of $G(3,5)$ are naturally identified with the maximal minors of a generic 3×5 matrix. Consider the homomorphism φ of the homogeneous coordinate ring of $G(3,5)$ to $\mathrm{Cox}(X_4)$, which sends these generic minors into the corresponding minors of the matrix M above. Since $\mathrm{Cox}(X_4)$ is generated by $\{x_E\}$, this homomorphism is surjective. By 3.6, φ respects the $\mathrm{Pic}(X_r)$-grading (in particular, φ respects the $\mathbb{Z}_{\geqslant 0}$-grading as well). The surjectivity of φ induces a closed embedding of $\mathbb{P}(X_4)$ into $G(3,5)$. Since both varieties are irreducible of dimension 6 (see 1.4), we obtain an isomorphism between $\mathbb{P}(X_4)$ and $G(3,5)$ as subvarieties of $\mathbb{P}^9$. Therefore φ is an isomorphism of the homogeneous coordinate ring of $G(3,5)$ and $\mathrm{Cox}(X_4)$. In particular, $\mathrm{Cox}(X_4)$ is defined by 5 quadratic Plücker relations. One of these relations is

$$M_{\{1,2,5\}} M_{\{3,4,5\}} - M_{\{1,3,5\}} M_{\{2,4,5\}} + M_{\{1,4,5\}} M_{\{2,3,5\}} = 0.$$

□

The article [**Ses78**] describes a $\Bbbk$-basis for the homogeneous coordinate ring of G/P in the case, when P is a maximal parabolic subgroup containing a Borel subgroup B such that the fundamental weight ϖ corresponding to P is minuscule (see 2.2). It also shows that this ring is always defined by quadratic relations.

The paper [**Lic82**] gives an explicit way to write the quadratic relations for the orbit of the highest weight vector for any representation of a semisimple Lie group. A more geometric approach to these quadratic equations is contained in the proof of Theorem 1.1 in [**LT79**]:

Proposition 4.2. *The orbit G/P_ϖ of the highest weight vector in the projective space $\mathbb{P}(V(\varpi))$ is the intersection of the second Veronese embedding of $\mathbb{P}(V(\varpi))$ with the subrepresentation $V(2\varpi)$ of the symmetric square $S^2V(\varpi)$. Moreover, these quadratic relations generate the ideal of $G/P_\varpi \subset \mathbb{P}(V(\varpi))$.*

We expect that the following general statement is true:

Conjecture 4.3. *The ideal of relations between the degree-1 generators of $\mathrm{Cox}(X_r)$ is generated by quadrics for $4 \leqslant r \leqslant 8$.*

For any exceptional curve $E \subset X$, we consider the open chart $U_E \subset \mathbb{P}^{N_r-1}$ defined by the condition $x_E \neq 0$. Thus we obtain an open covering of $\mathbb{P}(X_r)$ by N_r affine subsets $U_E \cap \mathbb{P}(X_r)$.

Proposition 4.4. *Let X_{r-1} be the Del Pezzo surface obtained by contracting E on X_r. Then there exists a natural isomorphism*

$$U_E \cap \mathbb{P}(X_r) \cong \mathbb{A}(X_{r-1}).$$

Proof. Let $\pi\colon X_r \to X_{r-1}$ be the contraction of E. Then we obtain the ring homomorphism $\pi^*\colon \mathrm{Cox}(X_{r-1}) \to \mathrm{Cox}(X_r)$. We shall show that the localization $\mathrm{Cox}(X_r)_{x_E}$ of the ring $\mathrm{Cox}(X_r)$ by the element x_E can be identified with the Laurent polynomial extension of $\pi^*\mathrm{Cox}(X_{r-1})$ by x_E; i.e., there exists a ring isomorphism

$$\mathrm{Cox}(X_r)_{x_E} \cong \pi^*\mathrm{Cox}(X_{r-1})[x_E, x_E^{-1}].$$

For simplicity, we assume that $[E] = l_r$ and $\{l_0, \ldots, l_{r-1}\}$ is the pullback of the standard basis in $\mathrm{Pic}(X_{r-1})$. Any divisor class

$$[D] = m_r l_r + \sum_{i=0}^{r-1} m_i l_i \in \mathrm{Pic}(X_r)$$

is uniquely represented as a sum $m_r[E] + [D']$ where

$$[D'] = \sum_{i=0}^{r-1} m_i l_i \in \pi^*(\mathrm{Pic}(X_{r-1})).$$

Using π^*, we identify the fields of rational functions $\mathscr{K}(X_{r-1})$ and $\mathscr{K}(X_r)$. This identification allows us to consider the vector space

$$H^0\left(X_r, \mathscr{O}\left(m_r l_r + \sum_{i=0}^{r-1} m_i l_i\right)\right)$$

as a subspace of

$$H^0\left(X_{r-1}, \mathscr{O}\left(\sum_{i=0}^{r-1} m_i l_i\right)\right) x_E^{m_r} \subset \pi^*\mathrm{Cox}(X_{r-1})[x_E, x_E^{-1}].$$

For fixed integers $m_0, m_1, \ldots, m_{r-1}$, the embedding of the vector spaces

$$H^0\left(X_r, \mathscr{O}\left(m_r l_r + \sum_{i=0}^{r-1} m_i l_i\right)\right) \hookrightarrow H^0\left(X_{r-1}, \mathscr{O}\left(\sum_{i=0}^{r-1} m_i l_i\right)\right) x_E^{m_r}$$

is an isomorphism for sufficiently large m_r. Moreover, this embedding of vector spaces respects the multiplications in $\mathrm{Cox}(X_r)$ and in $\pi^*\mathrm{Cox}(X_{r-1})[x_E, x_E^{-1}]$. Thus we obtain an embedding of rings

$$\mathrm{Cox}(X_r) \hookrightarrow \pi^*\mathrm{Cox}(X_{r-1})[x_E, x_E^{-1}].$$

On the other hand, $\pi^*\mathrm{Cox}(X_{r-1})[x_E, x_E^{-1}]$ is a subring of the localization $\mathrm{Cox}(X_r)_{x_E}$. Thus we get an isomorphism

$$\mathrm{Cox}(X_r)_{x_E} \cong \pi^*\mathrm{Cox}(X_{r-1})[x_E, x_E^{-1}].$$

Now we remark that the coordinate ring of the affine variety $U_E \cap \mathbb{P}(X_r)$ is degree-0 component of $\mathrm{Cox}(X_r)_{x_E}$. By the above isomorphism, this component is isomorphic to $\mathrm{Cox}(X_{r-1})$. □

Corollary 4.5. *The singular locus of the algebraic varieties $\mathbb{P}(X_r)$ and $\mathbb{A}(X_r)$ has codimension* 7.

Proof. Since $\mathbb{A}(X_3) \cong \mathbb{A}^6$, we obtain that $\mathbb{P}(X_4)$ is a smooth variety covered by 10 affine charts which are isomorphic to $\mathbb{A}^6$. Using the isomorphism $\mathbb{P}(X_4) \cong G(3,5)$ (see 4.1), we obtain that $\mathbb{A}(X_4)$ has an isolated singularity at 0. Therefore, the singular locus of $\mathbb{P}(X_5)$ consists of 16 isolated points. The singular locus of $\mathbb{P}(X_6)$ is 1-dimensional and the singular locus of $\mathbb{P}(X_7)$ is 2-dimensional, etc. □

Definition 4.6. A divisor class $[D]$ is called *a ruling* if it can be written as a sum of two classes of exceptional curves $[E_i] + [E_j]$ such that $(E_i, E_j) = 1$, or equivalently, if D satisfies the conditions $(D, D) = 0$, $(D, -K) = 2$. The invertible sheaf corresponding to a ruling determines a conic bundle morphism $X_r \to \mathbb{P}^1$.

Remark 4.7. Lemma 5.3 of **[FM02]** says that the Weyl group acts transitively on rulings.

Each ruling $[D]$ can be written in $r-1$ different ways as a sum of two classes of exceptional curves corresponding to degenerate fibers of the conic bundle $X_r \to \mathbb{P}^1$. Thus, we obtain $r-1$ distinguished sections in the 2-dimensional space $H^0(X_r, \mathcal{O}(D))$. If $r \geqslant 4$, then for each ruling $[D]$, we obtain in this way $r-3$ linearly independent quadratic relations between generators of $\mathrm{Cox}(X_r)$.

Remark 4.8. We note that $\mathrm{Pic}(X_4)$ has exactly 5 rulings. Each such a ruling defines a Plücker quadric (see the proof of 4.1).

We cannot expect in general that all quadratic relations among generators come from rulings. However, the following statement is true:

Theorem 4.9. *For $4 \leqslant r \leqslant 6$, the ring $\mathrm{Cox}(X_r)$ is defined by the radical of the ideal generated by the quadratic relations corresponding to rulings.*

Proof. Let $Z_r \subset \mathbb{A}^{N_r}$ be the affine subvariety defined by the quadratic relations coming from rulings. We want to show that $Z_r = \mathbb{A}(X_r)$ $(4 \leqslant r \leqslant 6)$.

For $r = 4$, the statement follows from 4.8.

Obviously, the zero $0 \in \mathbb{A}^{N_r}$ is common point of Z_r and $\mathbb{A}(X_r)$ for all r. Consider the affine open coverings of $Z_r \setminus \{0\}$ and $\mathbb{A}(X_r) \setminus \{0\}$ defined by affine open subsets $x_E \neq 0$, where E runs over all exceptional curves of X_r. Using induction on r and Proposition 4.4, we need only show that $Z_r \cap U_E = \mathbb{A}(X_r) \cap U_E$ for each exceptional curve. For this purpose, it is important to remark that the affine coordinate ring of $Z_r \cap U_E$ is generated by all elements x_F/x_E such that $(E, F) = 0$. For $r = 5, 6$, the last property follows from the fact that if $(E, E') > 0$ for two exceptional curves E, E' on X_r, then $(E, E') = 1$; i.e., $[E] + [E']$ is a ruling and there exists a ruling quadratic relation

$$x_E x_{E'} = \sum_i a_i X_{E_i} X_{E_i'},$$

where all exceptional curves E_i, E_i' do not intersect E. The last property shows that

$$Z_r \cap U_E \cong Z_{r-1} \times (\mathbb{A}^1 \setminus \{0\}).$$

It follows from the proof of 4.4 that

$$\mathbb{A}(X_r) \cap U_E \cong \mathbb{A}(X_{r-1}) \times (\mathbb{A}^1 \setminus \{0\}).$$

By induction, we have the equality $Z_{r-1} = \mathbb{A}(X_{r-1})$. This implies the equality $Z_r \cap U_E = \mathbb{A}(X_r) \cap U_E$ for each exceptional curve. Thus, $Z_r = \mathbb{A}(X_r)$. □

References

[Cox95] D. A. Cox – The homogeneous coordinate ring of a toric variety, *J. Algebraic Geom.* **4** (1995), no. 1, 17–50.

[CTS87] J.-L. Colliot-Thélène & J.-J. Sansuc – La descente sur les variétés rationnelles. II, *Duke Math. J.* **54** (1987), no. 2, 375–492.

[DP80] M. Demazure & H. C. Pinkham (eds.) – *Séminaire sur les Singularités des Surfaces*, Lecture Notes in Mathematics, vol. 777, Springer, Berlin, 1980.

[FM02] R. Friedman & J. W. Morgan – Exceptional groups and Del Pezzo surfaces, Symposium in Honor of C. H. Clemens (Salt Lake City, UT, 2000), Contemp. Math., vol. 312, Amer. Math. Soc., Providence, RI, 2002, 101–116.

[HK00] Y. Hu & S. Keel – Mori dream spaces and GIT, *Michigan Math. J.* **48** (2000), 331–348.

[HT03] B. Hassett & Y. Tschinkel – Universal torsors and Cox rings, this volume, 2003.

[Hum75] J. E. Humphreys – *Linear Algebraic Groups*, Springer-Verlag, New York, 1975.

[Kap93] M. M. Kapranov – Chow quotients of Grassmannians. I, I. M. Gel′fand Seminar, Adv. Soviet Math., vol. 16, Amer. Math. Soc., Providence, RI, 1993, 29–110.

[Leu00] N. Leung – *ADE*-bundle over rational surfaces, configuration of lines and rulings, `math.AG/0009192`, 2000.

[Lic82] W. Lichtenstein – A system of quadrics describing the orbit of the highest weight vector, *Proc. Amer. Math. Soc.* **84** (1982), no. 4, 605–608.

[LMS79] V. Lakshmibai, C. Musili & C. S. Seshadri – Geometry of G/P. III. Standard monomial theory for a quasi-minuscule P, *Proc. Indian Acad. Sci. Sect. A Math. Sci.* **88** (1979), no. 3, 93–177.

[LS86] V. Lakshmibai & C. S. Seshadri – Geometry of G/P. V, *J. Algebra* **100** (1986), no. 2, 462–557.

[LT79] G. Lancaster & J. Towber – Representation-functors and flag-algebras for the classical groups. I, *J. Algebra* **59** (1979), no. 1, 16–38.

[Man86] Y. I. Manin – *Cubic Forms*, second ed., North-Holland Mathematical Library, vol. 4, North-Holland Publishing Co., Amsterdam, 1986.

[Nag60] M. NAGATA – On rational surfaces. I. Irreducible curves of arithmetic genus 0 or 1, *Mem. Coll. Sci. Univ. Kyoto Ser. A Math.* **32** (1960), 351–370.

[Nag61] ———, On rational surfaces. II, *Mem. Coll. Sci. Univ. Kyoto Ser. A Math.* **33** (1960/1961), 271–293.

[Ram85] A. RAMANATHAN – Schubert varieties are arithmetically Cohen-Macaulay, *Invent. Math.* **80** (1985), no. 2, 283–294.

[Ses78] C. S. SESHADRI – Geometry of G/P. I. Theory of standard monomials for minuscule representations, C. P. Ramanujam—a tribute, Tata Inst. Fund. Res. Studies in Math., vol. 8, Springer, Berlin, 1978, 207–239.

[Sko93] A. N. SKOROBOGATOV – On a theorem of Enriques-Swinnerton-Dyer, *Ann. Fac. Sci. Toulouse Math. (6)* **2** (1993), no. 3, 429–440.

[Sko01] A. SKOROBOGATOV – *Torsors and Rational Points*, Cambridge Tracts in Mathematics, vol. 144, Cambridge University Press, Cambridge, 2001.

Arithmetic of Higher-dimensional Algebraic Varieties
(B. Poonen, Yu. Tschinkel, eds.), p. 105–120
Progress in Mathematics, Vol. 226, © 2004 Birkhäuser Boston, Cambridge, MA

COUNTING RATIONAL POINTS ON THREEFOLDS

Niklas Broberg
University of Durham, Science Laboratories, South Rd, Durham DH1 3LE, UK
E-mail : niklas.broberg@durham.ac.uk

Per Salberger
Department of Mathematics, Chalmers University of Technology, S-412 96, Göteborg, Sweden • *E-mail :* salberg@math.chalmers.se

Abstract. Let $X \subset \mathbb{P}^4$ be an irreducible hypersurface and $\varepsilon > 0$ be given. We show that there are $O(B^{3+\varepsilon})$, resp. $O(B^{55/18+\varepsilon})$, rational points on $\mathbb{P}^4$ lying on X when X is of degree $d \geqslant 4$, resp. $d = 3$. The implied constants depend only on d and ε.

1. Introduction

Let $\overline{\mathbb{Q}}$ be an algebraic closure of $\mathbb{Q}$ and $\mathbb{P}^n$ be projective n-space over $\mathbb{Q}$. That is, $\mathbb{P}^n$ is the set of one-dimensional linear subspaces of $\overline{\mathbb{Q}}^{n+1}$. A point on $\mathbb{P}^n$ is said to be rational if it represents a subspace of $\overline{\mathbb{Q}}^{n+1}$ generated by an element in $\mathbb{Q}^{n+1}$. The height of a rational point $p \in \mathbb{P}^n(\mathbb{Q})$ is given by

$$H(p) = \max(|Z_0|, \dots, |Z_n|),$$

Key words and phrases. Rational points, heights, threefolds.

While working on this paper, the first author was supported by the EC network Arithmetic Algebraic Geometry.

where $Z_0, \ldots, Z_n$ are relatively prime integers such that $p = [Z_0, \ldots, Z_n]$. For a Zariski closed subset X of $\mathbb{P}^n$, let $N(X, B)$ be the counting function

$$N(X,B) = \#\{p \in \mathbb{P}^n(\mathbb{Q}) \cap X \ : \ H(p) \leqslant B\},$$

where $\mathbb{P}^n(\mathbb{Q})$ is the set of rational points on $\mathbb{P}^n$. In the case where X is defined by an irreducible form $F(Z_0, \ldots, Z_n)$ of degree $d \geqslant 2$, Heath-Brown [**4**] conjectured that

$$N(X,B) = O_{n,d,\varepsilon}(B^{n-1+\varepsilon})$$

for any $\varepsilon > 0$. The implied constant should thus not depend on F, only on n, d, and ε. Such uniform estimates of $N(X, B)$ for hypersurfaces $X \subset \mathbb{P}^n$ are useful for estimating $N(X, B)$ for hypersurfaces $X \subset \mathbb{P}^m$, where $m > n$. In [**4**], Heath-Brown verified the conjecture for curves, surfaces, and for quadrics of any dimension. In [**1**], Browning proved the conjecture for non-singular hypersurfaces in $\mathbb{P}^4$ of degree at least four. In this paper we shall prove the following result.

Theorem 1. *Let $X \subset \mathbb{P}^4$ be a hypersurface defined by an irreducible form $F(Z_0, \ldots, Z_4)$ with coefficients in $\overline{\mathbb{Q}}$. Then the following holds for any $\varepsilon > 0$:*

$$N(X,B) = \begin{cases} O_{d,\varepsilon}(B^{3+\varepsilon}) & \text{if } d \geqslant 4 \\ O_{\varepsilon}(B^{55/18+\varepsilon}) & \text{if } d = 3. \end{cases}$$

2. Preliminaries

In this section we collect some known estimates of counting functions. We also state and prove some results that we use in the proof of Theorem 1. First note that:

- We follow the convention that a subvariety X of $\mathbb{P}^n$ is a closed subset that is not necessarily irreducible. A hypersurface of $\mathbb{P}^n$ is a subvariety of codimension one. All varieties are defined over $\overline{\mathbb{Q}}$.
- We shall often count rational points of bounded height on $\mathbb{P}^n$ which lie on X even if X is not defined over $\mathbb{Q}$. The same situation occurs for subvarieties of the dual space $\mathbb{P}^{n*}$ of $\mathbb{P}^n$ when we count hyperplanes Γ of $\mathbb{P}^n$ defined over $\mathbb{Q}$ for which $\Gamma \cap X$ is reducible.
- If $X \subset \mathbb{P}^n$ is a hypersurface, then we define the degree of X to be the degree of the corresponding reduced scheme. That is, we let $d = \deg(X)$ be the minimal degree among all forms defining X. This implies that an intersection $\Lambda \cap X$ with a linear subspace $\Lambda \subset \mathbb{P}^n$ not contained in X may have lower degree than X.

– Our calculations involve numerous constants. To avoid introducing the constants explicitly, we use the following notation.

Suppose that f_1 and f_2 are functions such that $f_i(B) \geqslant 0$ for all $B \geqslant 1$. We write

$$f_1(B) \ll_{p_1,\dots,p_k} f_2(B),$$

if there exists a positive constant C, depending only on the parameters $p_1, \dots, p_k$, such that $f_1(B) \leqslant C f_2(B)$ for all $B \geqslant 1$. We write

$$f_1(B) \asymp_{p_1,\dots,p_k} f_2(B),$$

if $f_1(B) \ll_{p_1,\dots,p_k} f_2(B)$ and $f_2(B) \ll_{p_1,\dots,p_k} f_1(B)$.

– The Grassmannian $\mathbb{G}(k, n)$ of k-dimensional linear subspaces of $\mathbb{P}^n$ is assumed to be embedded into projective space by the Plücker embedding. In particular, we identify $\mathbb{G}(n-1, n)$ with the dual projective space $\mathbb{P}^{n*}$. The height of a rational linear subspace $\Lambda \subset \mathbb{P}^n$ is by definition the height of its Plücker coordinates. According to [**6**, Chapter I, Corollary 5I], we have

$$H(\Lambda) \asymp_n \det(\mathbf{\Lambda}),$$

where

$$\mathbf{\Lambda} = \left\{ \mathbf{x} \in \mathbb{Z}^{n+1} \ : \ [\mathbf{x}] \in \Lambda \right\} \cup \{\mathbf{0}\}$$

is the lattice associated to Λ and $\det(\mathbf{\Lambda})$ is the volume of a fundamental domain of $\mathbf{\Lambda}$.

2.1. Results from the Geometry of Numbers. The following result is well known [**4**, Lemma 1(iii)]. It is a basic fact from the geometry of numbers, and it is one of the key tools in the proof of Theorem 1.

Lemma 2.1.1. *Let $\mathbf{\Lambda} \subset \mathbb{Z}^n$ be a lattice of dimension m. Then $\mathbf{\Lambda}$ has a basis $\mathbf{b}_1, \dots, \mathbf{b}_m$ such that if one writes $\mathbf{x} \in \mathbf{\Lambda}$ as $\mathbf{x} = \sum_j \lambda_j \mathbf{b}_j$, then*

$$|\lambda_j| \ll_n |\mathbf{x}| \, / \, |\mathbf{b}_j| \, .$$

Moreover one has

$$\det(\mathbf{\Lambda}) \asymp_n \prod_{j=1}^{m} |\mathbf{b}_j| \, .$$

The following result is a consequence of [**6**, Chapter I, Corollary 5J].

Lemma 2.1.2. *Suppose that $\mathbf{a}_1, \dots, \mathbf{a}_k$ are linearly independent n-dimensional vectors with integer components and let*

$$\mathbf{\Lambda} = \{ \mathbf{x} \in \mathbb{Z}^n \ : \ \mathbf{a}_1 \cdot \mathbf{x} = \cdots = \mathbf{a}_k \cdot \mathbf{x} = 0 \} \, .$$

Then $\mathbf{\Lambda} \subset \mathbb{Z}^n$ *is a lattice of dimension* $n-k$ *and*

$$\det(\mathbf{\Lambda}) \ll_n \prod_{j=1}^{k} |\mathbf{a}_j| .$$

2.2. Bad linear sections. The homogeneous ideal of a hypersurface $X \subset \mathbb{P}^n$ of degree d is generated by a single homogeneous polynomial $F(Z_0, \ldots, Z_n)$ of degree d. This means that hypersurfaces in $\mathbb{P}^n$ of degree d are parametrised by points in $\mathbb{P}(\overline{\mathbb{Q}}[Z_0, \ldots, Z_n]_d)$, where $\overline{\mathbb{Q}}[Z_0, \ldots, Z_n]_d$ is the vector space of homogeneous polynomials of degree d in $n+1$ variables. Let $V(F)$ denote the hypersurface in $\mathbb{P}^n$ given by the zero locus of $F \in \overline{\mathbb{Q}}[Z_0, \ldots, Z_n]_d$. The set of pairs $(\Lambda, F) \in \mathbb{G}(k,n) \times \mathbb{P}(\overline{\mathbb{Q}}[Z_0, \ldots, Z_n]_d)$ for which $\Lambda \cap V(F)$ is an irreducible variety of dimension $k-1$ and degree d is an open subset of $\mathbb{G}(k,n) \times \mathbb{P}(\overline{\mathbb{Q}}[Z_0, \ldots, Z_n]_d)$. We denote the complement of this open subset by $\Phi_{n,d,k}$.

The following result is well known but the proof is so short that we reproduce it here.

Lemma 2.2.1. *Let* $X \subset \mathbb{P}^n$ *be an irreducible hypersurface of degree* d *and dimension at least two. Then the set of hyperplanes* Γ *for which the linear section* $\Gamma \cap X$ *is reducible or of degree less than* d *is a proper closed subset of* $\mathbb{P}^{n*}$*. Furthermore, this closed subset is cut out by hypersurfaces of degrees bounded solely in terms of* n *and* d*. The number of required hypersurfaces is also bounded in terms of* n *and* d*.*

Proof. By choosing a basis of $\overline{\mathbb{Q}}[Z_0, \ldots, Z_n]_d$, we may identify $\mathbb{P}(\overline{\mathbb{Q}}[Z_0, \ldots, Z_n]_d)$ with $\mathbb{P}^N$ for some N. Suppose that the ideal of $\Phi_{n,d,n-1} \subset \mathbb{P}^{n*} \times \mathbb{P}^N$ is generated by bihomogeneous polynomials

$$G_i(Z_0, \ldots, Z_n; W_0, \ldots, W_N) \quad \text{for} \quad i = 1, 2, \ldots, m. \tag{2.1}$$

The set of hyperplanes Γ for which $\Gamma \cap X$ is reducible or of degree less than d is then the common zero locus of the polynomials (2.1), where $W_0, \ldots, W_N$ are the coefficients of any homogeneous polynomial generating the ideal of X. Since a general hyperplane section $\Gamma \cap X$ is irreducible [**3**, Proposition 18.10], all of the polynomials (2.1) cannot vanish identically on $\mathbb{P}^{n*}$. □

Lemma 2.2.2. *Let* $X \subset \mathbb{P}^4$ *be an irreducible hypersurface of degree* d *and let* $V \subset \mathbb{P}^{4*}$ *be the set of hyperplanes* Γ *with the following property. There is a point* $p \in \Gamma \cap X$ *such that for every two-plane* $\Lambda \subset \Gamma$ *that contains* p *either* Λ *is contained in* X *or* $\Lambda \cap X$ *contains an irreducible component of degree less than* d*. Then we have the following:*

(a) *V is a closed subset of $\mathbb{P}^{4*}$, and if $V \neq \mathbb{P}^{4*}$, then V is cut out by hypersurfaces of degrees bounded in terms of d. The number of required hypersurfaces is also bounded in terms of d.*

(b) *If $Y = \Gamma \cap X$ is an irreducible surface for some hyperplane $\Gamma \in V$, then Y is a cone over a plane curve.*

(c) *$V = \mathbb{P}^{4*}$ if and only if X is a cone over a plane curve with respect to a vertex line.*

Proof. As in the proof of Lemma 2.2.1, we identify $\mathbb{P}(\overline{\mathbb{Q}}[Z_0, \ldots, Z_4]_d)$ with $\mathbb{P}^N$. Let $\Psi \subset \mathbb{P}^{4*} \times \mathbb{P}^4 \times \mathbb{P}^N \times \mathbb{G}(2,4)$ be the set of four-tuples (Γ, p, F, Λ) such that $F(p) = 0$, $p \in \Lambda$, $\Lambda \subset \Gamma$, and $(\Lambda, F) \in \Phi_{4,d,2}$. Let π be the projection map from Ψ to $\mathbb{P}^{4*} \times \mathbb{P}^4 \times \mathbb{P}^N$. Then Ψ is a projective variety and the function

$$\lambda(q) = \dim(\pi^{-1}(q))$$

is an upper-semicontinuous function on the image $\pi(\Psi)$ [**3**, Corollary 11.13]. In particular,

$$\Omega = \{(\Gamma, p, F) \in \pi(\Psi) \; : \; \lambda(\Gamma, p, F) \geqslant 2\}$$

is a subvariety of $\mathbb{P}^{4*} \times \mathbb{P}^4 \times \mathbb{P}^N$. Now the set of two-planes $\Lambda \subset \mathbb{P}^4$ for which $p \in \Lambda$ and $\Lambda \subset \Gamma$ for some $(\Gamma, p) \in \mathbb{P}^{4*} \times \mathbb{P}^4$ is a two-dimensional linear subspace of $\mathbb{G}(2,4)$. The fiber $\pi^{-1}(\Gamma, p, F)$ is contained in this linear subspace. Hence, Ω is the set of triples (Γ, p, F) such that $F(p) = 0$ and such that $\Lambda \subset V(F)$ or $\Lambda \cap V(F)$ contains an irreducible component of degree less than d for every two-plane $\Lambda \subset \Gamma$ containing p. Let Σ be the projection of Ω on $\mathbb{P}^{4*} \times \mathbb{P}^N$, and let

$$G_i(Z_0, \ldots, Z_n; W_0, \ldots, W_N) \quad \text{for} \quad i = 1, 2, \ldots, m, \tag{2.2}$$

be bihomogeneous polynomials generating the ideal of Σ. If $W_0, \ldots, W_N$ are the coefficients of some homogeneous polynomial generating the ideal of X, then V is the common zero locus of the polynomials (2.2). This proves (a).

Next we consider (b). Let $Y = \Gamma \cap X$ be an irreducible hyperplane section and assume that p is a point of Y such that $\Lambda \cap X$ contains an irreducible component of degree less than d for every two-plane $\Lambda \subset \Gamma$ containing p. Let $\pi : \widetilde{Y} \to Y$ be the blow-up of Y at p. There is, then, a unique map $\psi : \widetilde{Y} \to \mathbb{P}^2$ which extends the projection map $Y \dashrightarrow \mathbb{P}^2$ from p to some two-plane $\mathbb{P}^2 \subset \Gamma$. If ψ is surjective, then $\psi^{-1}(L)$ is irreducible of degree d for a general line $L \subset \mathbb{P}^2$ [**2**, Theorem 1.1]. This contradicts the assumption that $\pi(\psi^{-1}(L))$ is reducible for every line $L \subset \mathbb{P}^2$. Hence, the map ψ is not surjective, so Y is cone over a plane curve with vertex p.

To prove (c) we assume that $V = \mathbb{P}^{4*}$ and consider the incidence correspondence $\Omega \subset \mathbb{P}^{4*} \times X$ consisting of all pairs (Γ, p) such that $\Lambda \subset X$ or $\Lambda \cap X$

contains an irreducible component of degree less than d for every two-plane $\Lambda \subset \Gamma$ containing p. It follows from the proof of (a) that Ω is a closed subset of $\mathbb{P}^{4*} \times X$. Since V is the projection of $\Omega \subset \mathbb{P}^{4*} \times X$ on the first factor, we have that the dimension of Ω is at least four. According to (b), we can then find a point $p \in X$ and a family of hyperplanes $\{\Gamma_\lambda\}$ through p such that $\Gamma_\lambda \cap X$ are all cones with the common vertex p. It follows that X is a cone with vertex p over $\Gamma \cap X$ for any hyperplane $\Gamma \subset \mathbb{P}^4$ which does not contain p. Let $\Gamma \subset \mathbb{P}^4$ be such that $\Gamma \cap X$ is a cone over a plane curve $C \subset X$ with vertex $q \in X$. Then X is a cone over C with two different vertices p and q. This proves the first implication of (c). The other one is immediate. □

2.3. Linear subspaces of hypersurfaces. In this section we state some elementary results about linear subspaces contained in a hypersurface $X \subset \mathbb{P}^n$. Let $F_k(X) \subset \mathbb{G}(k,n)$ denote the Fano variety of k-planes contained in the variety $X \subset \mathbb{P}^n$. It can be shown that the number of irreducible components and the dimensions of the irreducible components of $F_k(X)$ can be bounded in terms of n and d.

Lemma 2.3.1. *Let $X \subset \mathbb{P}^n$ be an irreducible hypersurface and assume that X is not a hyperplane. Then the dimension of $F_{n-2}(X)$ is at most one and $F_{n-2}(X)$ contains no lines.*

Proof. Suppose that Y is an irreducible component of $F_{n-2}(X)$ of dimension at least one. Then the variety $\bigcup_{\Lambda \in Y} \Lambda$ has dimension at least $n-1$ and is therefore equal to X. In particular, every point on X belongs to an $(n-2)$-plane Λ in Y. Consider the incidence correspondence

$$\Psi = \{(p, \Lambda) \in X \times Y \,:\, p \in \Lambda\}.$$

The fiber of Ψ over an $(n-2)$-plane Λ is irreducible of dimension $n-2$. The variety Ψ is therefore irreducible of dimension $\dim(Y)+n-2$ [**3**, Theorem 11.14]. Now if an $(n-2)$-plane $\Lambda \subset X$ contains the point p, then Λ must lie in the projective tangent space $\mathbb{T}_p(X)$ of X at p. For a non-singular point $p \in X$, the dimension of $X \cap \mathbb{T}_p(X)$ is $n-2$, so the fiber of Ψ over a general point of X is finite. The dimension of Ψ is thus at most $n-1$. Hence, the dimension of Y is at most one. To finish the proof we note that $\bigcup_{\Lambda \in Y} \Lambda$ is a hyperplane when Y is a line. Since X is irreducible, $F_{n-2}(X)$ cannot contain any lines. □

The following lemma is a modification of Example 19.11 on page 244 in [**3**].

Lemma 2.3.2. *Let $X \subset \mathbb{P}^n$ be the surface swept out by the lines parametrised by an irreducible curve $C \subset \mathbb{G}(1,n)$. Then the degree of X does not exceed the degree of C.*

Proof. The degree of X is by definition the cardinality of the intersection $\Lambda\cap X$ for a general $(n-2)$-plane $\Lambda\subset\mathbb{P}^n$. Assume that $\Lambda\cap X$ contains $\deg(X)$ points. Now every point of $\Lambda\cap X$ belongs to a line $L\in C$ that meets Λ. The locus of lines $L\in\mathbb{G}(1,n)$ that meet Λ is a hyperplane section $\Gamma\cap\mathbb{G}(1,n)$. If C is contained in Γ, then every line $L\in C$ meets Λ. That is, we have a regular map $C\to\Lambda$ given by $L\mapsto L\cap\Lambda$. But C is irreducible so the image of this map is irreducible. Hence, $\Lambda\cap X$ contains only one point so that the degree of X is one. If C is not contained in Γ, then there are at most $\deg(C)$ points in $\Gamma\cap C$. Hence, $\Lambda\cap X$ contains at most $\deg(C)$ points. □

2.4. Estimates for counting functions. It this section we list those known estimates for counting functions that we use in the proof of Theorem 1.

(E1) Let Λ be a k-dimensional linear subspace of $\mathbb{P}^n$. If Λ contains $k+1$ linearly independent rational points of height at most B, then Λ is defined over $\mathbb{Q}$ and

$$N(\Lambda,B)\ll_n\frac{B^{k+1}}{H(\Lambda)}.$$

To see this, let $\mathbf{b}_0,\ldots,\mathbf{b}_k$ be a basis of the lattice

$$\boldsymbol{\Lambda}=\{\mathbf{x}\in\mathbb{Z}^{n+1}\;:\;[\mathbf{x}]\in\Lambda\}\cup\{\mathbf{0}\}$$

with the properties stated in Lemma 2.1.1. Since Λ contains $k+1$ linearly independent rational points of height at most B, we must have $|\mathbf{b}_i|\ll_n B$. Hence,

$$N(\Lambda,B)\ll_n\frac{B^{k+1}}{|\mathbf{b}_0|\cdots|\mathbf{b}_k|}\asymp_n\frac{B^{k+1}}{H(\Lambda)}.$$

(E2) If $X\subset\mathbb{P}^n$ is an irreducible variety of degree d and dimension r, then

$$N(X,B)\ll_{n,d}B^{r+1}.$$

This is proved for hypersurfaces in [**4**, Theorem 1]. The general result follows by a standard projection argument (see for example the proof of Lemma 1 in [**1**]).

(E3) If $X\subset\mathbb{P}^n$ is an irreducible variety of degree $d\geqslant 2$ and dimension r, then

$$N(X,B)\ll_{n,d,\varepsilon}B^{r+1/d+\varepsilon},$$

for every $\varepsilon>0$ [**5**].

(E4) If $X\subset\mathbb{P}^n$ is an irreducible curve of degree d, then

$$N(X,B)\ll_{n,d,\varepsilon}B^{2/d+\varepsilon},$$

for every $\varepsilon>0$. This estimate is proved for plane curves in [**4**, Theorem 3]. As in (E2), the general estimate follows by a projection argument.

(E5) Let $\Lambda \subset \mathbb{P}^n$ be a two-plane which is defined over the rational numbers. If $X \subset \Lambda$ is a non-singular curve of degree $d \geqslant 2$, then

$$N(X,B) \ll_{n,d,\varepsilon} 1 + \frac{B^{2/d+\varepsilon}}{H(\Lambda)^{2/3d}},$$

for every $\varepsilon > 0$. This follows from Theorem 3 and Lemma 1(iii) of [**4**].

(E6) If $X \subset \mathbb{P}^n$ is an irreducible surface of degree $d \geqslant 2$, then

$$N(X,B) \ll_{n,d,\varepsilon} B^{2+\varepsilon},$$

for every $\varepsilon > 0$ [**1**, Lemma 1].

(E7) If $X \subset \mathbb{P}^n$ is a quadratic hypersurface of rank at least three, then

$$N(X,B) \ll_{n,\varepsilon} B^{n-1+\varepsilon},$$

for every $\varepsilon > 0$ [**4**, Theorem 2].

3. Proof of the main theorem

The idea of the proof is simple. We cover the set

$$\{p \in \mathbb{P}^4(\mathbb{Q}) \;:\; H(p) \leqslant B\}$$

by a finite collection I of linear subspaces $\Lambda \subset \mathbb{P}^4$, and put

$$\Sigma = \bigcup_{\Lambda \in I} (\Lambda \cap X).$$

We then have

$$N(X,B) = N(\Sigma,B) \quad \text{and} \quad \dim(X) > \dim(\Sigma).$$

We may thus apply the sharp estimates (E4) and (E6) from Section 2.4. To determine a suitable set I, we apply the results from Section 2.1.

Let $p = [Z_0, \ldots, Z_4]$ be a point of $\mathbb{P}^4$ such that $Z_0, \ldots, Z_4$ are relatively prime integers. According to [**4**, Lemma 1(i)], the set

$$\mathbf{\Lambda}_1 = \{(x_0, \ldots, x_4) \in \mathbb{Z}^5 \;:\; Z_0x_0 + \cdots + Z_4x_4 = 0\}$$

is a lattice of dimension four and

$$\det(\mathbf{\Lambda}_1) = \sqrt{Z_0^2 + \cdots + Z_4^2} \asymp H(p).$$

Lemma 2.1.1 states that there exists a basis $\mathbf{b}_1, \mathbf{b}_2, \mathbf{b}_3, \mathbf{b}_4$ of $\mathbf{\Lambda}_1$ such that

$$|\mathbf{b}_1|\,|\mathbf{b}_2|\,|\mathbf{b}_3|\,|\mathbf{b}_4| \asymp \det(\mathbf{\Lambda}_1).$$

Without loss of generality we may assume that

$$|\mathbf{b}_1|\,|\mathbf{b}_2| \ll \det(\mathbf{\Lambda}_1)^{1/2}.$$

Let

$$\mathbf{\Lambda}_2 = \left\{ \mathbf{x} \in \mathbb{Z}^5 \;:\; \mathbf{b}_1 \cdot \mathbf{x} = \mathbf{b}_2 \cdot \mathbf{x} = 0 \right\}$$

and apply Lemma 2.1.1 again to find a basis $\mathbf{x}_1, \mathbf{x}_2, \mathbf{x}_3$ of the lattice $\mathbf{\Lambda}_2$ such that

$$|\mathbf{x}_1|\,|\mathbf{x}_2|\,|\mathbf{x}_3| \asymp \det(\mathbf{\Lambda}_2) \ll |\mathbf{b}_1|\,|\mathbf{b}_2|\,. \tag{3.1}$$

The last inequality of (3.1) follows from Lemma 2.1.2. Finally let

$$\mathbf{\Lambda}_3 = \left\{ \mathbf{a} \in \mathbb{Z}^5 \;:\; \mathbf{a} \cdot \mathbf{x}_1 = \mathbf{a} \cdot \mathbf{x}_2 = \mathbf{a} \cdot \mathbf{x}_3 = 0 \right\}$$

and apply Lemma 2.1.1 to find a basis $\mathbf{a}_1, \mathbf{a}_2$ of $\mathbf{\Lambda}_3$ such that

$$|\mathbf{a}_1|\,|\mathbf{a}_2| \asymp \det(\mathbf{\Lambda}_3) \ll |\mathbf{x}_1|\,|\mathbf{x}_2|\,|\mathbf{x}_3|\,.$$

Then $(Z_0, \ldots, Z_4) \in \mathbf{\Lambda}_2$,

$$\mathbf{\Lambda}_2 = \left\{ \mathbf{x} \in \mathbb{Z}^5 \;:\; \mathbf{a}_1 \cdot \mathbf{x} = \mathbf{a}_2 \cdot \mathbf{x} = 0 \right\}, \tag{3.2}$$

and,

$$|\mathbf{a}_1|\,|\mathbf{a}_2| \asymp \det(\mathbf{\Lambda}_2) \ll H(p)^{1/2}.$$

Equality (3.2) follows from the fact that the dual of the dual lattice $\mathbf{\Lambda}_3$ of $\mathbf{\Lambda}_2$ is the lattice $\mathbf{\Lambda}_2$ itself. This shows that there exists a rational two-plane $\Lambda \subset \mathbb{P}^4$ containing p such that $H(\Lambda) \ll H(p)^{1/2}$. It also shows that $\Lambda = \Gamma_1 \cap \Gamma_2$ for some rational hyperplanes Γ_1, Γ_2 in $\mathbb{P}^4$ such that $H(\Gamma_1)H(\Gamma_2) \asymp H(\Lambda)$.

Let A be a positive constant and let $I \subset \mathbb{P}^{4*}(\mathbb{Q}) \times \mathbb{P}^{4*}(\mathbb{Q})$ be the set of pairs (Γ_1, Γ_2) of hyperplanes such that

(i) $\Gamma_1 \neq \Gamma_2$,
(ii) $H(\Gamma_1)H(\Gamma_2) \leqslant AH(\Gamma_1 \cap \Gamma_2)$, where $\Gamma_1 \cap \Gamma_2$ is considered as an element of $\mathbb{G}(2,4)$,
(iii) $H(\Gamma_1) \leqslant AB^{1/4}$, and
(iv) $H(\Gamma_2) \leqslant AB^{1/2}/H(\Gamma_1)$.

Provided that A is large enough, we have

$$\left\{ p \in \mathbb{P}^4(\mathbb{Q}) \;:\; H(p) \leqslant B \right\} \subset \bigcup_{(\Gamma_1,\Gamma_2)\in I} (\Gamma_1 \cap \Gamma_2).$$

From the discussion above it follows that we may choose A independently of B. This defines I and Σ.

The next step of the proof is to use the estimates from Section 2.4 to estimate $N(\Sigma, B)$. The set I can be partitioned into three subsets:

I_1 is the set of $(\Gamma_1, \Gamma_2) \in I$ such that $\Gamma_1 \cap X$ contains an irreducible component of degree less than d.

I_2 is the set of $(\Gamma_1, \Gamma_2) \in I$ such that $\Gamma_1 \cap X$ is irreducible of degree d but $\Gamma_1 \cap \Gamma_2 \cap X$ contains an irreducible component of degree less than d.

I_3 is the set of $(\Gamma_1, \Gamma_2) \in I$ such that $\Gamma_1 \cap \Gamma_2 \cap X$ is an irreducible curve of degree d.

Let

$$\Sigma_i = \bigcup_{(\Gamma_1,\Gamma_2)\in I_i} (\Gamma_1 \cap \Gamma_2 \cap X) \quad \text{for} \quad i = 1, 2, 3.$$

Then

$$N(\Sigma, B) \leqslant N(\Sigma_1, B) + N(\Sigma_2, B) + N(\Sigma_3, B).$$

3.1. Estimate of $N(\Sigma_1, B)$. Let J be the set of rational hyperplanes $\Gamma \subset \mathbb{P}^4$ such that $H(\Gamma) \leqslant AB^{1/4}$ and such that $\Gamma \cap X$ is reducible or of degree less than $d = \deg(X)$. Consider an irreducible component $Y \subset \Gamma \cap X$ for some $\Gamma \in J$. If Y is not a two-plane, then

$$N(Y, B) \ll_{d,\varepsilon} B^{2+\varepsilon}$$

according to (E6). If Y is a two-plane such that all points of height at most B on $\mathbb{P}^4(\mathbb{Q}) \cap Y$ lie on a line, then

$$N(Y, B) \ll B^2$$

according to (E2). Hence,

$$N(\Sigma_1, B) \ll_{d,\varepsilon} B^{2+\varepsilon} |J| + N'(X, B), \tag{3.3}$$

where $N'(X, B)$ is the number of rational points on $\mathbb{P}^4$ of height at most B lying on the union of all two-planes in X that contain three non-collinear rational points on $\mathbb{P}^4$ of height at most B. By Lemma 2.2.1 and (E2), the cardinality of J is $O_d(B)$ so the first term of (3.3) is $O_{d,\varepsilon}(B^{3+\varepsilon})$. By the following lemma, the second term is also $O_{d,\varepsilon}(B^{3+\varepsilon})$.

Lemma 3.1.1. *Let $N'(X, B)$ be the number of rational points on $\mathbb{P}^4$ of height at most B lying on the union of all two-planes in X that contain three non-collinear rational points on $\mathbb{P}^4$ of height at most B. Then,*

$$N'(X, B) \ll_{d,\varepsilon} B^{3+\varepsilon}.$$

Proof. If a two-plane $\Lambda \subset X$ contains three non-collinear rational points on $\mathbb{P}^4$ of height at most B, then $H(\Lambda) \leqslant A'B^3$ for some constant A'. Furthermore, Λ contains $O(B^3/H(\Lambda))$ rational points of height at most B according to (E1). Hence,

$$N'(X, B) \leqslant \sum_{\substack{\Lambda \in F_2(X)(\mathbb{Q}) \\ H(\Lambda) \leqslant A'B^3}} \frac{B^3}{H(\Lambda)},$$

where $F_2(X) \subset \mathbb{G}(2, 4)$ is the Fano variety of two-planes in X. By Lemma 2.3.1, the dimension of $F_2(X)$ is at most one and $F_2(X)$ contains no lines. The number of $\Lambda \in F_2(X)(\mathbb{Q})$ with $H(\Lambda) \ll T$ for some $T \geqslant 1$ is thus $O_{d,\varepsilon}(T^{1+\varepsilon})$ according to (E4). Hence,

$$\sum_{\substack{\Lambda \in F_2(X)(\mathbb{Q}) \\ T < H(\Lambda) \leqslant 2T}} \frac{B^3}{H(\Lambda)} \ll_{d,\varepsilon} \frac{B^3}{T} T^{1+\varepsilon} \ll B^{3+\varepsilon},$$

when $T \ll B^3$. By summing over dyadic intervals, we get

$$N'(X, B) \ll_{d,\varepsilon} B^{3+\varepsilon}.$$

□

3.2. Estimate of $N(\Sigma_2, B)$. Consider an irreducible component $Y \subset \Gamma_1 \cap \Gamma_2 \cap X$ for some $(\Gamma_1, \Gamma_2) \in I_2$. If the degree of Y is at least three, then

$$N(Y, B) \ll_{d,\varepsilon} B^{2/3+\varepsilon}$$

according to (E4). If the degree of Y is two, then

$$N(Y, B) \ll_{\varepsilon} 1 + \frac{B^{1+\varepsilon}}{H(\Gamma_1 \cap \Gamma_2)^{1/3}}$$

according to (E5). If Y is a line such that Y contains at most one rational point of height at most B, then $N(Y, B) \leqslant 1$. Hence,

$$\begin{aligned} N(\Sigma_2, B) \ll_{d,\varepsilon} & B^{2/3+\varepsilon} |I_2| \\ & + \sum_{(\Gamma_1,\Gamma_2) \in I_2} \frac{B^{1+\varepsilon}}{H(\Gamma_1 \cap \Gamma_2)^{1/3}} \qquad (3.4) \\ & + |I_2| + N(Z, B), \end{aligned}$$

where $Z \subset X$ is the union of all lines $L \subset \Sigma_2$ that contain two different rational points of height at most B. Note that if X is a cone over a plane curve, then $\Gamma_1 \cap \Gamma_2 \cap X$ is a union of lines for every pair $(\Gamma_1, \Gamma_2) \in I_2$. Thus, the first two terms of (3.4) do not appear in this case.

Lemma 3.2.1. *We have*

$$\#\{(\Gamma_1, \Gamma_2) \in I_2 \ : \ H(\Gamma_i) \leqslant T_i\} \ll_{d,\varepsilon} T_1^5 T_2^{10/3+\varepsilon} + T_1^{5-\eta} T_2^4,$$

where $\eta = 1$ unless X is a cone over a plane curve in which case $\eta = 0$.

Proof. Let $f : \Gamma_1 \to \mathbb{P}^3$ be an isomorphism for some hyperplane $\Gamma_1 \subset \mathbb{P}^4$ such that $\Gamma_1 \cap X$ is irreducible. Let

$$g : \mathbb{P}^{4*} \setminus \{\Gamma_1\} \to \mathbb{P}^{3*}$$

be the projection map induced by f^{-1}. By Lemma 2.2.1, the set of two-planes $\Lambda \subset \mathbb{P}^3$ for which $\Lambda \cap f^{-1}(X)$ contains an irreducible component of degree less than $d = \deg(X)$ is a proper closed subset $V \subset \mathbb{P}^{3*}$. The set of hyperplanes $\Gamma_2 \in \mathbb{P}^{4*} \setminus \{\Gamma_1\}$ for which $\Gamma_1 \cap \Gamma_2 \cap X$ is reducible is thus contained in the proper closed subset $W = \overline{g^{-1}(V)}$ of $\mathbb{P}^{4*}$. There are two cases to consider.

If V does not contain any two-planes, then W does not contain any hyperplanes. Hence, the number of $(\Gamma_1, \Gamma_2) \in I_2$ with Γ_1 fixed and $H(\Gamma_2) \leqslant T_2$ is $O_{d,\varepsilon}(T_2^{10/3+\varepsilon})$ by (E3) and (E7). Note that W is cut out by $O_d(1)$ hypersurfaces of degrees $O_d(1)$ since V is. The number of $\Gamma_1 \in \mathbb{P}^{4*}(\mathbb{Q})$ of height at most T_1 is $O(T_1^5)$. Hence, the number of pairs $(\Gamma_1, \Gamma_2) \in I_2$ such that V does not contain any two-planes and $H(\Gamma_i) \leqslant T_i$ is $O_{d,\varepsilon}(T_1^5 T_2^{10/3+\varepsilon})$.

If V contains a two-plane, then W contains a hyperplane. The best estimate for the number of $(\Gamma_1, \Gamma_2) \in I_2$ with Γ_1 fixed and $H(\Gamma_2) \leqslant T_2$ is therefore $O_d(T_2^4)$. This is the trivial estimate (E2). Lemma 2.2.2 states that the set of Γ_1 for which V contains a two-plane is contained in a hypersurface in $\mathbb{P}^{4*}$ of degree $O_d(1)$, provided that X is not a cone over a plane curve. In this case there are $O_d(T_1^4)$ such $\Gamma_1 \in \mathbb{P}^{4*}(\mathbb{Q})$ of height at most T_1, again according to (E2). In the general case we have $O(T_1^5)$ hyperplanes. Hence, the number of pairs $(\Gamma_1, \Gamma_2) \in I_2$ such that V contains a two-plane and $H(\Gamma_i) \leqslant T_i$ is $O_d(T_1^{5-\eta} T_2^4)$, where $\eta = 1$ unless X is a cone over a plane curve in which case $\eta = 0$. $\square$

We can use Lemma 3.2.1 to estimate the cardinality of I_2. The number of $(\Gamma_1, \Gamma_2) \in I_2$ with $T < H(\Gamma_1) \leqslant 2T$ for some $T \ll B^{1/4}$ is

$$\ll_{d,\varepsilon} \left(\frac{B^{1/2}}{T}\right)^{10/3+\varepsilon} T^5 + \left(\frac{B^{1/2}}{T}\right)^{5-\eta} T^4 \ll \begin{cases} B^{25/12+\varepsilon} & \text{if } \eta = 1, \\ B^{5/2+\varepsilon} & \text{if } \eta = 0. \end{cases}$$

By summing over dyadic intervals we get

$$|I_2| \ll_{d,\varepsilon} \begin{cases} B^{25/12+\varepsilon} & \text{if } \eta = 1, \\ B^{5/2+\varepsilon} & \text{if } \eta = 0. \end{cases}$$

Consequently, the first term of (3.4) is $O_{d,\varepsilon}(B^{3+\varepsilon})$. Recall that this term does not even appear when $\eta = 0$.

In order to estimate the second term of (3.4) we divide the ranges of both Γ_1 and Γ_2 into dyadic intervals. If $T_1 \ll B^{1/4}$ and $T_1T_2 \ll B^{1/2}$, then

$$\sum_{\substack{(\Gamma_1,\Gamma_2)\in I_2\\ T_i<H(\Gamma_i)\leqslant 2T_i}} \frac{B^{1+\varepsilon}}{H(\Gamma_1\cap\Gamma_2)^{1/3}} \ll_{d,\varepsilon} \frac{B^{1+\varepsilon}}{(T_1T_2)^{1/3}}\left(T_1^5T_2^{10/3+\varepsilon}+T_1^4T_2^4\right)$$
$$\ll B^{35/12+\varepsilon}.$$

Hence, the second term of (3.4) is $O_{d,\varepsilon}(B^{3+\varepsilon})$.

We have already seen that $|I_2| = O_{d,\varepsilon}(B^{3+\varepsilon})$, so it remains to estimate the very last term in (3.4). Let $Z_1(T)$ be the union of all lines $L\subset Z$ with $H(L)>T$, for some $T>0$, and let $Z_2(T)$ be the union of all lines $L\subset Z$ with $H(L)\leqslant T$. Then

$$N(Z,B)\leqslant N(Z_1(T),B)+N(Z_2(T),B), \tag{3.5}$$

for every $T>0$.

Lemma 3.2.2. *If $T\leqslant B^2$, then*

$$N(Z_1(T),B) \ll_{d,\varepsilon} B^{13/4}T^{2/d-1+\varepsilon}.$$

Proof. Let $J\subset\mathbb{P}^{4*}(\mathbb{Q})$ be the projection of $I_2\subset\mathbb{P}^{4*}(\mathbb{Q})\times\mathbb{P}^{4*}(\mathbb{Q})$ on the first factor. If $\Gamma_1\in J$, then $\Gamma_1\cap X$ is an irreducible surface of degree d. By Lemma 2.3.1, the dimension of $F_1(\Gamma_1\cap X)$ is at most one, and by Lemma 2.3.2, every one-dimensional irreducible component of $F_1(\Gamma_1\cap X)$ has degree at least d. There are thus $O_{d,\varepsilon}(R^{2/d+\varepsilon})$ rational lines $L\subset\Gamma_1\cap X$ of height at most R according to (E1). If the line L is contained in Z, then L contains two different rational points on $\mathbb{P}^4$ of height at most B. Hence,

$$\sum_{\substack{L\in F_1(\Gamma_1\cap X)(\mathbb{Q})\\ L\subset Z\\ R<H(L)\leqslant 2R}} N(L,B) \ll_{d,\varepsilon} \frac{B^2}{R}R^{2/d+\varepsilon} \ll B^2R^{2/d-1+\varepsilon} \tag{3.6}$$

according to (E1). The cardinality of J is $O(B^{5/4})$, so

$$\sum_{\substack{L\subset Z\\ R<H(L)\leqslant 2R}} N(L,B) \ll_{d,\varepsilon} B^{13/4}R^{2/d-1+\varepsilon}. \tag{3.7}$$

By summing over dyadic intervals, we get

$$N(Z_1(T),B) \ll_{d,\varepsilon} B^{13/4}T^{2/d-1+\varepsilon}.$$

Note that we only have to sum over $O(\log B)$ dyadic intervals since $H(L)\ll B^2$ for every line L that contains two rational points of height at most B. $\square$

Lemma 3.2.3. *If $T \leqslant B^{3/4}$, then*

$$N(Z_2(T), B) \ll_{d,\varepsilon} B^{3+\varepsilon} + B^{2+\varepsilon}T^{2/3+2/d+\varepsilon}.$$

Proof. For each line $L \subset Z_2(T)$ we may choose a rational hyperplane $\Gamma_L \subset \mathbb{P}^4$ containing L such that

$$H(\Gamma_L) \leqslant A'' H(L)^{1/3}$$

for some positive constant A''. We may also find a second hyperplane $\Gamma'_L \subset \mathbb{P}^4$ containing L such that

$$H(\Gamma'_L) \leqslant A'' H(L)^{2/3}/H(\Gamma_L).$$

This is a consequence of Lemma 2.1.1 and 2.1.2. Let $Z'_2(T)$ be the union of all lines $L \subset Z_2(T)$ for which $\Gamma_L \cap X$ are irreducible surfaces of degree d, and let $Z''_2(T) \subset Z_2(T)$ be the union of the other lines. The constant A'' does not depend on B so we may assume that $A'' \leqslant A$, where A is the constant introduced in the definition of the set I. Hence, $Z''_2(T) \subset \Sigma_1$, provided that $T \leqslant B^{3/4}$. We already know that $N(\Sigma_1, B) \ll_{d,\varepsilon} B^{3+\varepsilon}$.

To estimate $N(Z'_2(T), B)$ we note that (3.6) is valid for any $\Gamma_1 \in \mathbb{P}^{4*}(\mathbb{Q})$ for which $\Gamma_1 \cap X$ is an irreducible surface of degree d. Consequently, the contribution to $N(Z_2(T), B)$ from the lines $L \subset Z'_2(T)$ for which $R < H(L) \leqslant 2R$ and $\Gamma_L = \Gamma$, for some positive number R and some fixed $\Gamma \in \mathbb{P}^{4*}(\mathbb{Q})$, is $O_{d,\varepsilon}(B^2 R^{2/d-1+\varepsilon})$. The number of $\Gamma \in \mathbb{P}^{4*}(\mathbb{Q})$ with $H(\Gamma) \leqslant A''(2R)^{1/3}$ is $O(R^{5/3})$. Hence,

$$\sum_{\substack{L \subset Z'_2(T) \\ R < H(L) \leqslant 2R}} N(L, B) \ll_{d,\varepsilon} B^2 R^{2/3+2/d+\varepsilon}.$$

By summing over dyadic intervals we get the desired result. □

If we put $T = B^{3/4}$ in (3.5) and apply Lemma 3.2.2 and 3.2.3, then we get

$$N(Z, B) \ll_{d,\varepsilon} B^{3+\varepsilon} + B^{5/2+3/2d+\varepsilon}.$$

3.3. Estimate of $N(\Sigma_3, B)$. Consider those pairs $(\Gamma_1, \Gamma_2) \in I_3$ with $T < H(\Gamma_1) \leqslant 2T$ for some $T \ll B^{1/4}$. For each such pair we use Lemma 2.1.1 to find a basis $\mathbf{b}_0, \mathbf{b}_1, \mathbf{b}_2$ of the lattice

$$\{\mathbf{x} \in \mathbb{Z}^5 \ : \ [\mathbf{x}] \in \Gamma_1 \cap \Gamma_2\} \cup \{\mathbf{0}\}.$$

Without loss of generality, we may assume that

$$|\mathbf{b}_0| \leqslant |\mathbf{b}_1| \leqslant |\mathbf{b}_2|.$$

Let $\varphi : \Gamma_1 \cap \Gamma_2 \to \mathbb{P}^2$ be the map

$$[\lambda_0 \mathbf{b}_0 + \lambda_1 \mathbf{b}_1 + \lambda_2 \mathbf{b}_2] \mapsto [\lambda_0, \lambda_1, \lambda_2].$$

Then $H(\varphi(p)) \ll H(p)/|\mathbf{b}_0|$ for a rational point $p \in \Gamma_1 \cap \Gamma_2$, so

$$N(\Gamma_1 \cap \Gamma_2 \cap X, B) \ll_{d,\varepsilon} \left(\frac{B}{|\mathbf{b}_0|}\right)^{2/d+\varepsilon}$$

according to (E4). Now consider all bases $\mathbf{b}_0, \mathbf{b}_1, \mathbf{b}_2$ satisfying $C_i < |\mathbf{b}_i| \leqslant 2C_i$ for some positive numbers C_i with $C_0 \ll C_1 \ll C_2$. The set of $\Gamma \in \mathbb{P}^{4*}$ which contains the point $[\mathbf{b}_0]$ is a hyperplane Λ in $\mathbb{P}^{4*}$, and the number of $\Gamma \in \Lambda(\mathbb{Q})$ with $H(\Gamma) \leqslant R$ is $O(R^4/|\mathbf{b}_0|)$, provided that $R \gg |\mathbf{b}_0|$ [**4**, Lemma 1(v)]. Hence, the number of pairs $(\Gamma_1, \Gamma_2) \in I_3$ with $T < H(\Gamma_1) \leqslant 2T$, $C_i < |\mathbf{b}_i| \leqslant 2C_i$, $H(\Gamma_2) \ll (C_0C_1C_2)/T$ and $\mathbf{b}_0$ fixed is

$$\ll \left(\frac{((C_0C_1C_2)/T)^4}{C_0}\right)\left(\frac{T^4}{C_0}\right) = C_0^2C_1^4C_2^4.$$

The number of $\mathbf{b}_0$ with $|\mathbf{b}_0| \ll C_0$ is $O(C_0^5)$. Recall that $C_0C_1C_2 \ll B^{1/2}$. Hence,

$$\begin{aligned}\sum_{\substack{(\Gamma_1,\Gamma_2)\in I_3\\ T<H(\Gamma_1)\leqslant 2T\\ C_i<|\mathbf{b}_i|\leqslant 2C_i}} N(\Gamma_1 \cap \Gamma_2 \cap X, B) &\ll_{d,\varepsilon} C_0^{7-2/d}C_1^4C_2^4B^{2/d+\varepsilon}\\ &\ll (C_0C_1C_2)^{5-2/3d}B^{2/d+\varepsilon}\\ &\ll B^{5/2+5/3d+\varepsilon}.\end{aligned}$$

By summing over dyadic intervals we get

$$N(\Sigma_3, B) \ll_{d,\varepsilon} B^{5/2+5/3d+\varepsilon}.$$

3.4. Conclusion. We have shown that if $X \subset \mathbb{P}^4$ is an irreducible hypersurface of degree $d \geqslant 3$, then

$$N(X,B) \ll_{d,\varepsilon} B^{3+\varepsilon} + B^{\frac{5}{2}+\frac{3}{2d}+\varepsilon} + B^{\frac{5}{2}+\frac{5}{3d}+\varepsilon},$$

for every $\varepsilon > 0$. It is easy to see that this estimate implies Theorem 1.

References

[1] T. D. Browning – A note on the distribution of rational points on threefolds, *Q. J. Math.* **54** (2003), no. 1, 33–39.

[2] W. Fulton & R. Lazarsfeld – Connectivity and its applications in algebraic geometry, Algebraic Geometry (Chicago, Ill., 1980), Lecture Notes in Math., vol. 862, Springer, Berlin, 1981, 26–92.

[3] J. Harris – *Algebraic Geometry*, Graduate Texts in Mathematics, vol. 133, Springer-Verlag, New York, 1992.

[4] D. R. Heath-Brown – The density of rational points on curves and surfaces, *Ann. of Math. (2)* **155** (2002), no. 2, 553–595.

[5] J. Pila – Density of integral and rational points on varieties, *Astérisque* (1995), no. 228, 4, 183–187, Columbia University Number Theory Seminar (New York, 1992).

[6] W. M. Schmidt – *Diophantine Approximations and Diophantine Equations*, Lecture Notes in Mathematics, vol. 1467, Springer-Verlag, Berlin, 1991.

Arithmetic of Higher-dimensional Algebraic Varieties
(B. Poonen, Yu. Tschinkel, eds.), p. 121–134
Progress in Mathematics, Vol. 226, © 2004 Birkhäuser Boston, Cambridge, MA

REMARQUES SUR L'APPROXIMATION FAIBLE SUR UN CORPS DE FONCTIONS D'UNE VARIABLE

Jean-Louis Colliot-Thélène
CNRS, UMR 8628, Mathématiques, Bâtiment 425,
Université de Paris-Sud, F-91405 Orsay, France
E-mail : Jean-Louis.Colliot-Thelene@math.u-psud.fr

Philippe Gille
CNRS, UMR 8628, Mathématiques, Bâtiment 425,
Université de Paris-Sud, F-91405 Orsay, France
E-mail : Philippe.Gille@math.u-psud.fr

Résumé. On établit l'approximation faible pour les espaces homogènes de groupes linéaires connexes définis sur le corps des fonctions d'une courbe complexe. On en déduit la même propriété pour les variétés qui se ramènent à de tels espaces par fibrations successives. Ainsi l'approximation faible vaut pour une surface fibrée en coniques au-dessus d'une droite. On montre par contre que l'approximation faible peut être en défaut pour une surface d'Enriques.

1. Introduction

Soit C/k une courbe projective, lisse, connexe, définie sur un corps algébriquement clos k de caractéristique nulle. On note $F = k(C)$ le corps de fonctions de C et F_M le complété de F pour la valuation v_M définie par un point fermé M de C.

Mots clefs. Approximation faible, corps de fonctions d'une variable, espaces homogènes de groupes linéaires, surfaces de Del Pezzo, intersections de deux quadriques, surfaces d'Enriques.

Soit X/F une variété lisse géométriquement connexe. L'ensemble $X(F_M)$ des F_M-points est muni d'une topologie naturelle ([**Kne62**]). On dit que l'approximation faible vaut pour un ensemble fini S de points fermés de C si l'image diagonale de $X(F)$ dans $\prod_{M \in S} X(F_M)$ est dense pour la topologie produit. Si c'est le cas pour tout ensemble fini S de points fermés, on dit que l'approximation faible vaut pour la variété X/F.

Comme l'a rappelé B. Hassett dans son exposé à l'American Institute of Mathematics, à Palo Alto en décembre 2002, on se demande si toute F-variété lisse et géométriquement rationnellement connexe satisfait à l'approximation faible. Un produit de k-variétés projectives, lisses et rationnellement connexes, est encore une variété projective, lisse, rationnellement connexe ; on en déduit, par un argument standard utilisant la restriction des scalaires à la Weil, que pour répondre à la question sur $F = k(C)$ il suffit de le faire sur $F = k(\mathbf{P}^1)$.

Mais le seul résultat général connu à ce jour est un théorème de Kollár, Miyaoka et Mori (voir [**Kol96**] IV. 6.10), qui ne concerne qu'une version faible de l'approximation, et ce seulement pour les $M \in C$ où X/F a bonne réduction.

Dans cette note nous établissons cet énoncé pour les F-variétés géométriquement rationnellement connexes qui se ramènent par fibrations à des espaces homogènes de groupes linéaires connexes.

Nous montrons par ailleurs que sous la simple hypothèse d'annulation des $H^i(X, O_X)$ (pour $i \geqslant 1$) il peut y avoir défaut d'approximation faible. Notre exemple est une surface d'Enriques. L'outil utilisé est une loi de réciprocité et l'existence d'un revêtement non ramifié non trivial sur une telle surface.

On notera que des résultats particuliers sur l'approximation faible ont été obtenus sur des corps de fonctions plus compliqués que ceux considérés ici : corps de fonctions d'une variable sur le corps des réels ([**CT96**], [**Sch96**], [**Duc98**]), et corps de fonctions de deux variables sur un corps algébriquement clos ([**CTGP01**], [**CTGP03**]).

2. Rappels : Approximation faible

Soient K un corps et Ω un ensemble de valuations de K, distinctes deux à deux, discrètes de rang un. Pour tout $v \in \Omega$ soit K_v le complété de K en v. L'approximation faible pour les K-variétés lisses géométriquement connexes se définit comme dans l'introduction. On note $\mathbf{A}^n_K$ l'espace affine de dimension n sur K.

Le fait suivant est bien connu :

Proposition 2.1. *Soient K et Ω comme ci-dessus. Soient X et Y deux K-variétés lisses géométriquement connexes. S'il existe $n, m \in \mathbf{N}$ tels que les K-variétés $X \times_K \mathbf{A}^n_K$ et $Y \times_K \mathbf{A}^m_K$ soient K-birationnellement équivalentes, alors l'approximation faible vaut pour X si et seulement si elle vaut pour Y. En particulier, l'approximation faible vaut pour X si et seulement si elle vaut pour un ouvert non vide de X.*

Proposition 2.2. *Soient K et Ω comme ci-dessus. Soit $f : X \to Y$ un K-morphisme lisse de K-variétés lisses géométriquement connexes, à fibre générique géométriquement connexe. Si pour tout point $M \in Y(K)$ l'approximation faible vaut pour la F-variété fibre $X_M = f^{-1}(M)$, et si l'approximation faible vaut pour Y, alors elle vaut pour X.*

Démonstration. Soit $S \subset \Omega$ un ensemble fini, et supposons donné pour chaque $v \in S$ un point $P_v \in X(K_v)$. Soit $Q_v = f(P_v)$. Par le théorème des fonctions implicites, il existe un voisinage ouvert $\omega_v \subset Y(K_v)$ contenant Q_v, équipé d'une section analytique $\sigma_v : \omega_v \to X(K_v)$ de la projection $X(K_v) \to Y(K_v)$. L'approximation faible valant pour Y, on peut trouver un point F-rationnel $Q \in Y(K)$ tel que pour chaque $v \in S$ le point Q soit très proche de Q_v dans $Y(K_v)$ et qu'il appartienne à ω_v. Soit $Z = f^{-1}(Q)$ la fibre en Q. Alors $R_v = \sigma_v(Q) \in Z(K_v)$ est très proche de P_v dans $X(K_v)$. Par hypothèse, l'approximation faible vaut pour la F-variété Z. Ainsi il existe un point $P \in Z(K) \subset X(K)$ très proche de R_v dans $Z(K_v)$ et donc dans $X(K_v)$ pour chaque $v \in S$. Un tel point est très proche de chaque P_v pour $v \in S$. □

Remarque 2.3. On ne peut espérer appliquer cette proposition générale que lorsque l'on a déjà la propriété : toute fibre non vide de f au-dessus d'un point K-rationnel de Y possède un point K-rationnel. Sur un corps $K = k(C)$ du type considéré dans l'introduction, d'après Graber, Harris et Starr ([**GHS03**]) et d'après de Jong et Starr ([**dJSt03**]), c'est le cas si la fibre générique de f est birationnelle à une variété projective, lisse, géométriquement connexe et rationnellement connexe. Dans ce cas, pour les variétés qui se dévissent en variétés rationnellement connexes pour lesquelles l'approximation faible a déjà été établie, on obtient l'approximation faible.

3. Rappels : Cohomologie galoisienne des corps C_1

Le théorème suivant regroupe des résultats de Springer (cas des corps C_1, qui nous suffirait ici, les corps $F = k(C)$ et $F_M \simeq k((t)))$ étant des corps C_1) et de Steinberg (cas général).

Théorème 3.1. *Soit K un corps de caractéristique zéro et de dimension cohomologique 1.*

(a) Tout K-groupe réductif connexe est quasi-déployé.

(b) Tout espace homogène sous un K-groupe linéaire connexe possède un point rationnel.

(c) Soient G un K-groupe linéaire connexe et $H \subset G$ un sous-groupe fermé. La projection $G(K)$-équivariante $G(K) \to (G/H)(K)$, où les deux ensembles sont pointés par l'élément neutre de G et par son image, se prolonge en une suite exacte d'ensembles pointés

$$G(K) \to (G/H)(K) \to H^1(K,H) \to 1.$$

(d) Soit H un K-groupe linéaire, et soit $H^0 \subset H$ sa composante connexe, sous-groupe normal de H. L'homomorphisme quotient $H \to H/H^0$ induit une bijection $H^1(K,H) \to H^1(K,H/H^0)$.

Démonstration.

(a) Voir [**Ser94**] III.2.3 Thm. 1′ p. 139.

(b) Voir [**Ser94**] III.2.4 Cor. 1 p. 141. L'énoncé (b) peut se reformuler ainsi : pour tout F-espace homogène X sous un K-groupe linéaire connexe G, il existe un K-morphisme G-équivariant de G (vu comme espace principal homogène à gauche sous G) vers X.

(c) Combiner $H^1(K,G) = 0$ (qui est une reformulation de (b) dans le cas des espaces principaux homogènes) et [**Ser94**] I.5.4 Prop. 36 p. 47.

(d) Voir [**Ser94**] III.2.4 Cor. 3 p. 142.

□

4. Approximation faible pour les espaces homogènes de groupes linéaires connexes et pour les variétés qui s'y ramènent

Le théorème suivant devrait être considéré comme bien connu.

Théorème 4.1. *Soit $F = k(C)$ un corps de fonctions d'une variable sur un corps algébriquement clos de caractéristique zéro et soit G un F-groupe linéaire connexe. Alors G satisfait à l'approximation faible.*

Démonstration. Le groupe G/F est le produit semi-direct de son quotient réductif G_{red} par son radical unipotent R_uG qui est F-isomorphe (en tant que variété) à un espace affine. Par la Proposition 2.1, on est ramené au cas où G est un groupe réductif connexe. Comme le corps F est C_1, le groupe réductif

G est quasi-déployé. Soit B un sous-groupe de Borel de G_{red} et soit T un F-tore maximal de B. La F-variété sous-jacente au radical unipotent R_uB est un espace affine, il en est de même du radical unipotent R_uB^- du sous-groupe de Borel B^- opposé à B, et le groupe G contient un ouvert, "la grosse cellule", isomorphe au produit $(R_uB)\times_F T\times_F (R_uB)^-$ (**[SGA64]**, exp. XXII, Proposition 4.1.2 p. 172). Par la Proposition 2.1, on est ramené à établir l'approximation faible pour un F-tore T.

Par considération du groupe des caractères, on voit aisément que pour tout F-tore T (ici F pourrait être un corps quelconque), il existe une suite exacte de F-tores

$$1 \to R \to E \to T \to 1,$$

où E est un F-tore quasi-trivial, i.e. un produit $\prod_i R_{F_i/F}\mathbf{G}_m$ de restrictions à la Weil du groupe multiplicatif $\mathbf{G}_m$ pour diverses extensions de corps F_i/F, et où R est un F-tore. Le tore quasi-trivial E est un ouvert d'un espace affine sur F, il satisfait donc à l'approximation faible. Soit S un ensemble fini de points fermés de la courbe C. Comme les complétés $F_M, M \in S$, du corps $F = k(C)$ sont des corps C_1, les groupes $H^1(F_M, R)$ sont nuls. Ainsi l'application continue $\prod_{M\in S} E(K_M) \to \prod_{M\in S} T(K_M)$ est surjective. Comme $E(K)$ est dense dans $\prod_{M\in S} E(K_M)$, il s'en suit immédiatement que $T(K)$ est dense dans $\prod_{M\in S} T(K_M)$. □

Théorème 4.2. *Soit $F = k(C)$ un corps de fonctions d'une variable sur un corps algébriquement clos de caractéristique zéro et soit G un F-groupe linéaire. Pour tout ensemble fini S de points fermés de C, l'application diagonale*

$$H^1(F, G) \to \prod_{M\in S} H^1(F_M, G),$$

où le second produit est un ensemble fini, est surjective.

Démonstration. D'après le Théorème 3.1(d), il suffit d'établir le théorème lorsque G est un F-schéma en groupes finis.

Soient F_s une clôture séparable de F et $\mathcal{G} = \mathrm{Gal}(F_s/F)$. Soit M un point fermé de C. Soit $F_{M,s}$ une clôture séparable de $F_M \simeq k((t))$ et $F_s \subset F_{M,s}$ un plongement. Le groupe de Galois absolu $I_M = \mathrm{Gal}(F_{M,s}/F_M)$ est isomorphe à $\hat{\mathbf{Z}}$, le complété profini de $\mathbf{Z}$. On fixe un tel isomorphisme, i.e. on fixe un générateur profini $c_M \in I_M$. On fixe aussi, pour chaque $M \in C$, un plongement $F_s \subset F_{M,s}$. Ceci détermine un plongement $I_M \subset \mathcal{G}$.

On note $A = G(F_s) = G(F_{M,s})$. C'est un groupe fini muni d'une action de $\mathcal{G}$ et donc de I_M pour tout $M \in C$.

Quitte à agrandir S, on peut supposer que le F-schéma en groupe G provient d'un U-schéma en groupes fini étale sur U, où U désigne le complémentaire de

S dans C. En d'autres termes, on peut supposer que pour tout $M \notin S$, l'action de I_M sur A est triviale.

Soit H un sous-groupe fermé de $\mathscr{G}$. L'ensemble de cohomologie $H^1(H, A)$ est un quotient de l'ensemble $Z^1(H, A)$ des 1-cocycles continus de H à valeurs dans le groupe fini A. Cet ensemble s'identifie à l'ensemble des homomorphismes continus de H dans le produit semi-direct $A \cdot H$ qui sont des sections de la projection structurale $A \cdot H \to H$ ([**Ser94**], Chap. I, 5.1, Exercice 1 p. 43). Si $z_h \in A, h \in H$ est le 1-cocycle, la section est donnée par $h \mapsto z_h \cdot h$.

Si l'on se donne un système de générateurs topologiques de H, il existe au plus un 1-cocycle dans $Z^1(H, A)$ prenant des valeurs données sur ces générateurs.

En particulier, pour tout M, l'ensemble $Z^1(I_M, A)$ est fini (A étant fini). *A fortiori* $H^1(F_M, G) = H^1(I_M, A)$ est fini.

Soit $S \subset C$ un ensemble fini de points fermés de C. Pour établir le théorème, nous allons établir l'énoncé *a priori* plus fort que l'application naturelle de 1-cocycles continus

$$Z^1(\mathscr{G}, A) \to \prod_{M \in S} Z^1(I_M, A)$$

est surjective. Soit g le genre de la courbe C. Notons $S = \{M_1, \dots, M_s\}$. Choisissons un point fermé $M_{s+1} \in C \setminus S$. Notons $\mathscr{H} = \mathrm{Gal}(L/F)$ le groupe de Galois de la sous-extension maximale L/F de F_s/F non ramifiée en dehors de $S \cup M_{s+1}$. C'est un quotient de $\mathscr{G}$, et l'on a des inclusions induites $I_{M_j} \subset \mathscr{H}$ pour $j = 1, \dots, s+1$. Rappelons que pour chaque j, on a choisi un générateur topologique c_j du groupe d'inertie $I_j = I_{M_j} \simeq \hat{\mathbf{Z}}$. L'hypothèse initiale faite sur S garantit que l'inclusion naturelle $G(L) \subset G(F_s) = A$ est une égalité. On dispose donc d'une inclusion naturelle $Z^1(\mathscr{H}, A) \hookrightarrow Z^1(\mathscr{G}, A)$, et pour établir l'énoncé il suffit de montrer que l'application diagonale

$$Z^1(\mathscr{H}, A) \to \prod_{M \in S} Z^1(I_M, A)$$

est surjective.

D'après le théorème d'existence de Riemann (cf. [**Ser92**], §1.2), le groupe $\mathscr{H}$ est le groupe profini engendré par les générateurs

$$a_1, b_1, \dots, a_g, b_g, c_1, \dots, c_s, c_{s+1}$$

avec l'unique relation

$$[a_1, b_1] \cdots [a_g, b_g]\, c_1 c_2 \cdots c_s c_{s+1} = 1,$$

où pour tout $j = 1, \dots, s$, l'élément c_j est comme ci-dessus.

Un élément du produit $\prod_{M\in S} Z^1(I_M, A)$ détermine une famille $z_j = z_{M_j} \in A$, où z_j est la valeur du 1-cocycle sur le générateur topologique c_j de I_j. Pour tout $j = 1, \ldots, s$, on dispose donc de l'homomorphisme $I_j \to A \cdot I_j$ envoyant c_j sur $z_j \cdot c_j$. Cet homomorphisme induit un homomorphisme $\rho_j : I_j \to A \cdot \mathscr{H}$. Pour définir un homomorphisme continu $\rho : \mathscr{H} \to A \cdot \mathscr{H}$, il suffit de le définir sur les générateurs de $\mathscr{H}$, et de s'assurer que la relation ci-dessus est respectée. Définissons $\rho(a_i) = 1 \cdot a_i$ et $\rho(b_i) = 1 \cdot b_i$ pour tout $i = 1, \ldots, g$, puis $\rho(c_j) = z_j \cdot c_j$ pour $j = 1, \ldots, s$. Pour que la relation soit respectée, il suffit de choisir $\rho(c_{s+1}) \in A \cdot \mathscr{H}$ tel que

$$[a_1, b_1] \cdots [a_g, b_g] z_1 \cdot c_1 \cdots z_s \cdot c_s \cdot \rho(c_{s+1}) = 1$$

(par abus de langage, on note ici $a_i \cdot 1 = a_i$ et $b_i \cdot 1 = b_i$), ce qui compte tenu de la relation initiale dans $\mathscr{H}$ se traduit encore par

$$c_{s+1}^{-1} \cdot c_s^{-1} \cdots c_1^{-1} \cdot z_1 \cdot c_1 \cdots z_s \cdot c_s \cdot \rho(c_{s+1}) = 1.$$

Soit $u \in A \cdot \mathscr{H}$ tel que $\rho(c_{s+1}) = u \cdot c_{s+1}$. Un calcul immédiat (moins magique qu'il n'y paraît) montre que u appartient à $A \subset A \cdot \mathscr{H}$. On définit alors $z_{s+1} = u \in A$.

On dispose donc d'un homomorphisme continu envoyant $\mathscr{H}$ dans $A \cdot \mathscr{H}$, envoyant a_i sur $1 \cdot a_i$ et b_i sur $1 \cdot b_i$ pour $i = 1, \ldots, g$, et par ailleurs c_j sur $z_j \cdot c_j$ pour $j = 1, \ldots, s+1$. Cet homomorphisme définit une section de la projection $A \cdot \mathscr{H} \to \mathscr{H}$, comme on le voit sur les générateurs. Enfin sa restriction à chaque sous-groupe I_j pour $j = 1, \ldots, s$ a son image dans $A \cdot I_j$ et coïncide avec l'homomorphisme initial $I_j \to A \cdot I_j$. □

Théorème 4.3. *Soit $F = k(C)$ un corps de fonctions d'une variable sur un corps algébriquement clos de caractéristique zéro, et soit G un F-groupe linéaire connexe. L'approximation faible vaut pour tout espace homogène sous G.*

Démonstration. Soit X/F un tel espace homogène. D'après le Théorème 3.1(b), on peut écrire $X = G/H$, où $H \subset G$ est un F-sous-groupe fermé de G. D'après le Théorème 3.1(c), et le fait que tant le corps F que les corps F_M sont des corps C_1, on a un diagramme commutatif de suites exactes d'ensembles pointés

$$\begin{array}{ccccccc} G(F) & \longrightarrow & (G/H)(F) & \xrightarrow{\varphi} & H^1(F,H) & \longrightarrow & 1 \\ \downarrow & & \downarrow & & \downarrow & & \\ \prod\limits_{M\in S} G(F_M) & \longrightarrow & \prod\limits_{M\in S} (G/H)(F_M) & \xrightarrow{\varphi_M} & \prod\limits_{M\in S} H^1(F_M,H) & \longrightarrow & 1. \end{array}$$

D'après le Théorème 4.2, la flèche verticale de droite est surjective. Pour chaque $M \in S$, soit $x_M \in (G/H)(F_M)$. D'après le diagramme ci-dessus et ses propriétés, il existe $x \in (G/H)(F)$ qui a même image que la famille $\{x_M\}$ dans

$\prod_{M \in S} H^1(F_M, H)$. Il existe alors pour chaque $M \in S$ un élément $g_M \in G(F_M)$ tel que $x_M = g_M \cdot x$. Comme l'approximation faible vaut pour G (Théorème 4.1), on peut trouver $g \in G(F)$ arbitrairement proche de chaque $g_M \in G(F_M)$, et alors $g \cdot x \in (G/H)(F)$ est arbitrairement proche de chaque $x_M \in (G/H)(F_M)$ (chaque application $G(F_M) \to (G/H)(F_M)$ étant continue). □

Théorème 4.4. *Soit $F = k(C)$ le corps des fonctions d'une courbe projective et lisse C sur un corps k algébriquement clos de caractéristique zéro. Soit $f : X \to Y$ un F-morphisme dominant de F-variétés lisses géométriquement intègres, dont la fibre générique est géométriquement connexe et $F(Y)$-birationnelle à un espace homogène d'un groupe linéaire connexe G sur le corps des fonctions $F(Y)$ de Y. Si Y satisfait à l'approximation faible, alors X satisfait à l'approximation faible.*

Démonstration. Pour établir le théorème, on peut d'après la Proposition 2.1 restreindre Y à un ouvert et X à l'image réciproque de cet ouvert. On peut donc supposer que le $F(Y)$-groupe linéaire connexe est la restriction d'un Y-groupe linéaire fidèlement plat sur Y, à fibres connexes, soit G/Y, et que pour tout point $P \in Y(F)$ la fibre $f^{-1}(P)$ est lisse, géométriquement connexe et F-birationnelle à un F-espace homogène sous le F-groupe linéaire connexe fibre G_P. D'après le Théorème 4.3 et la Proposition 2.1, toute telle fibre satisfait à l'approximation faible. Il résulte alors des hypothèses et de la Proposition 2.2 que X satisfait à l'approximation faible. □

Les exemples non triviaux abondent (par exemple non trivial, on entend des exemples de F-variétés qui ne sont pas nécessairement F-birationnelles à un espace projectif). Les plus évidents sont les surfaces fibrées en coniques (de dimension au moins un) au-dessus de la droite projective $\mathbf{P}^1_F$, et plus généralement les fibrés en quadriques, resp. en variétés de Severi–Brauer au-dessus d'un espace projectif de dimension arbitraire. Dans les cas cités, la démonstration de l'approximation faible se fait bien sûr à moindres frais : outre l'élémentaire Proposition 2.2, on utilise le fait que toute F-quadrique lisse possède un F-point (cas particulier du théorème de Tsen remontant à Max Noether) et est donc F-birationnelle à un espace projectif sur F, resp. le fait que toute variété de Severi–Brauer sur F est F-isomorphe à un espace projectif sur F (théorème de Tsen et théorie élémentaire des variétés de Severi–Brauer, due à F. Châtelet).

Dégageons le :

Corollaire 4.5. *Soit $F = k(C)$ comme ci-dessus et soit X/F une surface de Del Pezzo de degré 4, c'est-à-dire une intersection complète lisse de deux quadriques dans $\mathbf{P}^4_F$. L'approximation faible vaut pour X.*

Démonstration. Comme F est un corps C_1, on a $X(F) \neq \emptyset$. Comme F est infini, il existe un point F-rationnel R non situé sur l'une quelconque des 16 droites (sur une clôture algébrique de F) contenues dans X ([**Man86**], Chap. IV, §30, Theorem 30.1 p. 162). En éclatant R, on obtient une surface cubique lisse $Y \subset \mathbf{P}^3_F$ qui contient une droite définie sur F (la courbe exceptionnelle image inverse de R). Le pinceau des 2-plans de $\mathbf{P}^3_F$ passant par cette droite définit sur Y une structure de surface fibrée en coniques au-dessus de $\mathbf{P}^1_F$, et l'on a vu ci-dessus que l'approximation faible vaut pour une telle surface. Par la Proposition 2.1, l'approximation faible vaut donc aussi pour X. □

Remarque 4.6.

(a) Toute F-surface projective, lisse, géométriquement rationnellement connexe est rationnelle, et donc F-birationnelle soit à une surface fibrée en coniques au-dessus de $\mathbf{P}^1_F$, soit à une surface de Del Pezzo de degré d, avec $1 \leqslant d \leqslant 9$. Toute F-surface de Del Pezzo de degré $d \geqslant 5$ est F-birationnelle à $\mathbf{P}^2_F$, donc satisfait à l'approximation faible. Nous venons de voir que cette dernière propriété vaut pour $d = 4$. La question de savoir si l'approximation faible vaut reste ouverte pour les F-surfaces de Del Pezzo de degré 3 (surfaces cubiques lisses), et *a fortiori* pour les F-surfaces de Del Pezzo de degré 2 et 1.

(b) En utilisant la Proposition 2.2, on déduit du Corollaire 4.5 *l'approximation faible pour toute F-variété X intersection complète lisse de deux quadriques dans $\mathbf{P}^n_F$, $n \geqslant 4$.* Pour $n \geqslant 5$, on peut aussi déduire l'approximation faible de [**CTSaSD87**], I, Theorem 3.27 p. 80. Pour $n \geqslant 6$, la situation est encore plus simple : dans ce cas, toute telle variété X est F-birationnelle à un espace projectif ([**CTSaSD87**], I, Theorem 3.2 p. 60 ; Theorem 3.4 p. 62).

(c) Soit $Z \subset \mathbf{P}^n_F$, $n \geqslant 3$, une F-hypersurface cubique géométriquement intègre, non conique. Supposons que Z possède un ensemble globalement F-rationnel d'au plus 3 points singuliers. Alors pour tout modèle lisse X de Z, l'approximation faible vaut. Nous ne donnons que le principe de la démonstration. Lorsque Z possède un point singulier F-rationnel, alors Z est F-birationnel à un espace projectif. Si Z possède un ensemble globalement F-rationnel de deux points singuliers, alors Z est F-birationnel à un espace projectif ou à une hypersurface affine d'équation

$$y^2 - az^2 = P(x_1, \dots, x_{n-2})$$

avec $a \in F^*$ et P un polynôme non nul (de degré total au plus 4) ([**CTSaSD87**], II, Prop. 9.8 p. 109). Cette variété est fibrée en coniques au-dessus d'un espace affine, son ouvert de lissité satisfait donc à l'approximation faible (Prop. 2.2). Lorsque Z possède un ensemble globalement F-rationnel de trois points singuliers, soit Z est F-rationnelle, soit Z admet une équation d'un type particulier ([**CTSal89**], Prop. 1.6 p. 522). La variété Z est alors fibrée au-dessus d'un espace projectif, la fibre générique étant une surface cubique géométriquement intègre, non conique, possédant trois points singuliers conjugués. Sur le corps F, toute telle surface est F-rationnelle ([**CoTs88**]). La Proposition 2.2 permet donc d'établir l'approximation faible.

5. Une surface d'Enriques qui ne satisfait pas à l'approximation faible

Soit $F = k(C)$ comme dans l'introduction. Nous commençons par décrire un mécanisme familier dans un cadre plus délicat, à savoir celui des corps de nombres.

La somme des degrés des diviseurs d'une fonction est nulle. Le complexe ainsi obtenu

$$F^* \to \oplus_{M \in C} \mathbf{Z} \to \mathbf{Z},$$

où la dernière flèche est la somme, induit pour tout entier $n > 0$ un complexe

$$F^*/F^{*n} \to \oplus_{M \in C} \mathbf{Z}/n \to \mathbf{Z}/n.$$

Pour tout $M \in C$, la flèche $F^*/F^{*n} \to \mathbf{Z}/n$ induite par l'application diviseur en M s'identifie à la flèche naturelle $F^*/F^{*n} \to F_M^*/F_M^{*n}$.

Soit X une F-variété projective, lisse, géométriquement connexe. Soit $F(X)$ son corps des fonctions et $f \in F(X)^*$ une fonction dont le diviseur est une puissance n-ième dans le groupe des diviseurs de X. Soit $U \subset X$ un ouvert non vide sur lequel f est inversible. L'équation $f = t^n$ définit sur l'ouvert U un μ_n-torseur qui, grâce à l'hypothèse sur le diviseur, s'étend en un μ_n-torseur $Y \to X$.

Pour tout corps L contenant F, à ce μ_n-torseur est associé une application d'évaluation $\varphi_L : X(L) \to H^1(L, \mu_n) = L^*/L^{*n}$ qui sur $U(L)$ n'est autre que l'application associant à $P \in U(L)$ la classe de $f(P)$ dans L^*/L^{*n}. Pour $M \in C$ point fermé et $L = F_M$, l'application $\varphi_{F_M} : X(F_M) \to F_M^*/F_M^{*n} = \mathbf{Z}/n$ est continue, i.e. localement constante. Par ailleurs, comme X/F est propre, un argument de bonne réduction montre que pour presque tout $M \in C$ (i.e. tout M sauf un nombre fini), l'application $\varphi_{F_M} : X(F_M) \to F_M^*/F_M^{*n} = \mathbf{Z}/n$ se factorise par $O_M^*/(O_M^*)^n = 1$, donc a une image nulle dans $\mathbf{Z}/n$. Ici O_M

désigne le complété de l'anneau local de C en M, qui est isomorphe à $k[[t]]$. Soit $S \subset C$ l'ensemble fini des points où φ_{F_M} n'est pas l'application constante de valeur $0 \in \mathbf{Z}/n$.

Le torseur $Y \to X$ définit donc une application $X(F) \to \oplus_{M\in S}\mathbf{Z}/n$ qui composée avec la somme $\oplus_{M\in S}\mathbf{Z}/n \to \mathbf{Z}/n$ donne zéro.

Nous pouvons alors conclure :

Proposition 5.1. *Dans la situation ci-dessus, soit, pour chaque $M \in S$, un point $P_M \in X(F_M)$. Si $\sum_{M\in S}\varphi_M(P_M) \neq 0 \in \mathbf{Z}/n$, alors la famille $\{P_M\} \in \prod_{M\in S} X(F_M)$ n'est pas dans l'adhérence de l'image de $X(F)$ dans le produit topologique $\prod_{M\in S} X(F_M)$.*

Remarque 5.2. L'idée de cette proposition n'est pas nouvelle, on a déjà utilisé des lois de réciprocités variées pour définir une obstruction à l'approximation faible dans divers contextes. L'obstruction la plus connue est l'obstruction de Brauer–Manin sur un corps de nombres, qui fait intervenir le groupe $H^2(\cdot,\mu_n)$. Toujours sur un corps de nombres, on peut aussi utiliser $H^1(\cdot,\mu_n)$ (voir [**Har00**]). Sur un corps de fonctions d'une variable sur les réels, voir [**CT96**] et [**Duc98**].

Soit $\mathbf{A}^1_{\mathbf{C}} = \mathrm{Spec}\mathbf{C}[t]$ la droite affine, et soit $\mathbf{A}^1_{\mathbf{C}} \subset C = \mathbf{P}^1_{\mathbf{C}}$ le plongement naturel. Notons $F = \mathbf{C}(\mathbf{P}^1) = \mathbf{C}(t)$ le corps des fonctions de $\mathbf{P}^1_{\mathbf{C}}$.

Soient $a, b, c, d, e \in \mathbf{C}$, et soient

$$c(u) = t(u-a)(u-b), \quad d(u) = et(u-c)(u-d).$$

Supposons que le polynôme $e \cdot c(u) \cdot d(u) \cdot (c(u) - d(u)) \cdot u(u-1) \in \mathbf{C}[u]$ est séparable de degré 8. Soit $U \subset \mathbf{A}^4_F$, avec coordonnées affines (x, y, z, u), la F-surface lisse définie par le système d'équations

$$x^2 - t(u-a)(u-b) = u(u-1)y^2 \neq 0, \quad x^2 - te(u-c)(u-d) = u(u-1)z^2 \neq 0. \qquad (*)$$

Soit X/F un modèle projectif et lisse, F-minimal, de la surface U. Il existe un ouvert non vide $V \subset U$ qu'on peut identifier à un ouvert de X.

Proposition 5.3. *La F-surface X est une surface d'Enriques, elle satisfait en particulier $H^1(X, O_X) = 0$ et $H^2(X, O_X) = 0$. La fonction rationnelle définie par $f = u(u-1)$ sur U a son diviseur sur X qui est un double.*

Démonstration. Que le diviseur de f sur tout modèle lisse soit un double est facile à établir par des calculs valuatifs (voir [**CTSkSD97**]). Pour le détail de la démonstration du fait que X est une surface d'Enriques, affirmé dans [**CTSkSD97**], voir [**Laf03**]. □

D'après ce qui a été rappelé ci-dessus, il existe un $\mathbf{Z}/2$-torseur Y au-dessus de X dont la restriction à V est obtenue en extrayant la racine carrée de la fonction f (l'espace total du torseur est une surface $K3$). Pour tout corps L contenant F, on a une application induite $\varphi_L : X(L) \to L^*/L^{*2}$, qui sur $V(L)$ n'est autre que l'application envoyant $P \in V(L)$ sur la classe de $f(P)$ dans L^*/L^{*2}.

Proposition 5.4. *Avec les notations ci-dessus, pour $M \neq 0, \infty \in \mathbf{P}^1$, l'image de l'application $\varphi_M : X(F_M) \to F_M^*/F_M^{*2} = \mathbf{Z}/2$ est nulle. Par ailleurs pour $M = 0$ et pour $M = \infty$, l'image de φ_M est tout le groupe $\mathbf{Z}/2$.*

Démonstration. Par continuité et par le théorème des fonctions implicites, qui garantit que $V(F_M)$ est dense dans $X(F_M)$, il suffit d'établir ces faits pour chacune des applications $U(F_M) \to F_M^*/F_M^{*2} = \mathbf{Z}/2$ définie par la fonction f.

Soient $M \in \mathbf{P}^1$, soit $v = v_M$ la valuation associée sur F. Soit (x, y, z, u) un point de $U(F_M)$. Si $v(u) < 0$, alors $v(u(u-1)) = 2v(u)$ est pair. Supposons $v(u) \geqslant 0$ et $v(u\cdot(u-1)) \geqslant 0$ impair. Supposons d'abord $v(t) = 0$. Des équations (*) il résulte alors $v(x^2) = v(t(u-a)(u-b)) = 0$ (car tous deux sont pairs), et $v(x^2 - t(u-a)(u-b)) > 0$. De même $v(x^2 - et(u-c)(u-d)) > 0$. On a alors $v(c(u) - d(u)) > 0$, ce qui est impossible. L'énoncé pour tout $v = v_M$ avec $M \neq 0, \infty$ est donc établi.

Soit $v = v_M$ avec $M = 0 \in \mathbf{A}_F^1$, et donc $v(t) = 1$. Il existe un point $(x, y, z, u) = (1/t, y, z, 1/t) \in U(F_M)$ avec $v(y) = v(z) = 0$, et donc $v(u\cdot(u-1))$ pair. Par ailleurs il existe un point $(x, y, z, u) = (0, y, z, t) \in U(F_M)$ avec $v(y) = v(z) = 0$. Pour un tel point, on a $v(u \cdot (u-1)) = 1$ impair.

Soit $v = v_M$ avec $M = \infty \in \mathbf{P}_F^1$, et donc $v(t) = -1$. Il existe des points de $U(F_M)$ avec $v(u \cdot (u-1)) = 0$ et $v(x) = v(y) = v(z) < 0$. Par ailleurs il existe des points de $U(F_M)$ avec $x = 0$, $u = 1/t$, $v(y) = v(z) = -1$ et donc $v(u \cdot (u-1)) = 1$. □

Théorème 5.5. *La surface d'Enriques X/F possède des points rationnels. L'image de l'application diagonale $X(F) \to X(F_0) \times X(F_\infty)$ n'est pas dense dans ce produit.*

Démonstration. Fixons $u = u_0 \in \mathbf{C}$ assez général. On obtient alors une F-courbe d'équations

$$\begin{aligned} x^2 - t(u_0 - a)(u_0 - b) &= u_0(u_0 - 1)y^2 \neq 0, \\ x^2 - et(u_0 - c)(u_0 - d) &= u_0(u_0 - 1)z^2 \neq 0, \end{aligned}$$

contenue dans U, et rencontrant V. Cette courbe admet une F-compactification lisse Γ donnée en coordonnées homogènes (X, Y, Z, T) par

$$\begin{aligned} X^2 - t(u_0 - a)(u_0 - b)T^2 &= u_0(u_0 - 1)Y^2, \\ X^2 - et(u_0 - c)(u_0 - d)T^2 &= u_0(u_0 - 1)Z^2. \end{aligned}$$

Sur cette courbe on trouve pour $T = 0$ des F-points lisses (à coordonnées dans $\mathbf{C}$). L'application rationnelle de la F-courbe lisse Γ vers la F-variété propre X est partout définie, on a donc $X(F) \neq \emptyset$.

Le reste de l'énoncé résulte de la combinaison des Propositions 5.1, 5.3 et 5.4. □

Remarque 5.6. Les équations concrètes de surfaces d'Enriques utilisées ci-dessus furent introduites dans **[CTSkSD97]**. Lafon **[Laf03]** utilise des formes tordues des équations (*) pour exhiber des exemples de surfaces d'Enriques sur $F = \mathbf{C}(t)$ et même sur $F = \mathbf{C}((t))$ sans point rationnel. Nous ne doutons pas que l'on puisse utiliser de telles équations (tordues) pour exhiber des contre-exemples au principe de Hasse (on a $X(F_M) \neq \emptyset$ pour tout $M \in \mathbf{P}^1$, mais $X(F) = \emptyset$) reposant sur la loi de réciprocité sur F^*/F^{*2} utilisée plus haut, mais cela demandera sans doute un peu d'acharnement.

Références

[CT96] J.-L. Colliot-Thélène – Groupes linéaires sur les corps de fonctions de courbes réelles, *J. reine angew. Math.* **474** (1996), 139–167.

[CTGP01] J.-L. Colliot-Thélène, P. Gille & R. Parimala – Arithmétique des groupes algébriques linéaires sur certains corps géométriques de dimension deux, *C. R. Acad. Sci. Paris* Sér. I Math. **333** (2001), no. 9, 827–832.

[CTGP03] J.-L. Colliot-Thélène, P. Gille & R. Parimala – Arithmetic of linear algebraic groups over two-dimensional fields, *Duke Math. J.* (2003), to appear.

[CTSal89] J.-L. Colliot-Thélène & P. Salberger – Arithmetic on some singular cubic hypersurfaces, *Proc. London Math. Soc. (3)* **58** (1989), no. 3, 519–549.

[CTSaSD87] J.-L. Colliot-Thélène, J.-J. Sansuc & P. Swinnerton-Dyer – Intersections of two quadrics and Châtelet surfaces I, *J. reine angew. Math.* **373** (1987), 37–107 ; II, *J. reine angew. Math.* **374** (1987), 72–168.

[CTSkSD97] J.-L. Colliot-Thélène, A. N. Skorobogatov & P. Swinnerton-Dyer – Double fibres and double covers : paucity of rational points, *Acta Arith.* **79** (1997), no. 2, 113–135.

[CoTs88] D. F. Coray & M. A. Tsfasman – Arithmetic on singular Del Pezzo surfaces, *Proc. London Math. Soc. (3)* **57** (1988), no. 1, 25–87.

[dJSt03] A. J. de Jong & J. Starr – Every rationally connected variety over the function field of a curve has a rational point, à paraître.

[Duc98] A. Ducros – L'obstruction de réciprocité à l'existence de points rationnels pour certaines variétés sur le corps des fonctions d'une courbe réelle, *J. reine angew. Math.* **504** (1998), 73–114.

[GHS03] T. Graber, J. Harris & J. Starr – Families of rationally connected varieties, *J. Amer. Math. Soc.*, **16** (2003), 57–67.

[Har00] D. Harari – Weak approximation and non-abelian fundamental groups, *Ann. Sci. École Norm. Sup. (4)* **33** (2000), no. 4, 467–484.

[Kne62] M. Kneser – Schwache Approximation in algebraischen Gruppen, *Colloque de Bruxelles* (1962), 41–52.

[Kol96] J. Kollár – *Rational curves on algebraic varieties*, Ergebnisse der Mathematik und ihrer Grenzgebiete. 3. Folge., Bd. 32, Springer-Verlag, Berlin, 1996.

[Laf03] G. Lafon – Une surface d'Enriques sans point sur $\mathbf{C}((t))$, prépublication.

[Man86] Y. I. Manin – *Cubic forms*, seconde éd., North-Holland Mathematical Library, vol. 4, North-Holland Publishing Co., Amsterdam, 1986.

[Sch96] C. Scheiderer – Hasse principles and approximation theorems for homogeneous spaces over fields of virtual cohomological dimension one, *Invent. Math.* **125** (1996), no. 2, 307–365.

[Ser92] J.-P. Serre – Revêtements de courbes algébriques, Séminaire Bourbaki (1991/92), Exp. No. 749 *Astérisque*, vol. 206 (1992) 167–182.

[Ser94] ———, *Cohomologie galoisienne*, cinquième éd., Lecture Notes in Mathematics, vol. 5, Springer-Verlag, Berlin, 1994.

[SGA64] *Schémas en groupes. III : Structure des schémas en groupes réductifs* – Séminaire de Géométrie Algébrique du Bois Marie 1962/64 (SGA 3). Dirigé par M. Demazure et A. Grothendieck. Lecture Notes in Mathematics, vol. 153, Springer-Verlag, Berlin, 1962/1964.

Arithmetic of Higher-dimensional Algebraic Varieties
(B. Poonen, Yu. Tschinkel, eds.), p. 135–140
Progress in Mathematics, Vol. 226, © 2004 Birkhäuser Boston, Cambridge, MA

K3 SURFACES OVER NUMBER FIELDS WITH GEOMETRIC PICARD NUMBER ONE

Jordan S. Ellenberg
Department of Mathematics, Princeton University, Princeton NJ 085440-1000, USA • *E-mail :* ellenber@math.princeton.edu

Abstract. If d is an even positive integer, the generic complex K3 surface of degree d has Picard number one. However, because $\bar{\mathbb{Q}}$ is countable it is not a priori obvious that there exists such a K3 surface defined over a number field. We prove that for every d, there exists a number field K and a K3 surface X/K which has geometric Picard number one.

1. Introduction

A long-standing question in the theory of rational points of algebraic surfaces is whether a K3 surface X over a number field K acquires a Zariski-dense set of L-rational points over some finite extension L/K. In this case, we say X has *potential density of rational points.* In case $X_{\mathbb{C}}$ has Picard rank greater than 1, Bogomolov and Tschinkel [**2**] have shown in many cases that X has potential density of rational points, using the existence of elliptic fibrations on X or large automorphism groups of X. By contrast, we do not know a single example of a K3 surface X/K with geometric Picard number 1 which can be shown to have

Key words and phrases. K3 surfaces, monodromy.

Partially supported by NSA Young Investigator Grant MDA905-02-1-0097.

potential density of rational points; nor is there an example which we can show *not* to have potential density of rational points. In fact, the situation is even worse; the moduli space of polarized K3 surfaces of a given degree contains a countable union of subvarieties, each parametrizing a family of K3 surfaces with geometric Picard number greater than 1. Since $\bar{\mathbb{Q}}$ is countable, it is not *a priori* obvious that these subvarieties don't cover the $\bar{\mathbb{Q}}$-points of the moduli space. In other words, it is a non-trivial fact that there exists a K3 surface over any number field with geometric Picard number 1!

In this note, we correct this slightly embarrassing situation by proving the following theorem:

Theorem 1. *Let d be an even positive integer. Then there exists a number field K and a polarized K3 surface X/K, of degree d, such that* $\operatorname{rank}\operatorname{Pic}(X_{\mathbb{C}}) = 1$.

The main idea is to use an argument of Serre on ℓ-adic groups to reduce the problem to proving the existence of K3 surfaces whose associated mod n Galois representations have large image for some finite n; we then use Hilbert's irreducibility theorem and global Torelli for K3's to complete the proof.

Acknowledgements: This note is the result of a conversation between the author, Brendan Hassett, and A.J. de Jong, which took place at the American Institute of Mathematics during the workshop, "Rational and integral points on higher-dimensional varieties." It should also be pointed out that the main idea, in case $d = 4$, is implicit in the final remark of [**3**].

2. Proofs

We begin by recalling some notations and basic facts regarding K3 surfaces. An element x of an abelian group L is called *primitive* if it is not contained in kL for any integer $k > 1$. Let X be a K3 surface over a number field K, and write $\bar{X}$ for $X \times_K \bar{K}$. The group $H^2(X_{\mathbb{C}}, \mathbb{Z})$ is isomorphic to $\mathbb{Z}^{22}$; the cup product on $H^2(X_{\mathbb{C}}, \mathbb{Z})$ is a quadratic form with signature $(3, 19)$, which we denote $\langle , \rangle$. A *polarized K3 surface* is a pair $(X, \mathscr{L})$, where X/K is a K3 surface and $\mathscr{L}$ is an ample line bundle on X. If X is a polarized K3, we let x be the class of $\mathscr{L}$ in $H^2(X_{\mathbb{C}}, \mathbb{Z})$; then the positive even integer $\langle x, x\rangle$ is called the *degree* of X. We denote by L_X the orthogonal complement of x in $H^2(X_{\mathbb{C}}, \mathbb{Z})$. Denote by Γ the group of isometries of $H^2(X_{\mathbb{C}}, \mathbb{Z})$ which fix x and which lie in the identity component of $\operatorname{Aut}(H^2(X_{\mathbb{C}}, \mathbb{R}))$. So Γ is an arithmetic subgroup of $SO(2, 19)(\mathbb{Q})$.

For each prime ℓ we denote by G_ℓ the group of linear transformations α of $L_X \otimes_{\mathbb{Z}} \mathbb{Z}_\ell$ such that there exists $\chi(\alpha) \in \mathbb{Z}_\ell^*$ satisfying

$$\langle \alpha x, \alpha x \rangle = \chi(\alpha)\langle x, x \rangle$$

for all $x \in L_X \otimes_{\mathbb{Z}} \mathbb{Z}_\ell$. There is a natural inclusion

$$\iota : \Gamma \to G_\ell$$

and we denote by H_ℓ the closure, in the ℓ-adic topology, of $\iota(\Gamma)$.

When a polarized K3 surface X is defined over a number field K, the inclusion

$$L_X \otimes_{\mathbb{Z}} \mathbb{Z}_\ell \subset H^2(\bar{X}, \mathbb{Z}_\ell)$$

induces a $\mathrm{Gal}(\bar{K}/K)$-module structure on $L_X \otimes_{\mathbb{Z}} \mathbb{Z}_\ell$; we denote by

$$\rho_X : \mathrm{Gal}(\bar{K}/K) \to G_\ell$$

the resulting ℓ-adic Galois representation.

We begin by showing that the desired statement about $\operatorname{Pic} X_{\mathbb{C}}$ follows if the image of ρ_X is large enough.

Lemma 2. *Let ℓ be a prime. Suppose $\rho_X(\mathrm{Gal}(\bar{K}/K))$ contains a finite-index subgroup of H_ℓ. Then* $\operatorname{rank} \operatorname{Pic} X_{\mathbb{C}} = 1$.

Proof. Suppose $\operatorname{rank} \operatorname{Pic}(X_{\mathbb{C}})$ is greater than 1; that is, there is divisor on $X_{\mathbb{C}}$ whose class is linearly independent from the class of the polarization. This divisor can be defined over some finite extension L/K. It follows that $\rho_X(\mathrm{Gal}(\bar{K}/L))$ is contained in the stabilizer of a line in $L_X \otimes_{\mathbb{Z}} \mathbb{Z}_\ell$. But this stabilizer does not contain a finite-index subgroup of H_ℓ. □

We also need a general lemma on linear ℓ-adic groups.

Lemma 3. *Let H be a closed subgroup of $\mathrm{GL}_m(\mathbb{Z}_\ell)$. Let $\Gamma_H(\ell^n)$ be the kernel of projection from H to $\mathrm{GL}_m(\mathbb{Z}/\ell^n\mathbb{Z})$. Then there exists an integer N such that no proper closed subgroup of H projects surjectively onto $H/\Gamma_H(\ell^N)$.*

Proof. Since H is a closed subgroup of $\mathrm{GL}_m(\mathbb{Z}_\ell)$, it is an analytic subgroup. In particular, there is a subspace $L \subset M_m(\mathbb{Q}_\ell)$ and a positive integer N such that, for all $n \geqslant N$, the group $\Gamma_H(\ell^n)$ is precisely the set of matrices $\exp(\lambda)$, where λ ranges over $\ell^n M_m(\mathbb{Z}_\ell) \cap L$. Thus, every element of $\Gamma_H(\ell^n)$ can be written as $\exp(\ell\lambda)$ for some $\lambda \in L$; in particular, for every $u \in \Gamma_H(\ell^n)$ there exists $v \in \Gamma_H(\ell^{n-1})$ with $v^\ell = u$. (See **[4]** for basic facts used here about ℓ-adic Lie groups.) We also require $N \geqslant 2$.

We now proceed as in **[6**, IV.3.4, Lemma 3], which proves the lemma in the case $H = \mathrm{SL}_2$. Suppose H_0 is a closed subgroup projecting surjectively onto $H/\Gamma_H(\ell^N)$. It suffices to prove that H_0 projects surjectively onto $H/\Gamma_H(\ell^n)$

for all $n > N$. We proceed by induction and assume H_0 projects surjectively onto $H/\Gamma_H(\ell^{n-1})$. We therefore need only show that, for all $x \in \Gamma_H(\ell^{n-1})$, there exists $h \in H_0$ with $h^{-1}x \in \Gamma_H(\ell^n)$. Since $n-1 \geqslant N$, there exists $y \in \Gamma_H(\ell^{n-2})$ such that $y^\ell = x$. We may write $y = 1 + \ell^{n-2}Y + \ell^{n-1}M_1$ for matrices $Y, M_1 \in M_m(\mathbb{Z}_\ell)$. By hypothesis, there exists $h' \in H_0$ such that $(h')^{-1}y \in \Gamma_H(\ell^{n-1})$. Then

$$h' = 1 + \ell^{n-2}Y + \ell^{n-1}M_2.$$

for some $M_2 \in \mathrm{GL}_m(\mathbb{Z}_\ell)$. So take

$$h = (h')^\ell = 1 + \ell^{n-1}Y + \ell^n M_2 + (1/2)(\ell)(\ell-1)\ell^{2n-3}Y^2 + \dots$$

which is congruent to x mod ℓ^n, since $n > N \geqslant 2$. □

The purpose of Lemma 3 is to reduce the problem of showing that an ℓ-adic representation has large image to the corresponding problem for a mod ℓ^N representation. Below we show how to use Hilbert irreducibility to produce K3 surfaces X such that ρ_X has large image mod ℓ^N, where $N > 0$ is an integer to be specified at the end.

Write L_d for the rank-21 lattice $\langle -d\rangle \oplus H \oplus H \oplus E_8 \oplus E_8$. Then L_X is isomorphic to L_d for any polarized K3 of degree d.

By a *level m* structure on a polarized K3 we mean a choice of isometry

$$\varphi : L_X/mL_X \cong L_d/mL_d.$$

We denote by $\Gamma(m)$ the kernel of the map $\Gamma \to \mathrm{GL}(L_d/mL_d)$. Choose a p large enough so that $\Gamma(p)$ is a torsion-free group. (It suffices to choose p larger than the order of any finite-order element of $GL(L_d)$.) If (X, φ) is a polarized K3 with level p structure, any automorphism $\alpha : X \to X$ preserving the polarization and φ must have finite order (because it preserves the polarization) and thus must act trivially on L_X (by the hypothesis on p). But then α is trivial by the Torelli theorem for K3's [**5**].

Let $\tilde{\mathscr{M}}/\mathbb{Q}$ be the moduli space of pairs (X, φ_p), where X is a polarized K3 surface of degree d and φ_p is a level p structure, with $p \neq \ell$. We can construct this moduli space by geometric invariant theory, as in the final remark of [**1**]. The fact that (X, φ_p) admits no nontrivial automorphisms implies that $\tilde{\mathscr{M}}$ is a *fine* moduli space. Now let $\tilde{\mathscr{M}}(\ell^N)$ be the space of pairs $(X, \varphi_p, \varphi_{\ell^N})$, where φ_{ℓ^N} is a level ℓ^N structure on X. Note that $\tilde{\mathscr{M}}$ and $\tilde{\mathscr{M}(\ell^N)}$ are not *a priori* connected.

Using again the Torelli theorem for K3 surfaces, we know that the analytic moduli space of polarized K3 surfaces of degree d is a quotient $\Gamma\backslash\Omega$, where Ω is a certain connected 19-dimensional domain of periods. (See [**1**, §3], noting that our Γ is an index-2 subgroup of Beauville's Γ_q.) It follows that $\Gamma(p)\backslash\Omega$ is a connected component of the analytification $\tilde{\mathscr{M}}^{an}$ of $\tilde{\mathscr{M}}$, and $\Gamma(p\ell^N)\backslash\Omega$ is a connected component of $\tilde{\mathscr{M}}(\ell^N)^{an}$. Denote by $\mathscr{M}$ and $\mathscr{M}(\ell^N)$ the connected components of $\tilde{\mathscr{M}}$ and $\tilde{\mathscr{M}}(\ell^N)$ corresponding to the quotients above; then, for some number field K, the map $\pi : \mathscr{M}(\ell^N) \to \mathscr{M}$ is a Galois cover of varieties over K with Galois group $\Gamma(p)/\Gamma(p\ell^N)$. Denote this finite group by $\bar{\Gamma}$.

Now let $p : \mathscr{M} \to \mathbb{P}^{19}$ be a generically finite map of degree n. Then the composition $p \circ \pi$ expresses the function field $K(\mathscr{M}(\ell^N))$ as a finite extension of $K(\mathbb{P}^{19})$. Let U be a Galois cover of $\mathbb{P}^{19}$ whose function field is the Galois closure of $K(\mathscr{M}(\ell^N))/K(\mathbb{P}^{19})$. Then the Galois group G of $K(U)/K(\mathbb{P}^{19})$ is naturally contained in the wreath product W of $\bar{\Gamma}$ with S_n. The group W fits in an exact sequence

$$1 \to \bar{\Gamma}^n \to W \to S_n \to 1$$

and the intersection of G with a Cartesian factor of $\bar{\Gamma}^n$ is the full group $\bar{\Gamma}$, since $\bar{\Gamma}$ is the Galois group of the cover π.

Now, by the Hilbert irreducibility theorem, there is a Zariski-dense subset of $\mathbb{P}^{19}(K)$ consisting of points x such that the Galois group of $(p\circ\pi)^{-1}(x)$ over x is the full group G. Let x be such a point, and let y be a $\bar{\mathbb{Q}}$-point of $\mathscr{M}$ lying over x. Then $y \in \mathscr{M}(L)$ for some number field L, and the Galois group of $\pi^{-1}(y)$ over y is the full group $\bar{\Gamma}$. If X/L is the K3 surface corresponding to the point y, the map

$$\mathrm{Gal}(\bar{\mathbb{Q}}/L) \to GL_2(L_X \otimes_{\mathbb{Z}} (\mathbb{Z}/\ell^N\mathbb{Z}))$$

given by the Galois action on $H^2_{et}(X, \mathbb{Z}/\ell^N\mathbb{Z})$ has image $\bar{\Gamma}$. Now apply Lemma 3, taking H to be the closure in the ℓ-adic topology of the image of $\Gamma(p)$ in $GL_2(L_X \otimes_{\mathbb{Z}} \mathbb{Z}_\ell)$. We conclude that, having chosen N large enough, we can find a degree d polarized K3 surface X over a number field L such that the image of ρ_X contains H, which is a finite-index subgroup of H_ℓ. Now X has geometric Picard number 1 by Lemma 2.

Remark 4. Lemmas 2 and 3, in principle, should allow one to write down a K3 of any desired degree which has geometric Picard number 1. One would first compute suitable values of ℓ and N, as Lemma 3 guarantees we can. It remains to write down a K3 surface X such that the representation of Galois on $H^2_{et}(X, \mathbb{Z}/\ell^N Z)$ is as large as possible. In case $d = 4$, this computation is precisely the one suggested in the final remark of **[3]**. In order to make this computation more tractable, it might be a good idea to restrict to a family of quartic surfaces whose monodromy group Γ_0 is smaller than Γ, but which still doesn't have any stabilizers of points in L_X as finite-index subgroups.

References

[1] A. BEAUVILLE – Application aux espaces de modules, *Astérisque* (1985), no. 126, 141–152, Geometry of $K3$ surfaces: moduli and periods (Palaiseau, 1981/1982).

[2] F. A. BOGOMOLOV & Y. TSCHINKEL – Density of rational points on elliptic $K3$ surfaces, *Asian J. Math.* **4** (2000), no. 2, 351–368.

[3] T. EKEDAHL – An effective version of Hilbert's irreducibility theorem, Séminaire de Théorie des Nombres, Paris 1988–1989, Prog. Math., vol. 91, Birkhäuser Boston, Boston, MA, 1990, 241–249.

[4] R. HOOKE – Linear p-adic groups and their Lie algebras, *Ann. of Math. (2)* **43** (1942), 641–655.

[5] E. LOOIJENGA & C. PETERS – Torelli theorems for Kähler $K3$ surfaces, *Compositio Math.* **42** (1980/81), no. 2, 145–186.

[6] J.-P. SERRE – *Abelian l-adic representations and elliptic curves*, second ed., Advanced Book Classics, Addison-Wesley Publishing Company Advanced Book Program, Redwood City, CA, 1989.

Arithmetic of Higher-dimensional Algebraic Varieties
(B. POONEN, YU. TSCHINKEL, eds.), p. 141–147
Progress in Mathematics, Vol. 226, © 2004 Birkhäuser Boston, Cambridge, MA

JUMPS IN MORDELL–WEIL RANK AND ARITHMETIC SURJECTIVITY

Tom Graber
UC Berkeley, CA 94720-3840, USA • *E-mail :* graber@math.berkeley.edu

Joseph Harris
Harvard University, Cambridge MA 02138, USA
E-mail : harris@math.harvard.edu

Barry Mazur
Harvard University, Cambridge MA 02138, USA
E-mail : mazur@math.harvard.edu

Jason Starr
MIT, Cambridge MA 02139, USA • *E-mail :* jstarr@math.mit.edu

Abstract. We ask the question: If a pencil of curves of genus one defined over $\mathbf{Q}$ admits no section, can we find a number field $L/\mathbf{Q}$ and a member of that pencil defined over L having no L-rational points?

In our work establishing a "converse" to the theorem of Graber–Harris–Starr (see **[GHS03]**, **[GHMS03b]**, **[GHMS03a]**) the four of us had contemplated arithmetic issues related to our converse theorem. One of us (B. M.) presented a bit of this material (some linked *questions* together with a tiny piece of evidence in support of an affirmative answer to them) at the AIM workshop in Problems on Rational Points on Algebraic Varieties held in Palo Alto in December 2002. We record this in the text below. We are most thankful to the organizers for setting up such a wonderful engaging conference – an ideal format for discussion of these, and more general, topics.

Key words and phrases. Elliptic curves, Mordell–Weil rank, Families of curves of genus one, Sections.

1. Arithmetic surjectivity

Recall that a (smooth, proper) morphism $f : X \to B$ over a number field K is said to be **arithmetically surjective** if for all number field extensions L/K, the mapping on L-rational points, $f(L) : X(L) \to B(L)$, is surjective.

Denote by $C(f, L) \subset B(L)$ the complement of the image of $f(L)$, i.e., the set of L-valued points in the base whose f-fibers possess *no* L-rational points. So, f is arithmetically surjective if and only if for any number field extension L/K the complement of the image of L-valued points, $C(f, L)$, is empty.

We begin with a qualitative question.

Question 1. *Let $f : X \to B$ be a pencil of (projective, smooth) curves of genus one over a number field K. Are the following two conditions equivalent?*

(1) *The morphism f is arithmetically surjective.*
(2) *The morphism f admits a section over K.*

Here is a convenient way of restating this question.

Consider B an open subscheme of $\mathbf{P}^1$ over K and $X \to B$ a (flat) family of (projective, smooth) curves of genus 1 defined over K. Letting $E \to B$ be the jacobian (i.e., Pic^0) of the family, we have that $X \to B$ is a torsor over $E \to B$ and is represented therefore by some class $h \in H^1(B; E)$. This cohomology group is torsion, so let n denote the order of the element h. Invoking the Kummer sequence of group schemes over B,

$$0 \to E[n] \to E \to E \to 0,$$

where the mapping $E \to E$ is multiplication by n, we see that there is an element $\tilde{h} \in H^1(B; E[n])$ which maps to $h \in H^1(B; E)$. Choose such a $\tilde{h}$. The family $X \to B$ is determined up to isomorphism by the element $h \in H^1(B; E)$ and hence also by $\tilde{h} \in H^1(B; E[n])$; to reflect this fact we may refer to the family as $X_h \to B$, or $X_{\tilde{h}} \to B$.

Observation. *For $A \to S$, an abelian scheme over a $\mathbf{Q}$-scheme S and a class $\tilde{h} \in H^1(S; A[n])$, for some integer n, let $h \in H^1(B; A)$ be the image of $\tilde{h}$ under the natural mapping induced from the inclusion $A[n] \hookrightarrow A$, and let $f : X_h \to S$ be the associated A-torsor. The long exact sequence on cohomology obtained* *from the Kummer sequence*

$$0 \to A[n] \to A \to A \to 0$$

immediately gives the equivalence of the following statements:

- *The family $X_h \to S$ admits no section (over S).*
- *The cohomology class $h \in H^1(S; A)$ is not zero.*
- *The cohomology class $\tilde{h} \in H^1(S; A[n])$ is not in the image of $H^0(S; A) = A(S)$ under the connecting homomorphism of the long exact sequence on cohomology induced by the Kummer sequence.*

***Question 2* (for a given natural number n).** *Let $E \to B$ be an abelian scheme of dimension one over B, an open nonempty subscheme in $\mathbf{P}^1$ defined over a number field K. Is it the case that if a class $\tilde{h} \in H^1(B; E[n])$ has the property that the associated E-torsor $X_h \to B$ admits no section, then there exists a number field extension L/K and an L-valued point of B, $\beta =$ Spec $L \to B$ such that the fiber over β of the family $X_{\tilde{h}} \to B$ possesses no L-rational point?*

An affirmative answer to **Question 2** for all $n > 1$ and all $E \to B$ is the same as an affirmative answer to **Question 1.**

As for quantitative questions, fix the number field K, and for real numbers C, let b run through the K-rational points of height $< C$ in the base B, and ask for upper bounds, as C tends to $+\infty$, of the number of fibers X_b in the family $X \to B$ that have K-rational points. Here is one formulation of such a question, keeping the notation $\tilde{h}$ and h of **Question 2.**

***Question 3* (for a given natural number n).** *Let $E \to B$ be a smooth family of elliptic curves over B, an open nonempty subscheme in $\mathbf{P}^1$ over K. Is there a positive number e such that for any $\tilde{h} \in H^1(B; E[n])$ not in the image of $H^0(B; E) = E(B)$,*

$$\text{card}\{b \in B(K) \mid \text{height}(b) < C \text{ and } b^*(h) \neq 0\} \; > \; C^e$$

for sufficiently large C?

For a K-rational point $b \in B$ denote by $E(b)$ the Mordell–Weil group $E_b(K)$. By a **jump point** $b =$ Spec $K \to B$ (for a family of elliptic curves $E \to B$, over the number field K) let us mean a K-rational point of B such that the induced mapping on Mordell–Weil groups $E(B) \to E(b) := E_b(K)$ has cokernel of positive rank. Denote by $\mathscr{J}(E/B, K)$ the set of jump points of $E \to B$, over the number field K.

The remainder of this article consists of a fragmentary result related to **Question 3**, connecting it to the rarity of jump points. More specifically, for the E-torsors $f : X_h \to B$ to be examined in the proposition below, the set of jump points $\mathscr{J}(E/B, K) \subset B(K)$ will shown to be "small" and the union of the sets $\mathscr{J}(E/B, K)$ and $C(f, K)$ will shown to be all of $B(K)$. This will lead us to a query about a specific pencil of elliptic curves, constructed by Cassels and Schinzel, **[CS82]**, where every rational point in the base is a jump point.

2. Quadratic twist families

Let

$$E_1 : \quad y^2 = x^3 + ax + b$$

be an elliptic curve over the number field K. Let

$$B := \mathbf{P}^1 - \{0, \infty\} = \mathrm{Spec}(K[t, t^{-1}]).$$

By the **quadratic twist family** $E \to B$ attached to E_1 over K we mean the pencil of elliptic curves

$$E_t : \quad ty^2 = x^3 + ax + b.$$

Proposition. *Let $K = \mathbf{Q}$. Let E_1 be any elliptic curve over $\mathbf{Q}$ with none of its points of order two rational over $\mathbf{Q}$. Let $E \to B$ be its associated quadratic twist family. Then* **Question 3** *for $n = 2$ has an affirmative answer and we can take $e = 1 - \varepsilon$ for any positive ε.*

Corollary. *Under the same hypotheses,* **Question 2** *for $n = 2$ has an affirmative answer.*

Proof of the proposition. Let $\bar{\mathbf{Q}}$ be an algebraic closure of $\mathbf{Q}$. For X any $\mathbf{Q}$-scheme put

$$\bar{X} := X \otimes_{\mathrm{Spec}(\mathbf{Q})} \mathrm{Spec}(\bar{\mathbf{Q}}),$$

the scheme obtained by extending scalars to $\bar{\mathbf{Q}}$. Put $G := \mathrm{Gal}(\bar{\mathbf{Q}}/\mathbf{Q})$. We will be dealing with cohomology of the Kummer sequence

$$0 \to E[2] \to E \to E \to 0$$

over the $\mathbf{Q}$-scheme B and the restriction of that cohomology to rational points $b \in B$. We are in a particularly nice situation because the group scheme $E[2]$

over B is isomorphic to the pullback via $B \to \mathrm{Spec}(\mathbf{Q})$ of the group scheme $E_1[2]$ over $\mathrm{Spec}(\mathbf{Q})$.

The following result surely is somewhere in the literature; we give its proof at the end of this section.

Lemma. $E(\bar{B}) = E[2](\bar{B})$.

By the Hochschild–Serre spectral sequence we have an exact sequence

$$0 \to H^1(G; H^0(\bar{B}; E[2])) \to H^1(B; E[2]) \to H^1(\bar{B}; E[2])^G.$$

Since the group scheme $E[2]$ over B comes by pullback from the group scheme $E_1[2]$ over $\mathbf{Q}$ we see that

$$H^1(G; H^0(\bar{B}; E[2])) = H^1(G; E_1[2]).$$

Also,

$$H^1(\bar{B}; E[2])^G = \mathrm{Hom}(H_1(\bar{B}, \mathbf{Z}/2\mathbf{Z}), E[2])^G = \mathrm{Hom}(\mathbf{Z}/2\mathbf{Z}, E[2])^G = E_1[2]^G = 0,$$

the first equality coming from the Universal Coefficient Theorem, the second equality a consequence of the fact that $H_1(\bar{B}; \mathbf{Z}/2\mathbf{Z})$ is canonically $\mu_2 \cong \mathbf{Z}/2\mathbf{Z}$, and the last equality following from our hypothesis that there is no nontrivial 2-torsion in the Mordell–Weil group $E_1(\mathbf{Q})$.

Therefore we get that the natural mapping

$$H^1(G, E_1[2]) \to H^1(B; E[2])$$

is an isomorphism, which tells us that for any $\mathbf{Q}$-rational point $b \in B$, the induced morphism on cohomology,

$$H^1(B; E[2]) \to H^1(b; E[2])$$

is an isomorphism.

From the above discussion and our hypotheses we have that $E(B) = 0$. It follows that for any nontrivial $\tilde{h} \in H^1(B; E[2])$ the specialization of its image $h \in H^1(B; E)$ to a $\mathbf{Q}$-rational point b of B, $b^*(h) \in H^1(b; E)$, is nontrivial if the Mordell–Weil group $E_b(\mathbf{Q}) = E(b)$ has rank zero; equivalently, if b is *not* a jump point for the quadratic twist family. We have shown, in other words, that

$$\mathscr{J}(E, B) \cup C(f, K) = B(K)$$

where $f : X_h \to B$ is the torsor associated to h.

To conclude the proof of our proposition, then, we must control the number of jump points; specifically it suffices to show that for any positive ε,

$$\mathrm{card}\{b \in B(\mathbf{Q}) \mid \mathrm{height}(b) < C \text{ and } E(b) \text{ finite}\} > C^{1-\varepsilon}$$

for C sufficiently large.

For this, we use the fact that E_1 is modular. Let f be the newform of weight two which is attached to E_1. Perelli and Pomykała [**PP97**] have estimates on the number of nonzero Fourier coefficients of the modular form F of weight $3/2$ associated to such a newform f via the Shimura lift. By classical results of Waldspurger [**Wal81**], they then get that for any positive ε

$$\mathrm{card}\{b \in B(\mathbf{Q}) \mid \mathrm{height}(b) < C \text{ and analytic rank of } E_b \text{ is zero}\} \gg C^{1-\varepsilon},$$

for C sufficiently large, and they conclude the same estimate for arithmetic rank using the results of Kolyvagin [**Kol90**]. The result in [**PP97**] refined an earlier estimate due to Iwaniec [**Iwa90**], and it has, in turn, been improved by Ono and Skinner [**OS98**] who replace the $C^{1-\varepsilon}$ above by $C/\log(C)$. We are thankful to Brian Conrey and Yuri Tschinkel for pointing out these references to us. □

Proof of the lemma. Consider $\bar{B}' \to \bar{B}$ the double cover ramified at 0 and ∞ (i.e., extract a square root of the parameter t) and note that the pullback $\bar{E}' = \bar{E} \otimes_{\bar{B}} \bar{B}'$ is a product, $\bar{E}' = \bar{E}_1 \times \bar{B}'$. A section $\sigma : \bar{B} \to \bar{E}$ gives us a section $\sigma' : \bar{B}' \to \bar{E}'$ such that $\sigma' \cdot i = j \cdot \sigma'$ where $i : \bar{B}' \to \bar{B}'$ is the involution of $\bar{B}'$ as double cover of $\bar{B}$ and $j : \bar{E}' = \bar{E}_1 \times \bar{B}' \to \bar{E}_1 \times \bar{B}'$ is given by $j(e_1, b') = (-e_1, i(b'))$ for $e_1 \in \bar{E}_1$ and $b' \in \bar{B}'$. Since there are no nonconstant morphisms from $\bar{B}'$ to $\bar{E}_1$, σ' is a constant section, i.e., $\sigma'(b') = (e_1, b')$ for a fixed $e_1 \in \bar{E}_1$ and all $b' \in \bar{B}'$. Therefore e_1 is in $\bar{E}_1[2]$ proving the lemma. □

3. The Cassels–Schinzel example

Since jump points seem to be relevant to arithmetic surjectivity questions, it may be useful to recall a classic example due to Cassels–Schinzel [**CS82**] of a pencil of elliptic curves over $\mathbf{Q}$ where *every* $\mathbf{Q}$-rational point of the parameter space B is a jump point. Their example is a certain pencil E_t of twists of the elliptic curve $X_0(32)$ over $\mathbf{Q}$ where the Mordell–Weil group of its sections (over $\mathbf{Q}$) is equal to the Klein 4-group (consisting of four sections in the 2-torsion subgroup $E_t[2]$) and yet for each $t_0 \in \mathbf{Q}$ the elliptic curve E_{t_0} has odd, hence nonzero, (analytic) rank (the game in finding such examples is to "work the equations" to guarantee that odd parity happens for all rational t_0's). Of course, the "rank" computed is the analytic rank, but by a recent result of

Nekovár, it is the same parity as the Selmer rank. To pass from this "parity of the Selmer rank" to the parity of Mordell–Weil rank we must invoke the Shafarevich–Tate conjecture (at least when the Mordell–Weil rank is > 1 for then we can't invoke Heegner points and Kolyvagin's theorem). Here is the explicit example:

$$E_t : \quad y^2 = x(x^2 - (7 + 7t^4)^2).$$

It would be interesting to get asymptotic results (which are either like or unlike the estimates given in the proposition above) for torsors over such an example. It would also be interesting to get asymptotic results over $\mathbf{Q}$ for pencils of plane cubics possessing no sections.

References

[CS82] J. W. S. Cassels & A. Schinzel – Selmer's conjecture and families of elliptic curves, *Bull. London Math. Soc.* **14** (1982), no. 4, 345–348.

[GHMS03a] T. Graber, J. Harris, B. Mazur & J. Starr – Arithmetic questions related to rationally connected varieties, 2003, to appear.

[GHMS03b] ______, Rational connectivity and sections of families over curves, *Ann. Sci. École Norm. Sup. (4)* (2003), to appear.

[GHS03] T. Graber, J. Harris & J. Starr – Families of rationally connected varieties, *J. Amer. Math. Soc.* **16** (2003), no. 1, 57–67 (electronic).

[Iwa90] H. Iwaniec – On the order of vanishing of modular L-functions at the critical point, *Sém. Théor. Nombres Bordeaux (2)* **2** (1990), no. 2, 365–376.

[Kol90] V. A. Kolyvagin – Euler systems, The Grothendieck Festschrift, Vol. II, Prog. Math., vol. 87, Birkhäuser Boston, Boston, MA, 1990, 435–483.

[OS98] K. Ono & C. Skinner – Non-vanishing of quadratic twists of modular L-functions, *Invent. Math.* **134** (1998), no. 3, 651–660.

[PP97] A. Perelli & J. Pomykała – Averages of twisted elliptic L-functions, *Acta Arith.* **80** (1997), no. 2, 149–163.

[Wal81] J.-L. Waldspurger – Sur les coefficients de Fourier des formes modulaires de poids demi-entier, *J. Math. Pures Appl. (9)* **60** (1981), no. 4, 375–484.

Arithmetic of Higher-dimensional Algebraic Varieties
(B. Poonen, Yu. Tschinkel, eds.), p. 149–173
Progress in Mathematics, Vol. 226, © 2004 Birkhäuser Boston, Cambridge, MA

UNIVERSAL TORSORS AND COX RINGS

Brendan Hassett
Department of Mathematics, Rice University, Houston, TX 77005, USA
E-mail : hassett@rice.edu

Yuri Tschinkel
Department of Mathematics, Princeton University, Fine Hall, Washington Road, Princeton, NJ 08544-1000, USA • *E-mail :* ytschink@math.princeton.edu

Abstract. We study the equations of universal torsors on rational surfaces.

1. Introduction

The study of surfaces over nonclosed fields k leads naturally to certain auxiliary varieties, called *universal torsors*. The proofs of the Hasse principle and weak approximation for certain Del Pezzo surfaces required a very detailed knowledge of the projective geometry, in fact, explicit equations, for these torsors **[7]**, **[9]**, **[8]**, **[22]**, **[24]**, **[25]**. More recently, Salberger proposed using universal torsors to count rational points of bounded height, obtaining the first sharp upper bounds on split Del Pezzo surfaces of degree 5 and asymptotics on split toric varieties over $\mathbb{Q}$ **[23]**. This approach was further developed in the work of Peyre, de la Bretéche, and Heath-Brown **[20]**, **[21]**, **[3]**, **[15]**.

Key words and phrases. Cubic surfaces, torsors, Cox rings.

The first author was partially supported by the Sloan Foundation and by NSF Grants 0196187 and 0134259. The second author was partially supported by NSF Grant 0100277.

Colliot-Thélène and Sansuc have given a general formalism for writing down equations for these torsors. We briefly sketch their method: Let X be a smooth projective variety and $\{D_j\}_{j\in J}$ a finite set of irreducible divisors on X such that $U := X \setminus \cup_{j\in J} D_j$ has trivial Picard group. In practice, one usually chooses generators of the effective cone of X, e.g., the lines on the Del Pezzo surface. Consider the resulting exact sequence:

$$0 \longrightarrow \bar{k}[U]^*/\bar{k}^* \longrightarrow \oplus_{j\in J}\mathbb{Z}D_j \longrightarrow \mathrm{Pic}(X_{\bar{k}}) \longrightarrow 0.$$

Applying $\mathrm{Hom}(-,\mathbb{G}_m)$, one obtains an exact sequence of tori

$$1 \longrightarrow T(X) \longrightarrow T \longrightarrow R \longrightarrow 1,$$

where the first term is the *Néron–Severi torus* of X. Suppose we have a collection of rational functions, invertible on U, which form a basis for the relations among the $\{D_j\}_{j\in J}$. These can be interpreted as a section $U \to R \times U$, and thus naturally induce a $T(X)$-torsor over U, which canonically extends to the universal torsor over X. In practice, this extension can be made explicit, yielding equations for the universal torsor.

However, when the cone generated by $\{D_j\}_{j\in J}$ is simplicial, there are *no* relations and this method gives little information. In this paper, we outline an alternative approach to the construction of universal torsors and illustrate it in specific examples where the effective cone of X is simplicial.

We will work with varieties X such that the Picard and the Néron–Severi groups of X coincide and such that the ring

$$\mathrm{Cox}(X) := \bigoplus_{L\in \mathrm{Pic}(X)} \Gamma(X, L),$$

is finitely generated. This ring admits a natural action of the Néron–Severi torus and the corresponding affine variety is a natural embedding of the universal torsor of X. The challenge is to actually compute $\mathrm{Cox}(X)$ in specific examples; Cox has shown that it is a polynomial ring precisely when X is toric, provided the natural multigrading on this ring is compatible with the $\mathrm{Pic}(X)$-grading **[10]** (see also **[11]** for some recent related results).

Here is a roadmap of the paper: In Section 2 we introduce Cox rings and discuss their general properties. Finding generators for the Cox ring entails embedding the universal torsor into affine space, which yields embeddings of our original variety into toric quotients of this affine space. We have collected several useful facts about toric varieties in Section 3. Section 4 is devoted to a detailed analysis of the unique cubic surface S with an isolated singularity of type $\mathbf{E}_6$. We compute the (simplicial) effective cone of its minimal desingularization $\tilde{S}$, and produce 10 distinguished sections in $\mathrm{Cox}(\tilde{S})$. These satisfy a unique equation and we show the universal torsor naturally embeds in the corresponding hypersurface in $\mathbb{A}^{10}$. More precisely, we get a homomorphism from the coordinate ring of $\mathbb{A}^{10}$ to $\mathrm{Cox}(\tilde{S})$ and the main point is to prove its surjectivity. Here we use an embedding of $\tilde{S}$ into a simplicial toric threefold Y, a quotient of $\mathbb{A}^{10}$ under the action of the Néron–Severi torus so that $\mathrm{Cox}(Y)$ is the polynomial ring over the above 10 generators. The induced restriction map on the level of Picard groups is an isomorphism respecting the moving cones. We conclude surjectivity for each degree by finding an appropriate birational projective model of Y and using vanishing results on it. Finally, in Section 5 we write down equations for the universal torsors (the Cox rings) of a split and a nonsplit cubic surface with an isolated singularity of type $\mathbf{D}_4$.

For an account of the Cox rings of smooth Del Pezzo surfaces, we refer the reader to the paper of Batyrev and Popov in this volume [**2**].

Acknowledgements: The results of this paper have been reported at the American Institute of Mathematics conference "Rational and integral points on higher dimensional varieties". We benefited from the comments of the other participants, in particular, V. Batyrev and J.L. Colliot-Thélène. We also thank S. Keel for several helpful discussions about Cox rings and M. Thaddeus for advice about the geometric invariant theory of toric varieties.

2. Generalities on the Cox ring

For any finite subset Ξ of a real vector space, let $\mathrm{Cone}(\Xi)$ denote the closed cone generated by Ξ.

Let X be a normal projective variety of dimension n over an algebraically closed field k of characteristic zero. Let $\mathrm{A}_{n-1}(X)$ and $\mathrm{N}_{n-1}(X)$ denote Weil divisors on X up to linear and numerical equivalence, respectively. Let $\mathrm{A}_1(X)$ and $\mathrm{N}_1(X)$ denote the equivalence classes of curves. Let $\mathrm{NE}_{n-1}(X) \subset \mathrm{N}_{n-1}(X)_{\mathbb{R}}$ denote the cone of (pseudo)effective divisors, i.e., the smallest real closed cone containing all the effective divisors of X. Let $\mathrm{NE}_1(X) \subset \mathrm{N}_1(X)_{\mathbb{R}}$ denote the cone of effective curves and $\mathrm{NM}^1(X) \subset \mathrm{N}_{n-1}(X)_{\mathbb{R}}$ the cone of nef Cartier divisors, which is dual to the cone of effective curves. By Kleiman's criterion, this is the smallest real closed cone containing all ample divisors of X.

Let $L_1, \ldots, L_r$ be invertible sheaves on X. For $\nu = (n_1, \ldots, n_r) \in \mathbb{N}^r$ write

$$L^\nu := L_1^{\otimes n_1} \otimes \ldots \otimes L_r^{\otimes n_r}.$$

Consider the ring

$$R(X, L_1, \ldots, L_r) := \bigoplus_{\nu \in \mathbb{N}^r} \Gamma(X, L^\nu),$$

which need not be finitely generated in general.

By definition, an invertible sheaf L on X is *semiample* if L^N is globally generated for some $N > 0$:

Proposition 2.1. *([**16**], Lemma 2.8) If the invertible sheaves $L_1, \ldots, L_r$ are semiample then $R(X, L_1, \ldots, L_r)$ is finitely generated.*

Remark 2.2. If the L_i are ample then, after replacing each L_i by a large multiple, $R(X, L_1, \ldots, L_r)$ is generated by

$$\Gamma(X, L_1) \otimes \ldots \otimes \Gamma(X, L_r).$$

However, this is not generally the case if the L_i are only semiample (despite the assertion in the second part of Lemma 2.8 of [**16**]). Indeed, let $X \to \mathbb{P}^1 \times \mathbb{P}^1$ be a double cover and L_1 and L_2 be the pull-backs of the polarizations on the $\mathbb{P}^1$'s to X. For suitably large n_1 and n_2, $L_1^{n_1} \otimes L_2^{n_2}$ is very ample and its sections embed X. However,

$$\Gamma(X, L_1^{n_1}) \otimes \Gamma(X, L_2^{n_2}) \simeq \Gamma(\mathbb{P}^1, \mathcal{O}_{\mathbb{P}^1}(n_1)) \otimes \Gamma(\mathbb{P}^1, \mathcal{O}_{\mathbb{P}^1}(n_2)),$$

and any morphism induced by these sections factors through $\mathbb{P}^1 \times \mathbb{P}^1$.

Proposition 2.3. *Let $L_1, \ldots, L_r$ be a set of invertible sheaves on X such that L_j is generated by sections $s_{j,0}, \ldots, s_{j,d_j}$. Assume that the induced morphism $X \to \prod_j \mathbb{P}^{d_j}$ is birational into its image. Then the ring generated by the $s_{j,k}$'s has the same fraction field as $R(X, L_1, \ldots, L_r)$.*

Proof. Both rings have fraction field $k(X)(t_1, \ldots, t_r)$, where t_j is a nonzero section of L_j. □

Definition 2.4. [**16**] *Let X be a nonsingular projective variety so that* Pic(X) *is a free abelian group of rank r. The* Cox ring *for X is defined*

$$\mathrm{Cox}(X) := R(X, L_1, \ldots, L_r),$$

where $L_1, \ldots, L_r$ are lines bundles so that

(1) *the L_i form a $\mathbb{Z}$-basis of* Pic(X)*;*
(2) *the cone* Cone$(\{L_1, \ldots, L_r\})$ *contains* $\mathrm{NE}_{n-1}(X)$.

This ring is naturally graded by $\mathrm{Pic}(X)$*: for* $\nu \in \mathrm{Pic}(X)$ *the* ν*-graded piece is denoted* $\mathrm{Cox}(X)_\nu$.

Proposition 2.5. [**16**] *The ring* $\mathrm{Cox}(X)$ *does not depend on the choice of generators for* $\mathrm{Pic}(X)$.

Proof. Consider two sets of generators $L_1, \ldots, L_r$ and $M_1, \ldots, M_r$. Since the cones $\mathrm{Cone}(\{L_i\})$ and $\mathrm{Cone}(\{M_i\})$ contain all the effective divisors, the nonzero graded pieces of both $R(X, L_1, \ldots, L_r)$ and $R(X, M_1, \ldots, M_r)$ are indexed by the effective divisor classes in $\mathrm{Pic}(X)$. Choose isomorphisms

$$M_j \simeq L^{(a_{1j}, \ldots, a_{rj})}, \quad i = 1, \ldots, r, A = (a_{ij})$$

which naturally induce isomorphisms

$$\Gamma(M^\nu) \simeq \Gamma(L^{A\nu}), \quad A\nu = (a_{11}\nu_1 + \ldots + a_{1r}\nu_r, \ldots, a_{r1}\nu_1 + \ldots + a_{rr}\nu_r).$$

Thus we find $R(X, L_1, \ldots, L_r) \simeq R(X, M_1, \ldots, M_r)$. □

As $\mathrm{Cox}(X)$ is graded by $\mathrm{Pic}(X)$, a free abelian group of rank r, the torus

$$T(X) := \mathrm{Hom}(\mathrm{Pic}(X), \mathbb{G}_m)$$

acts on $\mathrm{Cox}(X)$. Indeed, each $\nu \in \mathrm{Pic}(X)$ naturally yields a character χ_ν of $T(X)$, and the action is given by

$$t \cdot \xi = \chi_\nu(t)\xi, \quad \xi \in \mathrm{Cox}(X)_\nu, t \in T(X).$$

Thus the isomorphism constructed in Proposition 2.5 is *not* canonical: Two such isomorphisms differ by the action of an element of $T(X)$. It is precisely this ambiguity that makes descending the universal torsor to nonclosed fields an interesting question.

The following conjecture is a special case of 2.14 of [**16**]:

Conjecture 2.6 **(Finiteness of Cox ring).** Let X be a log Fano variety. Then $\mathrm{Cox}(X)$ is finitely generated.

Remark 2.7. Note that if $\mathrm{Cox}(X)$ is finitely generated it follows trivially that $\mathrm{NE}_{n-1}(X)$ is finitely generated. Moreover, the nef cone $\mathrm{NM}^1(X)$ is also finitely generated.

Indeed, the nef cone corresponds to one of the chambers in the group of characters of $T(X)$ governed by the stability conditions for points $v \in \mathrm{Spec}(\mathrm{Cox}(X))$. These chambers are bounded by finitely many hyperplanes (see Theorem 0.2.3 in [**12**] for more details).

It has been conjectured by Batyrev [1] that the pseudo-effective cone of a Fano variety is finitely generated. However, the finiteness of the Cox ring is not a formal consequence of the finiteness of the pseudo-effective cone.

Example 2.8. Let $p_1, \ldots, p_9 \in H \subset \mathbb{P}^3$ be nine distinct coplanar points given as a complete intersection of two generic cubic curves in the hyperplane H, and let X be the blow-up of $\mathbb{P}^3$ at these points. Then $\mathrm{NE}^1(X)$ is finitely generated but $\mathrm{Cox}(X)$ is not. Indeed, X is an equivariant compactification of the additive group $\mathbb{G}_a^3$, acting by translation on the affine space $\mathbb{P}^3 - H$. The group action can be used to show that $\mathrm{NE}^1(X)$ is generated by the boundary components (see [14]). Similarly, one can show that the cone $\mathrm{NE}_1(X)$ is generated by classes of curves in the boundary components, e.g., the proper transform $\tilde{H} \subset X$ of H. It is well known that $\mathrm{NE}_1(\tilde{H})$ is infinite [18] §1.23(4): The pencil of cubic plane curves with base locus $p_1, \ldots, p_9$ induces an elliptic fibration,

$$\tilde{H} \rightarrow \mathbb{P}^1,$$

for which the nine exceptional curves of $\tilde{H} \rightarrow H$ are sections. Addition in the group law gives an infinite number of sections, which are also (-1)-curves and generators of $\mathrm{NE}_1(\tilde{H})$. These are also generators of $\mathrm{NE}_1(X)$, since the sections (other than the nine exceptional curves) intersect $\tilde{H}$ negatively. It follows that $\mathrm{NE}_1(X)$ and $\mathrm{NM}^1(X)$ are not finitely generated and hence $\mathrm{Cox}(X)$ is not finitely generated (see Remark 2.7).

Proposition 2.9. *Let X be a nonsingular projective variety whose anticanonical divisor $-K_X$ is nef and big. Suppose that D is a nef divisor on X. Then $H^i(X, \mathcal{O}_X(D)) = 0$ for each $i > 0$ and D is semiample.*

Proof. The first assertion is a consequence of Kawamata–Viehweg vanishing [18] §2.5. The second is a special case of the Kawamata Basepoint-freeness Theorem [18] §3.2. □

Proposition 2.9 largely determines the Hilbert function of the Cox ring:

Corollary 2.10. *Retain the assumptions of Proposition 2.9. Then for nef classes ν we have*

$$\dim \mathrm{Cox}(X)_\nu = \chi(\mathcal{O}_X(\nu)).$$

Remark 2.11. In practice, this will help us to find generators of $\mathrm{Cox}(X)$.

3. Generalities on toric varieties

We recall quotient constructions of toric varieties, following Brion–Procesi [**4**], Cox [**10**], and Thaddeus [**26**].

Let $T \simeq \mathbb{G}_m^r$ be a torus with character group $\mathfrak{X}^*(T)$. Suppose that T acts faithfully on the polynomial ring $k[x_1, \dots, x_{n+r}]$ by the formula

$$t(x_j) = \chi_j(t)x_j, \quad t \in T,$$

where $\{\chi_1, \dots, \chi_{n+r}\} \subset \mathfrak{X}^*(T)$. Define M as the kernel of the surjective morphism

$$\chi := (\chi_1, \dots, \chi_{n+r}) : \mathbb{Z}^{n+r} \to \mathfrak{X}^*(T).$$

We interpret M as the character group of the quotient torus $\mathbb{G}_m^{n+r}/T$. Set $N = \mathrm{Hom}(M, \mathbb{Z})$ so that dualizing gives

$$(\mathbb{Z}^{n+r})^* \to N \to 0.$$

Let $e_1, \dots, e_{n+r}$ and $e_1^*, \dots, e_{n+r}^*$ denote the coordinate vectors in $\mathbb{Z}^{n+r}$ and $(\mathbb{Z}^{n+r})^*$; let $\bar{e}_1^*, \dots, \bar{e}_{n+r}^* \in N$ denote the images of the e_i^* in N. Concretely, the $\bar{e}_i^*$ are the columns of the $n \times (n+r)$ matrix of dependence relations among the χ_j.

Consider a toric variety X of dimension n associated with a fan having one-skeleton $\{\bar{e}_1^*, \dots, \bar{e}_{n+r}^*\}$. In particular, we assume that none of $\bar{e}_i^*$ is zero or a positive multiple of any of the others. The variety X is a categorical quotient of an invariant open subset $U \subset \mathbb{A}^{n+r}$ under the action of T described above (see [**10**] 2.1). Elements $\nu \in \mathfrak{X}^*(T)$ classify T-linearized invertible sheaves $\mathscr{L}_\nu$ on $\mathbb{A}^{n+r}$ and

$$\Gamma(\mathbb{A}^{n+r}, \mathscr{L}_\nu) \simeq k[x_1, \dots, x_{n+r}]_\nu.$$

We have $\mathrm{A}_{n-1}(X) \simeq \mathfrak{X}^*(T)$ and we can identify

$$\Gamma(\mathscr{O}_X(D)) \simeq k[x_1, \dots, x_n]_{\nu(D)},$$

where $\nu(D) \in \mathfrak{X}^*(T)$ is associated with the divisor class of D. The variables x_i are associated with the irreducible torus-invariant divisors D_i on X (see [**13**] §3.4), and the cone of effective divisors $\mathrm{NE}_{n-1}(X)$ is generated by $\{D_1, \dots, D_{n+r}\}$. Geometrically, the effective cone in $\mathfrak{X}^*(T)$ is the image of the standard simplicial cone generated by $e_1, \dots, e_{n+r}$ under the projection homomorphism $\chi : \mathbb{Z}^{n+r} \to \mathfrak{X}^*(T)$.

Recall that the *moving cone*

$$\mathrm{Mov}(X) \subset \mathrm{NE}_{n-1}(X)$$

is defined as the smallest closed subcone containing the effective divisors on X without fixed components.

Proposition 3.1. *Retaining the notation and assumptions above,*

$$\mathrm{Mov}(X) = \bigcap_{i=1,\dots,n+r} \mathrm{Cone}(\chi_1, \dots, \chi_{i-1}, \chi_{i+1}, \dots, \chi_{n+r})$$

and has nonempty interior.

Proof. The fixed components of $\Gamma(X, \mathscr{O}_X(D))$ are necessarily invariant under the torus action, hence are taken from $\{D_1, \dots, D_{n+r}\}$. Moreover, D_i is fixed in each $\Gamma(X, \mathscr{O}_X(dD)), d > 0$ if and only if x_i divides each element of $k[x_1, \dots, x_{n+r}]_{d\nu(D)}$. This is the case exactly when

$$\nu(D) \in \mathrm{Cone}(\chi_1, \dots, \chi_{n+r}) - \mathrm{Cone}(\chi_1, \dots, \chi_{i-1}, \chi_{i+1}, \dots, \chi_{n+r}).$$

Suppose that the interior of the moving cone is empty. After permuting indices there are two possibilities: Either $\mathrm{Cone}(\chi_2, \dots, \chi_{n+r})$ has no interior, or the cones $\mathrm{Cone}(\chi_2, \dots, \chi_{n+r})$ and $\mathrm{Cone}(\chi_1, \chi_3, \dots, \chi_{n+r})$ have nonempty interiors but meet in a cone with positive codimension. As the T-action is faithful, the χ_i span $\mathfrak{X}^*(T)$. In the first case, $\chi_2, \dots, \chi_{n+r}$ span a codimension-one subspace of $\mathfrak{X}^*(T)$ that does not contain χ_1, so that each dependence relation

$$c_1\chi_1 + \dots + c_{n+r}\chi_{n+r} = 0$$

has $c_1 = 0$. This translates into $\bar{e}_1^* = 0$, a contradiction. In the second case, $\chi_3, \dots, \chi_{n+r}$ span a hyperplane, and χ_1 and χ_2 are on opposite sides of this hyperplane. Putting the dependence relations among the χ_i in row echelon form, we obtain a unique relation with nonzero first and second entries, and these two entries are both positive. This translates into the proportionality of $\bar{e}_1^*$ and $\bar{e}_2^*$. □

We now seek to characterize the projective toric n-folds X with one-skeleton $\{\bar{e}_1^*, \dots, \bar{e}_{n+r}^*\}$. These are realized as Geometric Invariant Theory quotients $\mathbb{A}^{n+r}//T$ associated with the various linearizations of our T-action. We consider the graded ring

$$R := \sum_{d \geq 0} \Gamma(\mathbb{A}^{n+r}, \mathscr{L}_{d\nu}) = \sum_{d \geq 0} k[x_1, \dots, x_{n+r}]_{d\nu}.$$

Proposition 3.2 **(see [26] §2,3).** *With the notation above, set* $X := \mathrm{Proj}(R)$.

(1) *X is projective over k if and only if 0 is not contained in the convex hull of $\{\chi_1, \dots, \chi_{n+r}\}$.*
(2) *X is toric of dimension n if ν is in the interior of*

$$\mathrm{Cone}(\chi_1, \dots, \chi_{n+r}).$$

(3) *In this case, the one-skeleton of* X *is contained in* $\{\bar{e}_1^*, \ldots, \bar{e}_{n+r}^*\}$. *Equality holds if* ν *is in the interior of the moving cone*

$$\bigcap_{i=1,\ldots,n+r} \operatorname{Cone}(\chi_1, \ldots, \chi_{i-1}, \chi_{i+1}, \ldots, \chi_{n+r}).$$

Remark 3.3. Our proof will show that X may still be of dimension n even when ν is contained in a facet of

$$\operatorname{Cone}(\chi_1, \ldots, \chi_{n+r}).$$

Similary, the one-skeleton of X may still be $\{\bar{e}_1^*, \ldots, \bar{e}_{n+r}^*\}$ even when ν is contained in a facet of

$$\bigcap_{i=1,\ldots,n+r} \operatorname{Cone}(\chi_1, \ldots, \chi_{i-1}, \chi_{i+1}, \ldots, \chi_{n+r}).$$

Proof. The monomials which appear in R are in one-to-one correspondence to solutions of

$$a_1\chi_1 + \ldots + a_{n+r}\chi_{n+r} = d\nu, \quad a_i \in \mathbb{Z}_{\geq 0}.$$

In geometric terms, the monomials appearing in R coincide with the elements of $\mathbb{Z}^{n+r}$ in the cone

$$\chi^{-1}(\operatorname{Cone}(\nu)) \cap \operatorname{Cone}(e_1, \ldots, e_{n+r}).$$

By Gordan's Lemma in convex geometry, R is generated as a k-algebra by a finite set of monomials $x^{m_1}, \ldots, x^{m_s}$. The monomials appearing in the dth graded piece R_d coincide with elements of $\mathbb{Z}^{n+r}$ in the polytope

$$P_{d\nu} := \chi_{\mathbb{R}}^{-1}(d\nu) \cap \operatorname{Cone}(e_1, \ldots, e_{n+r}).$$

Note that $\chi^{-1}(d\nu)$ is a translate of M.

For the first part, recall that $\operatorname{Proj}(R)$ is projective over $\operatorname{Spec}(R_0)$, where R_0 is the degree-zero part. Now 0 is in the convex hull of $\{\chi_1, \ldots, \chi_{n+r}\}$ if and only if there are nonconstant elements of R of degree zero. Our hypothesis just says that $R_0 = k$ and thus is equivalent to the projectivity of X over k.

As for the second part, T acts on R by homotheties and thus acts trivially on $\operatorname{Proj}(R)$, so we have an induced action of $\mathbb{G}_m^{n+r}/T$ on $\operatorname{Proj}(R)$. We claim this action is faithful, so the quotient is toric of dimension n. Let $\mu_1, \ldots, \mu_n$ be generators for $M = \mathfrak{X}^*(\mathbb{G}_m^{n+r}/T)$. Choose $v \in \mathbb{Z}^{n+r}$ in the interior of $\operatorname{Cone}(e_1, \ldots, e_{n+r})$ so that $\chi_{\mathbb{R}}(\operatorname{Cone}(v)) = \operatorname{Cone}(\nu)$. Replacing v by a suitably large integral multiple, we may assume each $v + \mu_i, i = 1, \ldots, n$, is in $\operatorname{Cone}(e_1, \ldots, e_{n+r})$. If $\chi(v) = d\nu$ then R_d contains a set of generators for M, so the induced representation of $\mathbb{G}_m^{n+r}/T$ on R_d is faithful.

For the third part, we extract the fan classifying X from $P_{d\nu}$, following [13] §1.5 and 3.4: For each face Q of $P_{d\nu}$, consider the cone

$$\sigma_Q = \{v \in N_{\mathbb{R}} : \langle u, v\rangle \leqslant \langle u', v\rangle \text{ for all } u \in Q, u' \in P_{d\nu}\}.$$

This assignment is inclusion reversing, so the one-dimensional cones of the fan correspond to facets of $P_{d\nu}$. Moreover, each facet of $P_{d\nu}$ is induced by one of the facets of $\mathrm{Cone}(e_1, \ldots, e_{n+r})$. The corresponding one-dimensional cone in $N_{\mathbb{R}}$ is spanned by $\bar{e}_i^*$. It remains to verify that each facet of $\mathrm{Cone}(e_1, \ldots, e_{n+r})$ actually induces a facet of $P_{d\nu}$. The hypothesis that ν is in the moving cone means that $P_{d\nu}$ intersects each of the $\mathrm{Cone}(e_1, \ldots, e_{i-1}, e_{i+1}, \ldots, e_{n+r})$. If ν is in the interior of the moving cone, then the intersection of $P_{d\nu}$ with $\mathrm{Cone}(e_1, \ldots, e_{i-1}, e_{i+1}, \ldots, e_{n+r})$ meets the relative interior of this cone, hence this cone induces a facet of $P_{d\nu}$. □

Proposition 3.2 yields the following nice consequence:

Proposition 3.4. *Let X be a complete toric variety and ν a divisor class in the interior of* $\mathrm{Mov}(X)$. *Then there exists a projective toric variety Y_ν, with the same one-skeleton as X, and polarized by ν.*

For generic T-linearized invertible sheaves on $\mathbb{A}^{n+r}$, all semistable points are actually stable; hence Y_ν is a simplicial toric variety for generic ν (see [4] 1.2 and [10] 2.1). For the special values ν_0, contained in the walls of the chamber decomposition of [26], this fails to be the case. However, for each special ν_0, there exists a generic ν so that $\mathrm{Cone}(\nu)$ is very close to $\mathrm{Cone}(\nu_0)$ and there is a projective, torus-equivariant morphism $Y_\nu \to Y_{\nu_0}$ [26] 3.11. The polarization associated to ν_0 pulls back to Y_ν, so we obtain the following:

Proposition 3.5. *Let X be a complete toric variety and ν_0 a divisor class in the moving cone of X. Then there exists a simplicial projective toric variety Y, with the same one-skeleton as X, so that ν_0 is semiample on Y.*

Of course, ν_0 is big when it is in the interior of the effective cone.

4. The $\mathbf{E}_6$ cubic surface

By definition, the $\mathbf{E}_6$ *cubic surface* is given by the homogeneous equation

$$S = \{(w, x, y, z) : xy^2 + yw^2 + z^3 = 0\} \subset \mathbb{P}^3. \tag{1}$$

We recall some elementary properties (see [5] for more details on singular cubic surfaces):

Proposition 4.1.

(1) *The surface S has a single singularity at the point $p := (0, 1, 0, 0)$, of type* $\mathbf{E}_6$.
(2) *S is the unique cubic surface with this property, up to projectivity.*
(3) *S contains a unique line, satisfying the equations $y = z = 0$.*

Any smooth cubic surface may be represented as the blow-up of $\mathbb{P}^2$ at six points in 'general position'. There is an analogous property of the $\mathbf{E}_6$ cubic surface:

Proposition 4.2. *The $\mathbf{E}_6$ cubic surface S is the closure of the image of $\mathbb{P}^2$ under the linear series*

$$w = a^2c \quad x = -(ac^2 + b^3) \quad y = a^3 \quad z = a^2b,$$

where

$$\Gamma(\mathbb{P}^2, \mathscr{O}_{\mathbb{P}^2}(1)) = \langle a, b, c\rangle.$$

This map is the inverse of the projection of S from the double point p. The affine open subset

$$\mathbb{A}^2 := \{a \neq 0\} \subset \mathbb{P}^2$$

is mapped isomorphically onto $S - \ell$. In particular, $S \setminus \ell \simeq \mathbb{A}^2$, so the $\mathbf{E}_6$ cubic surface is a compactification of $\mathbb{A}^2$.

Remark 4.3. Note that S is not an *equivariant* compactification of $\mathbb{G}_a^2$, so the general theory of [**6**] does not apply.

Indeed, if S were an equivariant compactification of $\mathbb{G}_a^2$ then the projection from p would be $\mathbb{G}_a^2$-equivariant (see [**14**]). Therefore, the map $\mathbb{P}^2 \dashrightarrow S$ given above has to be $\mathbb{G}_a^2$-equivariant. The only $\mathbb{G}_a^2$-action on $\mathbb{P}^2$ under which a line is invariant is the standard translation action [**14**]. However, the linear series above is not invariant under the standard translation action

$$b \mapsto b + \beta a \quad c \mapsto c + \gamma a.$$

We now compute the effective cone of the minimal resolution $\varphi_\ell : \tilde{S} \to S$. Let $\ell \subset \tilde{S}$ be the proper transform of the line mentioned in Proposition 4.1.

Proposition 4.4. *The Picard group $\mathrm{Pic}(\tilde{S})$ is a free abelian group of rank seven, generated by ℓ and the exceptional curves of φ_ℓ. For a suitable ordering $\{F_1, F_2, F_3, F_4, F_5, F_6\}$ of the exceptional curves, the intersection pairing takes*

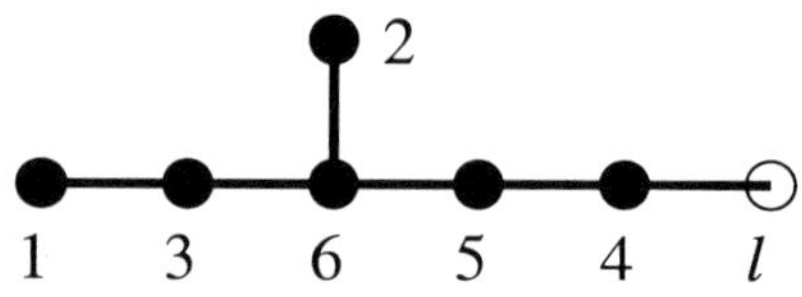

FIGURE 1. Dynkin diagram of $\mathbf{E}_6$

the form

(2)

	F_1	F_2	F_3	ℓ	F_4	F_5	F_6
F_1	-2	0	1	0	0	0	0
F_2	0	-2	0	0	0	0	1
F_3	1	0	-2	0	0	0	1
ℓ	0	0	0	-1	1	0	0
F_4	0	0	0	1	-2	1	0
F_5	0	0	0	0	1	-2	1
F_6	0	1	1	0	0	1	-2

Proposition 4.5. *The effective cone* $\mathrm{NE}(\tilde{S})$ *is simplicial and generated by* $\Phi := \{F_1, F_2, F_3, \ell, F_4, F_5, F_6\}$*. Each nef divisor is contained in the monoid generated by the divisors*

$$\begin{array}{rcl} A_1 & = & F_2 + F_3 + 2\ell + 2F_4 + 2F_5 + 2F_6 \\ A_2 & = & F_1 + F_2 + 2F_3 + 3\ell + 3F_4 + 3F_5 + 3F_6 \\ A_3 & = & F_1 + 2F_2 + 2F_3 + 4\ell + 4F_4 + 4F_5 + 4F_6 \\ A_\ell & = & 2F_1 + 3F_2 + 4F_3 + 3\ell + 4F_4 + 5F_5 + 6F_6 \\ A_4 & = & 2F_1 + 3F_2 + 4F_3 + 4\ell + 4F_4 + 5F_5 + 6F_6 \\ A_5 & = & 2F_1 + 3F_2 + 4F_3 + 5\ell + 5F_4 + 5F_5 + 6F_6 \\ A_6 & = & 2F_1 + 3F_2 + 4F_3 + 6\ell + 6F_4 + 6F_5 + 6F_6 \end{array}$$

Moreover A_ℓ *is the anticanonical class* $-K_{\tilde{S}}$ *and its sections induce the resolution morphism* $\varphi_\ell : \tilde{S} \to S$.

Proof. The intersection form in terms of $A := \{A_1, \dots, \}$ is

(3)

	A_1	A_2	A_3	A_ℓ	A_4	A_5	A_6
A_1	0	1	1	2	2	2	2
A_2	1	1	2	3	3	3	3
A_3	1	2	2	4	4	4	4
A_ℓ	2	3	4	3	4	5	6
A_4	2	3	4	4	4	5	6
A_5	2	3	4	5	5	5	6
A_6	2	3	4	6	6	6	6

This is the inverse of the intersection matrix (2) written in terms of the basis Φ, so the A_i generate the dual to $\mathrm{Cone}(\Phi)$. Observe that all the entries of matrix (3) are nonnegative and

$$\mathrm{Cone}(A) \subset \mathrm{Cone}(\Phi).$$

Suppose that D is an effective divisor on $\tilde{S}$. We write D as a sum of the fixed components contained in $\{F_1, \dots, F_6, \ell\}$ and the parts moving relative to Φ:

$$D = M_\Phi + a_1F_1 + \dots + a_6F_6 + a_\ell \ell, \quad a_1, \dots, a_6, a_\ell \geq 0.$$

A priori, M_Φ may have fixed components, but they are not contained in Φ (however, see Lemma 4.6). It follows that M_Φ intersects each element of Φ nonnegatively, i.e., it is contained in $\mathrm{Cone}(A)$ and thus in $\mathrm{Cone}(\Phi)$. We conclude that $D \in \mathrm{Cone}(\Phi)$. Since $A_1, \dots, A_6, A_\ell$ generate $\mathrm{N}_1(\tilde{S})$ over $\mathbb{Z}$ each nef divisor can be written as a nonnegative linear combination of these divisors.

To see that A_ℓ is the anticanonical divisor, we apply adjunction

$$K_{\tilde{S}}F_i = 0, i = 1, \dots, 6 \quad K_{\tilde{S}}\ell = -1.$$

Nondegeneracy of the intersection form implies $A_\ell = -K_{\tilde{S}}$. Since S has rational double points, the resolution map φ_ℓ is crepant, i.e., $\varphi_\ell^* K_S = K_{\tilde{S}}$. Thus

$$\Gamma(A_\ell) = \Gamma(-K_{\tilde{S}}) = \Gamma(\varphi_\ell^* K_S) = \Gamma(\varphi_\ell^* \mathscr{O}_S(+1))$$

so the sections of A_ℓ induce φ_ℓ. □

Choose nonzero sections $\xi_1, \dots, \xi_\ell$ generating $\Gamma(F_1), \dots, \Gamma(\ell)$:

$$\Gamma(F_1) = \langle \xi_1 \rangle, \dots, \Gamma(F_6) = \langle \xi_6 \rangle, \Gamma(\ell) = \langle \xi_\ell \rangle.$$

These are canonical up to scalar multiplication. Each effective divisor

$$D = b_1F_1 + b_2F_2 + b_3F_3 + b_\ell \ell + b_4F_4 + b_5F_5 + b_6F_6$$

has a distinguished nonzero section

$$\xi^{(b_1,b_2,b_3,b_\ell,b_4,b_5,b_6)} := \xi_1^{b_1} \dots \xi_6^{b_6} \xi_\ell^{b_\ell}.$$

The distinguished section of A_j is denoted $\xi^{\alpha(j)}$. Note that we have an injective ring homomorphism

$$k[\xi_1, \ldots, \xi_6, \xi_\ell] \to \mathrm{Cox}(\tilde{S}). \tag{4}$$

There is a partial order on the monoid of effective divisors of $\tilde{S}$: $D_1 \prec D_2$ if $D_2 - D_1$ is effective. The restriction of this order to the generators of the nef cone is illustrated in the diagram below:

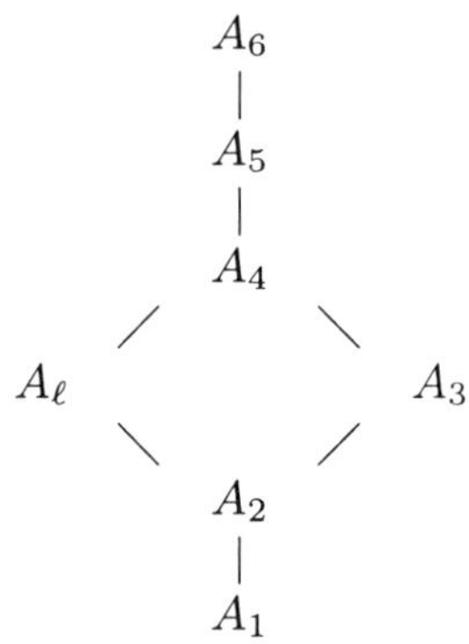

Whenever $D_1 \prec D_2$ we have an inclusion

$$\Gamma(D_1) \hookrightarrow \Gamma(D_2)$$

which is natural up to scalar multiplication: Indeed, express

$$D_1 - D_2 = b_1 F_1 + b_2 F_2 + \ldots + b_6 F_6 + b_\ell \ell, \quad b_j \geq 0$$

so we have

$$\begin{array}{ccc} s_1 & \mapsto & \xi^{(b_1, b_2, b_3, b_\ell, b_4, b_5, b_6)} s_1 \\ \Gamma(D_1) & \hookrightarrow & \Gamma(D_2). \end{array}$$

The homomorphism (4) is not surjective, and we now look for generators of $\mathrm{Cox}(\tilde{S})$ beyond the ξ_j. Consider the subring

$$\mathrm{Cox}_a(\tilde{S}) = \bigoplus_{\nu \in \mathrm{NM}(\tilde{S})} \mathrm{Cox}(\tilde{S})_\nu$$

obtained by restricting to degrees corresponding to nef classes on $\tilde{S}$. The following lemma implies that any homogeneous element $\mathsf{s}_D \in \mathrm{Cox}(\tilde{S})$ can be written in the form

$$\mathsf{s}_D = m_D \xi_1^{b_1} \cdots \xi_6^{b_6} \xi_\ell^{b_\ell}$$

with nonnegative exponents and $m_D \in \mathrm{Cox}_a(\tilde{S})$.

Lemma 4.6. *Let D be an effective divisor on $\tilde{S}$ with fixed part F_D and moving part M_D. Then F_D is supported in $\{F_1, \ldots, F_6, \ell\}$, and M_D is a linear combination of $A_1, \ldots, A_6, A_\ell$ with nonnegative coefficients.*

Proof. Clearly M_D is nef, so the description of the nef divisors in Proposition 4.5 gives the expression in terms of the A_i. Proposition 2.9 shows M_D is semiample with vanishing higher cohomology; the last part of Proposition 4.5 gives the requisite positivity of the anticanonical class.

Let F be a fixed component of D not supported in $\{F_1, \ldots, F_6, \ell\}$. To arrive at a contradiction, we need to show that $h^0(M_D + F) > h^0(M_D)$. Since M_D has vanishing higher cohomology and

$$h^2(F + M_D) = h^0(K - F - M_D) = 0$$

it suffices to show that

$$\chi(F + M_D) > \chi(M_D).$$

By Riemann–Roch, it suffices to show that

$$F^2 + 2M_D F - K_{\tilde{S}} F > 0$$

or, equivalently,

$$F^2 + K_{\tilde{S}} F + 2M_D F - 2K_{\tilde{S}} F = 2g(F) - 2 + 2M_D F - 2K_{\tilde{S}} F > 0.$$

Since F is irreducible, $g(F) \geq 0$ and $M_D F \geq 0$ and $-K_{\tilde{S}} F > 1$, as M_D is nef and $-K_{\tilde{S}}$ is nonpositive only along the exceptional curves and has degree 1 only on the line ℓ (see Proposition 4.1). □

Corollary 2.10 gives the dimensions of the graded pieces of $\mathrm{Cox}_a(\tilde{S})$. We focus first on the generators of the nef cone, introducing sections $\tau_j \in \Gamma(A_j)$ as needed to achieve the prescribed dimensions:

$$\begin{aligned}
\Gamma(A_1) &= \left\langle \xi^{\alpha(1)}, \tau_1 \right\rangle \\
\Gamma(A_2) &= \left\langle \xi^{\alpha(2)}, \xi^{\alpha(2)-\alpha(1)}\tau_1, \tau_2 \right\rangle \\
\Gamma(A_\ell) &= \left\langle \xi^{\alpha(\ell)}, \xi^{\alpha(\ell)-\alpha(1)}\tau_1, \xi^{\alpha(\ell)-\alpha(2)}\tau_2, \tau_\ell \right\rangle
\end{aligned}$$

The sections of A_ℓ induce $\varphi_\ell : \tilde{S} \to S \subset \mathbb{P}^3$ by Proposition 4.5, and can be identified with the coordinates w, x, y, z of Equation (1). Since $A_1 \prec A_2 \prec A_\ell$, we have

$$\Gamma(A_1) \hookrightarrow \Gamma(A_2) \hookrightarrow \Gamma(A_\ell).$$

We can identify $\Gamma(A_1) = \langle y, z \rangle$; these correspond to projecting S from the line $\ell = \{y = z = 0\}$ and induce a conic bundle structure

$$\varphi_1 : \tilde{S} \to \mathbb{P}^1.$$

We have $\Gamma(A_2) = \langle x, y, z\rangle$; these correspond to projecting S from the singularity $p = \{w = y = z = 0\}$ and induce the blow-up realization

$$\varphi_2 : \tilde{S} \to \mathbb{P}^2.$$

Therefore, we may choose τ_1, τ_2, and τ_ℓ so that

$$y = \xi^{\alpha(\ell)} \quad w = \xi^{\alpha(\ell)-\alpha(2)}\tau_2 \quad z = \xi^{\alpha(\ell)-\alpha(1)}\tau_1 \quad x = \tau_\ell.$$

We obtain the following induced sections for A_3, A_4, A_5, and A_6:

$$\begin{aligned}
\Gamma(A_3) &= \left\langle \xi^{\alpha(3)}, \xi^{\alpha(3)-\alpha(1)}\tau_1, \xi^{\alpha(3)-\alpha(2)}\tau_2, \xi^{\alpha(3)-2\alpha(1)}\tau_1^2 \right\rangle \\
\Gamma(A_4) &= \left\langle \xi^{\alpha(4)}, \xi^{\alpha(4)-\alpha(1)}\tau_1, \xi^{\alpha(4)-\alpha(2)}\tau_2, \xi^{\alpha(4)-\alpha(\ell)}\tau_\ell, \xi^{\alpha(4)-2\alpha(1)}\tau_1^2 \right\rangle \\
\Gamma(A_5) &= \left\langle \xi^{\alpha(5)}, \xi^{\alpha(5)-\alpha(1)}\tau_1, \xi^{\alpha(5)-\alpha(2)}\tau_2, \xi^{\alpha(5)-\alpha(\ell)}\tau_\ell, \xi^{\alpha(5)-2\alpha(1)}\tau_1^2, \right. \\
&\qquad \left. \xi^{\alpha(5)-\alpha(1)-\alpha(2)}\tau_1\tau_2 \right\rangle \\
\Gamma(A_6) &= \left\langle \xi^{\alpha(6)}, \xi^{\alpha(6)-\alpha(1)}\tau_1, \xi^{\alpha(6)-\alpha(2)}\tau_2, \xi^{\alpha(6)-\alpha(\ell)}\tau_\ell, \xi^{\alpha(6)-2\alpha(1)}\tau_1^2, \right. \\
&\qquad \left. \xi^{\alpha(6)-\alpha(1)-\alpha(2)}\tau_1\tau_2, \xi^{\alpha(6)-2\alpha(2)}\tau_2^2, \xi^{\alpha(6)-3\alpha(1)}\tau_1^3 \right\rangle
\end{aligned}$$

Equation (1) gives the relation

$$\tau_\ell \xi^{2\alpha(\ell)} + \tau_2^2 \xi^{3\alpha(\ell)-2\alpha(2)} + \tau_1^3 \xi^{3\alpha(\ell)-3\alpha(1)} = 0.$$

Dividing by a suitable monomial ξ^β, we obtain

$$\tau_\ell \xi_\ell^3 \xi_4^2 \xi_5 + \tau_2^2 \xi_2 + \tau_1^3 \xi_1^2 \xi_3 = 0,$$

a dependence relation in $\Gamma(A_6)$. This is the only such relation: Any other relation, after multiplying through by ξ^β, yields a cubic form vanishing on $S \subset \mathbb{P}^3$, but equation (1) is the only such form. It follows that the sections given above for $A_1, \ldots, A_5$ form bases for $\Gamma(A_1), \ldots, \Gamma(A_5)$.

Since

$$A_3 \prec A_4 \prec A_5 \prec A_6 \prec 2A_\ell$$

we have

$$\begin{aligned}
\Gamma(A_3) &\hookrightarrow \Gamma(A_4) \hookrightarrow \Gamma(A_5) \hookrightarrow \Gamma(A_6) \\
&\hookrightarrow \Gamma(2A_\ell) = \left\langle w^2, wx, wy, x^2, xy, xz, y^2, yz, z^2 \right\rangle
\end{aligned}$$

and identifications

$$\begin{aligned}\Gamma(A_3) &= \langle y^2, yz, wy, z^2\rangle \\ \Gamma(A_4) &= \langle y^2, yz, wy, xy, z^2\rangle \\ \Gamma(A_5) &= \langle y^2, yz, wy, xy, z^2, wz\rangle \\ \Gamma(A_6) &= \langle y^2, yz, wy, xy, z^2, wz, w^2\rangle .\end{aligned}$$

The sections of A_3 induce a morphism

$$\varphi_3 : \tilde{S} \to \mathbb{P}^3$$

onto a quadric surface with a single ordinary double point. The sections of A_4 induce a morphism

$$\varphi_4 : \tilde{S} \to \mathbb{P}^4$$

with image a quartic Del Pezzo surface with a rational double point of type $\mathbf{D}_5$. The sections of A_5 induce a morphism

$$\varphi_5 : \tilde{S} \to \mathbb{P}^5$$

with image a quintic Del Pezzo surface with a rational double point of type $\mathbf{A}_4$. The sections of A_6 induce a morphism

$$\varphi_6 : \tilde{S} \to \mathbb{P}^6$$

with image a sextic Del Pezzo surface with two rational double points, of types $\mathbf{A}_1$ and $\mathbf{A}_2$.

We summarize this analysis in the following proposition

Proposition 4.7. *Every section of $A_j, j = 1, 2, 3, \ell, 4, 5, 6$, can be expressed as a polynomial in $\xi_1, \ldots, \xi_6, \xi_\ell, \tau_1, \tau_2, \tau_6$. The only dependence relation among them is*

$$\tau_\ell \xi_\ell^3 \xi_4^2 \xi_5 + \tau_2^2 \xi_2 + \tau_1^3 \xi_1^2 \xi_3 = 0$$

in $\Gamma(A_6)$. Each A_j is globally generated and induces a morphism

$$\varphi_j : \tilde{S} \to \mathbb{P}^{\chi-1}, \quad \chi = \chi(\mathcal{O}_{\tilde{S}}(A_j)).$$

The remainder of this section is devoted to proving the following:

Theorem 4.8. *The homomorphism*

$$\varrho : k[\xi_1, \ldots, \xi_6, \xi_\ell, \tau_1, \tau_2, \tau_\ell]/\langle \tau_\ell \xi_\ell^3 \xi_4^2 \xi_5 + \tau_2^2 \xi_2 + \tau_1^3 \xi_1^2 \xi_3 \rangle \to \mathrm{Cox}(\tilde{S})$$

is an isomorphism.

If ϱ were not injective, its kernel would have nontrivial elements in degree $\nu = dA_\ell$, for some d sufficiently large. These translate into homogeneous polynomials of degree d vanishing on $S \subset \mathbb{P}^3$. All such polynomials are multiples of the cubic form defining S, which itself is a multiple of the relation we already have.

It remains to show that ϱ is surjective. By Proposition 4.5, Lemma 4.6 and the analysis of the sections of the A_i, it suffices to prove:

Proposition 4.9. *ϱ is surjective in degrees corresponding to nef divisor classes of $\tilde{S}$.*

Lemma 4.10. *For any positive integers $c_1, c_2, c_3, c_\ell, c_4, c_5, c_6$, the image of*

$$\Gamma(A_1)^{\otimes c_1} \otimes \ldots \otimes \Gamma(A_\ell)^{\otimes c_\ell} \otimes \ldots \otimes \Gamma(A_6)^{\otimes c_6} \longrightarrow \Gamma(c_1 A_1 + \ldots + c_6 A_6)$$

is a linear series embedding $\tilde{S}$.

Proof. Proposition 4.7 says that each A_j is globally generated, so if the image of

$$\Gamma(A_1) \otimes \ldots \otimes \Gamma(A_6) \longrightarrow \Gamma(A_1 + \ldots + A_6)$$

embed $\tilde{S}$ then the general result follows. We use the standard criterion: a linear series gives an embedding iff any length-two subscheme $\Sigma \subset \tilde{S}$ imposes two independent conditions on the linear series.

First, suppose the support of Σ is not contained in the exceptional locus of $\varphi_\ell : \tilde{S} \to S$, i.e., the curves $F_1, F_2, F_3, F_4, F_5, F_6$. Then φ_ℓ maps Σ to a subscheme of length two, which imposes independent conditions on $\Gamma(A_\ell)$, and thus independent conditions on the linear series in question. Second, suppose that $\Sigma \subset F_j$ for some j (resp. $\Sigma \subset \ell$). Since $A_j \cdot F_j = 1$ (resp. $A_\ell \cdot \ell = 1$), φ_j maps F_j (resp. ℓ) isomorphically onto a line. It follows that Σ imposes independent conditions on $\Gamma(A_j)$. Third, suppose that Σ is reduced with support in F_i and F_j, but is not contained in either F_i or F_j. Consider the chain of rational curves containing F_i and F_j (see Figure 4.) There exists a curve F_k in this chain so that $\varphi_k(F_i) \neq \varphi_k(F_j)$, so Σ imposes independent conditions on $\Gamma(A_k)$. Fourth, suppose that Σ is nonreduced and supported in F_j but not contained in any F_i or ℓ. The morphism φ_j ramifies at points where F_j meets one of the other exceptional curves, and the kernel of the tangent morphism $d\varphi_j$ consists of the tangent vectors to the curves contracted by φ_j. It follows that $\varphi_j(\Sigma)$ has length two and imposes independent conditions on $\Gamma(A_j)$. □

The polynomial ring

$$k[\xi_1, \ldots, \xi_6, \xi_\ell, \tau_1, \tau_2, \tau_\ell]$$

is graded by the Néron–Severi group of $\tilde{S}$

$$\deg(\xi_j) = F_j, j = 1, \dots, 6 \ \deg(\xi_\ell) = \ell \ \deg(\tau_j) = A_j, j = 1, 2, \ell.$$

This gives an action of the Néron–Severi torus $T(\tilde{S})$ on the corresponding affine space $\mathbb{A}^{10}$.

We consider the projective toric varieties that arise as quotients of $\mathbb{A}^{10}$ by $T(\tilde{S})$. As sketched in §3, these varieties have the one-skeleton

$$x_1 = (0,1,2), x_2 = (1,1,3), x_3 = (1,2,4), x_\ell = (2,3,3), x_4 = (2,3,4)$$
$$x_5 = (2,3,5), x_6 = (2,3,6), t_1 = (-1,0,0), t_2 = (0,-1,0), t_\ell = (0,0,-1)$$

where the x_j correspond to the ξ_j and the t_j correspond to the τ_j.

Lemma 4.11. *Let X be a toric threefold with one-skeleton $\{x_1, \dots, t_\ell\}$ and divisor class-group $\mathfrak{X}^*(T(\tilde{S})) = \mathrm{N}_1(\tilde{S})$. Then*

$$\mathrm{Mov}(X) = \mathrm{Cone}(A_1, \dots, A_6, A_\ell).$$

Proof. Proposition 3.1 reduces this to computing the intersection of the cones generated by subsets of

$$\{F_1, \dots, F_6, \ell, A_1, A_2, A_\ell\}$$

with nine elements. Since A_1, A_2, A_ℓ are effective combinations of the classes $F_1, \dots, F_6$, and ℓ, it suffices to compute

$$\mathrm{Cone}(F_1, \dots, \hat{\ell}, \dots, A_\ell) \bigcap \left(\bigcap_{i=1,\dots,6} \mathrm{Cone}(F_1, \dots, \hat{F_i}, \dots, A_\ell) \right).$$

This intersection obviously contains A_1, A_2, and A_ℓ, and it is a straightforward computation to show that it also contains A_3, A_4, A_5, A_6. For the reverse inclusion, suppose that D is contained in the intersection. Considering D as a divisor on $\tilde{S}$, we see that

$$D \cdot F_1, \dots, D \cdot F_6, D \cdot \ell$$

are all nonnegative. Thus D is an effective sum of A_j by Proposition 4.5. □

Combining Lemmas 4.11 and 4.10 with Propositions 3.2 and 4.7, we obtain the following:

Proposition 4.12. *Let ν be an ample divisor on $\tilde{S}$. Then there exists a projective toric variety Y_ν with one-skeleton $\{x_1, \dots, t_\ell\}$ and polarization ν, and an embedding $\tilde{S} \hookrightarrow Y_\nu$ with the following properties:*

(1) *the divisor class group of Y_ν is isomorphic to the divisor class group of $\tilde{S}$ so that the moving cone of Y_ν is identified with the nef cone of $\tilde{S}$;*

(2) *the equation for* $\tilde{S}$ *in the Cox ring of* Y_ν *is*

$$\tau_\ell \xi_\ell^3 \xi_4^2 \xi_5 + \tau_2^2 \xi_2 + \tau_1^3 \xi_1^2 \xi_3 = 0$$

and $[\tilde{S}] = A_6$ *in the divisor class group of* Y_ν*;*

(3) $\mathrm{Cox}(Y_\nu) = k[\xi_1, \ldots, \tau_\ell]$ *and is mapped isomorphically to the image of the homomorphism* ϱ*.*

For each toric variety Y_ν, we can consider the exact sequence of sheaves

$$0 \to I_{\tilde{S}} \to \mathscr{O}_{Y_\nu} \to \mathscr{O}_{\tilde{S}} \to 0,$$

where $I_{\tilde{S}} \simeq \mathscr{O}_{Y_\nu}(-A_6)$ is the ideal sheaf of $\tilde{S}$. Given an element θ in the divisor class group of Y_ν, we can twist to obtain

$$0 \to I_{\tilde{S}}(\theta) \to \mathscr{O}_{Y_\nu}(\theta) \to \mathscr{O}_{\tilde{S}}(\theta) \to 0.$$

We should make precise what we mean by the twist $\mathscr{F}(\theta)$ of a coherent sheaf $\mathscr{F}$ on Y_ν: Realize $\mathscr{F}$ as the sheafification of a graded module F over $\mathrm{Cox}(Y_\nu)$ (which exists by **[19]** Theorem 1.1, **[10]** Proposition 3.1), shift F by θ, and then resheafify the shifted module to obtain $\mathscr{F}(\theta)$. Twisting respects exact sequences **[10]** 3.1.

The anticanonical divisor of a toric variety is the sum of the invariant divisors **[13]** p. 89, so

$$-K_{Y_\nu} = F_1 + \ldots + F_6 + \ell + A_1 + A_2 + A_\ell = A_\ell + A_6$$

and we can rewrite our exact sequence as

$$0 \to \mathscr{O}_{Y_\nu}(K_{Y_\nu} + A_\ell + \theta) \to \mathscr{O}_{Y_\nu}(\theta) \to \mathscr{O}_{\tilde{S}}(\theta) \to 0.$$

Suppose that θ corresponds to a nef class on $\tilde{S}$; we shall prove that ϱ is surjective in degree θ, thus proving Proposition 4.9 and Theorem 4.8. Since

$$\Gamma(\mathscr{O}_{Y_\nu}(\theta)) \simeq k[\xi_1, \ldots, \xi_6, \xi_\ell, \tau_1, \tau_2, \tau_\ell]_\theta$$

it suffices to show that

$$H^1(\mathscr{O}_{Y_\nu}(K_{Y_\nu} + A_\ell + \theta)) = 0.$$

We apply Proposition 3.5, with $\nu_0 = A_\ell + \theta$, to get a *simplicial* toric variety Y_ν on which $A_\ell + \theta$ is nef. As A_ℓ is in the interior of the effective cone of Y_ν, $A_\ell + \theta$ is also big. Note that Y_ν has finite-quotient singularities, which are log terminal **[18]** §5.2. The desired vanishing follows from Theorem 2.17 of **[17]**. Alternately, we could apply Theorem 0.1 of **[19]**, which applies in arbitrary characteristic and obviates the need to pass to a simplicial model.

5. $\mathbf{D}_4$ cubic surface

The strategy of the previous section can applied to other surfaces as well. Here we illustrate it in the case of a cubic surface given by the homogeneous equation

$$S = \{(x_1, x_2, x_3, w) \,:\, w(x_1 + x_2 + x_3)^2 = x_1x_2x_3\} \subset \mathbb{P}^3.$$

We summarize its properties:

(1) S has a single singularity at the point $(0, 0, 0, 1)$ of type $\mathbf{D}_4$.

(2) S contains 6 lines with the equations

$$\begin{array}{ll} \ell_1' := \{w = x_1 = 0\} & m_1' := \{x_1 = x_2 + x_3 = 0\} \\ \ell_2' := \{w = x_2 = 0\} & m_2' := \{x_2 = x_1 + x_3 = 0\} \\ \ell_3' := \{w = x_3 = 0\} & m_3' := \{x_3 = x_1 + x_2 = 0\} \end{array}$$

(3) S is the closure of the image of $\mathbb{P}^2$ under the linear series

$$x_1 = u_1(u_1 + u_2 + u_3)^2,\ x_2 = u_2(u_1 + u_2 + u_3)^2,\ x_3 = u_3(u_1 + u_2 + u_3)^2,$$

$$w = u_1u_2u_3,$$

where $\langle u_1, u_2, u_3\rangle = \Gamma(\mathbb{P}^2, \mathscr{O}_{\mathbb{P}^2}(1))$.

Remark 5.1. There are two isomorphism classes of cubic surfaces with a $\mathbf{D}_4$ singularity **[5]** Lemma 4. The other class is

$$w(x_1 + x_2 + x_3)^2 = x_1x_2(-x_1 - x_2);$$

it is obtained from S by substituting

$$(w, x_1, x_2, x_3) \mapsto (t^{-2}w, x_1, x_2, tx_3 + (t - 1)x_1 + (t - 1)x_2)$$

and letting $t \to 0$ in the resulting equation.

We can distinguish S and S_0 geometrically: In S, the three lines not containing p do not share a common point. In S_0, the analogous lines

$$\{w = x_1 = 0\}, \{w = x_2 = 0\}, \{w = x_1 + x_2 = 0\} \subset S_0$$

are coincident at $w = x_1 = x_2 = 0$.

Let $\beta : \tilde{S} \to S$ denote the minimal desingularization of S and

$$\ell_1, \ell_2, \ell_3, m_1, m_2, m_3$$

the strict transforms of the lines. The rational map $S \dashrightarrow \mathbb{P}^2$ induces a morphism $\tilde{S} \to \mathbb{P}^2$ and let L denote the pullback of the hyperplane class. Let E_0, E_1, E_2, E_3 be the exceptional divisors of β, ordered so that we have the following intersection matrix:

$$
\begin{array}{c|ccccccc}
 & L & E_1 & E_2 & E_3 & m_1 & m_2 & m_3 \\
\hline
L & 1 & 0 & 0 & 0 & 0 & 0 & 0 \\
E_1 & 0 & -2 & 0 & 0 & 1 & 0 & 0 \\
E_2 & 0 & 0 & -2 & 0 & 0 & 1 & 0 \\
E_3 & 0 & 0 & 0 & -2 & 0 & 0 & 1 \\
m_1 & 0 & 1 & 0 & 0 & -1 & 0 & 0 \\
m_2 & 0 & 0 & 1 & 0 & 0 & -1 & 0 \\
m_3 & 0 & 0 & 0 & 1 & 0 & 0 & -1
\end{array}. \tag{5}
$$

This is a rank seven unimodular matrix; since the Picard group of $\tilde{S}$ has rank seven, it is generated by $L, E_1, E_2, E_3, m_1, m_2, m_3$. In particular, we have

$$E_0 = L - (E_1 + E_2 + E_3 + m_1 + m_2 + m_3) \quad \text{and} \quad \ell_j = L - E_j - 2m_j.$$

The anticanonical class is given by

$$-K_{\tilde{S}} = 3L - (E_1 + E_2 + E_3) - 2(m_1 + m_2 + m_3) = \ell_1 + \ell_2 + \ell_3.$$

Proposition 5.2. *The effective cone* $\mathrm{NE}(\tilde{S})$ *is generated by*

$$\Xi := \{E_0, E_1, E_2, E_3, m_j, \ell_j\}.$$

Proof. Each effective divisor D can be expressed as a sum

$$D = M_\Xi + b_{E_0}E_0 + b_{E_1}E_1 + \ldots + b_{\ell_3}\ell_3,$$

with nonnegative coefficients, where M_Ξ intersects each of the elements in Ξ nonnegatively and thus is in the dual cone to Cone(Ξ). Direct computation shows that the dual to Cone(Ξ) has generators

$$L, L - E_i - m_i, 2L - E_i - 2m_i, 2L - E_i - E_j - 2m_i - 2m_j,$$

$$2L - E_i - E_j - m_i - 2m_j.$$

Each of these is contained in Cone(Ξ):

$$
\begin{array}{rcl}
L & = & \ell_i + E_i + 2m_i, \\
2L - E_i - 2m_i & = & 2\ell_i + E_i + 2m_i, \\
2L - E_i - E_j - m_i - 2m_j & = & \ell_i + \ell_j + m_i, \\
L - E_i - m_i & = & \ell_i + m_i, \\
2L - E_i - E_j - 2m_i - 2m_j & = & \ell_i + \ell_j.
\end{array}
$$

It follows that M_Ξ and D are sums of elements in Ξ with nonnegative coefficients. □

Each of the divisors m_i, ℓ_i and E_i has a distinguished nonzero section (up to a constant), denoted μ_i, λ_i and η_i, respectively. We have

$$\{\lambda_i \eta_i \mu_i^2, \eta_0\eta_1\eta_2\eta_3\mu_1\mu_2\mu_3\} \subset \Gamma(L),$$

and we may identify

$$u_i = \lambda_i \eta_i \mu_i^2 \quad \text{and} \quad u_1 + u_2 + u_3 = \eta_0\eta_1\eta_2\eta_3\mu_1\mu_2\mu_3$$

after suitably normalizing the μ_i, λ_i, and η_i. The dependence relation among the sections in $\Gamma(L)$ translates into

$$\lambda_1\eta_1\mu_1^2 + \lambda_2\eta_2\mu_2^2 + \lambda_3\eta_3\mu_3^2 = \eta_0\eta_1\eta_2\eta_3\mu_1\mu_2\mu_3. \tag{6}$$

An argument similar to the one given at the end of Section 4 proves that the natural homomorphism

$$k[\eta_0, \ldots, \eta_3, \mu_i, \lambda_i]/\langle \lambda_1\eta_1\mu_1^2 + \lambda_2\eta_2\mu_2^2 + \lambda_3\eta_3\mu_3^2 - \eta_0\eta_1\eta_2\eta_3\mu_1\mu_2\mu_3 \rangle \to \mathrm{Cox}(\tilde{S})$$

is an isomorphism.

The cubic surface S admits an $\mathfrak{S}_3$-action on the coordinates x_1, x_2, x_3. In particular, it admits nonsplit forms over nonclosed ground fields. They can be expressed as follows: Let K/k be a cubic extension with Galois closure E/k. Fix a basis $\{\gamma, \gamma', \gamma''\}$ for K over k so that elements $Y \in K$ can be represented as

$$Y = y\gamma + y'\gamma' + y''\gamma''$$

with $y, y', y'' \in k$. Choose $\sigma \in \mathrm{Gal}(E/k)$ so that σ and σ^2 are coset representatives $\mathrm{Gal}(E/k)$ modulo $\mathrm{Gal}(E/K)$. Then

$$w \cdot \mathrm{Tr}_{K/k}(Y)^2 - \mathrm{N}_{K/k}(Y)$$

is isomorphic, over E, to S:

$$\begin{array}{rcl} x_1 & = & Y = y\gamma + y'\gamma' + y''\gamma'' \\ x_2 & = & \sigma(Y) = y\sigma(\gamma) + y'\sigma(\gamma') + y''\sigma(\gamma'') \\ x_3 & = & \sigma^2(Y) = y\sigma^2(\gamma) + y'\sigma^2(\gamma') + y''\sigma^2(\gamma'') \end{array} .$$

Assigning elements $U, V, W \in K$ to η_1, μ_1 and λ_1, respectively, the torsor equation (6) takes the form

$$\mathrm{Tr}_{K/k}(UV^2W) = \eta_0 \mathrm{N}_{K/k}(UV).$$

References

[1] V. V. Batyrev – The cone of effective divisors of threefolds, *Proceedings of the International Conference on Algebra, Part 3 (Novosibirsk, 1989)* (Providence, RI), Contemp. Math., vol. 131, Amer. Math. Soc., 1992, 337–352.

[2] V. V. Batyrev & O. N. Popov – The Cox ring of a Del Pezzo surface, this volume.

[3] R. de la Bretèche – Nombre de points de hauteur bornée sur les surfaces de del Pezzo de degré 5, *Duke Math. J.* **113** (2002), no. 3, 421–464.

[4] M. Brion & C. Procesi – Action d'un tore dans une variété projective, Operator algebras, unitary representations, enveloping algebras, and invariant theory (Paris, 1989), Prog. Math., vol. 92, Birkhäuser Boston, Boston, MA, 1990, 509–539.

[5] J. W. Bruce & C. T. C. Wall – On the classification of cubic surfaces, *J. London Math. Soc. (2)* **19** (1979), no. 2, 245–256.

[6] A. Chambert-Loir & Y. Tschinkel – On the distribution of points of bounded height on equivariant compactifications of vector groups, *Invent. Math.* **148** (2002), no. 2, 421–452.

[7] J.-L. Colliot-Thélène & J.-J. Sansuc – La descente sur les variétés rationnelles. II, *Duke Math. J.* **54** (1987), no. 2, 375–492.

[8] J.-L. Colliot-Thélène, J.-J. Sansuc & P. Swinnerton-Dyer – Intersections of two quadrics and Châtelet surfaces. I, *J. Reine Angew. Math.* **373** (1987), 37–107.

[9] ______, Intersections of two quadrics and Châtelet surfaces. II, *J. Reine Angew. Math.* **374** (1987), 72–168.

[10] D. A. Cox – The homogeneous coordinate ring of a toric variety, *J. Algebraic Geom.* **4** (1995), no. 1, 17–50.

[11] D. A. Cox, R. Krasauskas & M. Mustata – Universal rational parametrizations and toric varieties, 2003, `ArXiV:math.AG/0303316`.

[12] I. V. Dolgachev & Y. Hu – Variation of geometric invariant theory quotients, *Inst. Hautes Études Sci. Publ. Math.* (1998), no. 87, 5–56.

[13] W. Fulton – *Introduction to toric varieties*, Annals of Mathematics Studies, vol. 131, Princeton University Press, Princeton, NJ, 1993.

[14] B. Hassett & Y. Tschinkel – Geometry of equivariant compactifications of $\mathbf{G}_a^n$, *Internat. Math. Res. Notices* (1999), no. 22, 1211–1230.

[15] D. R. Heath-Brown – The density of rational points on Cayley's cubic surface, to appear.

[16] Y. Hu & S. Keel – Mori dream spaces and GIT, *Michigan Math. J.* **48** (2000), 331–348.

[17] J. Kollár – Singularities of pairs, Algebraic geometry—Santa Cruz 1995, Proc. Sympos. Pure Math., vol. 62, Amer. Math. Soc., Providence, RI, 1997, 221–287.

[18] J. Kollár & S. Mori – *Birational geometry of algebraic varieties*, Cambridge Tracts in Mathematics, vol. 134, Cambridge University Press, Cambridge, 1998.

[19] M. MUSTAŢĂ – Vanishing theorems on toric varieties, *Tohoku Math. J. (2)* **54** (2002), no. 3, 451–470.

[20] E. PEYRE – Terme principal de la fonction zêta des hauteurs et torseurs universels, *Astérisque* (1998), no. 251, 259–298.

[21] ______, Torseurs universels et méthode du cercle, Rational points on algebraic varieties, Prog. Math., vol. 199, Birkhäuser, Basel, 2001, 221–274.

[22] P. SALBERGER & A. N. SKOROBOGATOV – Weak approximation for surfaces defined by two quadratic forms, *Duke Math. J.* **63** (1991), no. 2, 517–536.

[23] P. SALBERGER – Tamagawa measures on universal torsors and points of bounded height on Fano varieties, *Astérisque* (1998), no. 251, 91–258.

[24] A. SKOROBOGATOV – *Torsors and rational points*, Cambridge Tracts in Mathematics, vol. 144, Cambridge University Press, Cambridge, 2001.

[25] A. N. SKOROBOGATOV – On a theorem of Enriques–Swinnerton-Dyer, *Ann. Fac. Sci. Toulouse Math. (6)* **2** (1993), no. 3, 429–440.

[26] M. THADDEUS – Toric quotients and flips, Topology, geometry and field theory, World Sci. Publishing, River Edge, NJ, 1994, 193–213.

Arithmetic of Higher-dimensional Algebraic Varieties
(B. Poonen, Yu. Tschinkel, eds.), p. 175–184
Progress in Mathematics, Vol. 226, © 2004 Birkhäuser Boston, Cambridge, MA

RANDOM DIOPHANTINE EQUATIONS

Bjorn Poonen

Department of Mathematics, UC Berkeley, CA 94720-3840, USA • *E-mail :* `poonen@math.berkeley.edu`
URL : `http://www.math.berkeley.edu/~poonen`

José Felipe Voloch

Department of Mathematics, University of Texas, Austin, TX 78712, USA • *E-mail :* `voloch@math.utexas.edu`
URL : `http://www.ma.utexas.edu/users/voloch/`

(with appendices by Jean-Louis Colliot-Thélène and Nicholas M. Katz)

Abstract. Consider hypersurfaces of fixed degree d in a fixed projective space $\mathbf{P}^n$ over $\mathbb{Q}$. We present a conjecture about the fraction of these that have rational points, and present evidence for the conjecture, including a proof that a positive fraction of the hypersurfaces have points over every completion of $\mathbb{Q}$, provided that $n, d \geq 2$ and $(n, d) \neq (2, 2)$. Generalizations to number fields are discussed. One of our proofs uses a result of Colliot-Thélène, proved in an appendix, that there is no Brauer–Manin obstruction to the Hasse principle for smooth complete intersections of dimension ≥ 3 in projective space over number fields. Colliot-Thélène's proof, in turn, uses a consequence of the Weak Lefschetz Theorem proved in an appendix by Katz.

Key words and phrases. Diophantine equation, Hasse principle, Brauer–Manin obstruction, hypersurface, complete intersection, Weak Lefschetz Theorem, cubic surface.

The first author was partially supported by a Packard Fellowship.

1. Introduction

The main result of this paper is that, in a precise sense, a positive proportion of all hypersurfaces in $\mathbf{P}^n$ of degree d defined over $\mathbf{Q}$ are everywhere locally solvable, provided that $n, d \geq 2$ and $(n, d) \neq (2, 2)$. This result is motivated by a conjecture discussed in detail below about the proportion of hypersurfaces as above that are globally solvable, i.e., have a rational point.

Acknowledgements: This paper developed during a workshop sponsored by the American Institute of Mathematics. We thank Jean-Louis Colliot-Thélène and Nick Katz for allowing us to include their results in the appendices to this paper. We thank also Jean-Louis Colliot-Thélène, Jordan Ellenberg, Brendan Hassett, Roger Heath-Brown, Bill McCallum, and Trevor Wooley for many conversations which helped shape Conjecture 2.2.

2. A conjecture

Fix $n, d \geq 2$. Let $\mathbf{Z}[x_0, \ldots, x_n]_d$ denote the set of homogeneous polynomials in $\mathbf{Z}[x_0, \ldots, x_n]$ of degree d. Let $m = \binom{n+d}{d}$ denote the number of monomials in $x_0, \ldots, x_n$ of degree d. Define the height $h(f)$ of $f \in \mathbf{Z}[x_0, \ldots, x_n]_d$ as the maximum of the absolute values of the coefficients of f. Let $M_{\mathbf{Q}}$ be the set of places of $\mathbf{Q}$, and let $\mathbf{Q}_v$ be the completion of $\mathbf{Q}$ at the place v. Define

$$\begin{aligned} N_{\mathrm{tot}}(H) &= \#\{\, f \in \mathbf{Z}[x_0, \ldots, x_n]_d : h(f) \leqslant H \,\} = (2\lfloor H \rfloor + 1)^m, \\ N(H) &= \#\{\, f \in \mathbf{Z}[x_0, \ldots, x_n]_d : h(f) \leqslant H, \\ &\qquad \text{and } \exists x \in \mathbf{Z}^{n+1} \setminus \{0\} \text{ with } f(x) = 0 \,\}, \\ N_{\mathrm{loc}}(H) &= \#\{\, f \in \mathbf{Z}[x_0, \ldots, x_n]_d : h(f) \leqslant H, \\ &\qquad \text{and } \forall v \in M_{\mathbf{Q}}, \exists x \in \mathbf{Q}_v^{n+1} \setminus \{0\} \text{ with } f(x) = 0 \,\}. \end{aligned}$$

The limit of $N(H)/N_{\mathrm{tot}}(H)$ as $H \to \infty$, if it exists, will be called the proportion of globally solvable hypersurfaces. Similarly, the limit of $N_{\mathrm{loc}}(H)/N_{\mathrm{tot}}(H)$ will be called the proportion of locally solvable hypersurfaces.

Remark 2.1.

1. Restricting the set of f's to those such that $f = 0$ defines a smooth geometrically integral hypersurface in $\mathbf{P}^n$ does not change the values of these limits, since the f's that violate these conditions correspond to integer points on a Zariski closed subset of positive codimension in some affine space.

2. A standard argument involving Möbius inversion shows that the values of the limits do not change if in all our counts we restrict to f's whose coefficients are coprime.
3. We chose to work over $\mathbf{Q}$ to keep statements and proofs simple. In Section 4, we sketch the changes that would be needed to generalize the results to other number fields.

Conjecture 2.2.

(i) *If $d > n+1$, then $N(H)/N_{\text{tot}}(H) \to 0$.*

(ii) *If $d < n+1$ and $(d,n) \neq (2,2)$, then $N(H)/N_{\text{tot}}(H) \to c$ for some real $c > 0$ (depending on d and n). Moreover, $c = \prod_{v \in M_{\mathbf{Q}}} c_v$, where c_v is the proportion of polynomials in $\mathbf{Z}[x_0, \ldots, x_n]_d$ with a nontrivial zero over $\mathbf{Q}_v$.*

In the case $d = n+1$, we do not know what to expect. As a special case, if you write down a plane cubic, how likely is it to have a rational point?

Remark 2.3. Each local proportion c_v exists, since if we define

$$\mathbf{Z}_v = \{\, x \in \mathbf{Q}_v : |x|_v \leq 1 \,\}$$

and normalize Haar measure (Lebesgue measure if $v = \infty$) on the space $\mathbf{Z}_v^m$ parametrizing homogeneous polynomials of degree d in $x_0, \ldots, x_n$ with coefficients in $\mathbf{Z}_v$, then c_v is the measure of the v-adically closed subset of $\mathbf{Z}_v^m$ corresponding to homogeneous polynomials with a nontrivial zero over $\mathbf{Q}_v$.

3. Motivation and evidence

To motivate part (i) of Conjecture 2.2, consider the set of $f \in \mathbf{Z}[x_0, \ldots, x_n]_d$ of height at most H having a given zero $a \in \mathbf{Z}^{n+1} \setminus \{0\}$ with coprime coordinates. This forms a hyperplane in the parameter space $\mathbf{Z}^m$, and contains $c(a)H^{m-1}/\varphi(a) + O(H^{m-2})$ integer points of height at most H, where $c(a)$ is the $(m-1)$-dimensional volume of the part of the hyperplane inside $[-1,1]^m$, and $\varphi(a)$ is the covolume of the lattice of integer points lying on the hyperplane. Lemma 3.1 below shows that $\varphi(a)$ equals the norm of the vector b formed by the monomials of degree d in the coordinates of a. If we ignore the error term, then we get that $N(H) \lesssim H^{m-1} \sum_a c(a)/\varphi(a)$, where a ranges over $\mathbf{Z}^{n+1} \setminus \{0\}$. Now $c(a)$ is bounded, and it is easy to show that $\sum 1/\varphi(a)$ converges precisely when $d > n+1$, so our heuristic predicts $N(H) = O(H^{m-1})$. Since $N_{\text{tot}}(H) \sim (2H)^m$, this leads to the first part of the conjecture.

Lemma 3.1. *Let b be a vector in $\mathbf{R}^m$ with coprime integer coordinates and norm $|b| = \sqrt{M}$. The covolume of the lattice $\Lambda = \{ x \in \mathbf{Z}^m : \langle x, b \rangle = 0 \}$ is $\sqrt{M}$.*

Proof. The lattice $\Gamma = \{ x \in \mathbf{Z}^m : \langle x, b \rangle \equiv 0 \pmod{M} \}$ is the inverse image of $M\mathbf{Z}$ under the surjection $\mathbf{Z}^m \to \mathbf{Z}$ mapping x to $\langle x, b \rangle$, so Γ has covolume M in $\mathbf{R}^m$. On the other hand, Γ is the orthogonal direct sum of Λ and $\mathbf{Z}b$, so the covolume of Λ is $M/|b| = \sqrt{M}$. □

Our next proposition will be conditional on cases of the following very general conjecture.

***Conjecture 3.2* (Colliot-Thélène).** *Let X be a smooth, proper, geometrically integral variety over a number field k. Suppose that X is (geometrically) rationally connected. Then the Brauer–Manin obstruction to the Hasse principle for X is the only obstruction.*

Remark 3.3. Conjecture 3.2 has a long history. In the special case of rational surfaces, it appeared as Question (k1) on page 233 of **[CTS80]** (a paper later developed as **[CTS87]**), based on evidence eventually published in the papers **[CTCS80]** and **[CTS82]**. Theoretical evidence and some numerical evidence have been gathered since then in the case of rational surfaces. The conjecture was generalized to (geometrically) unirational and Fano varieties on the first page of **[CTSD94]**. The full version of Conjecture 3.2 was raised as a question in lectures by Colliot-Thélène at the Institut Henri Poincaré in the Spring 1999, and was repeated in print on page 3 of **[CT03]**.

Proposition 3.4. *Assume Conjecture 3.2. If $2 \le d \le n$, then*

$$\frac{N_{\text{loc}}(H) - N(H)}{N_{\text{tot}}(H)} \to 0 \quad \textit{as} \quad H \to \infty.$$

Proof. By Remark 2.1, we may restrict attention to f's for which $f = 0$ defines a smooth, geometrically integral hypersurface X in $\mathbf{P}^n$. The assumption $d \le n$ implies that X is Fano, hence rationally connected (see Theorem V.2.13 of **[Kol96]**). If $n \ge 4$, then by Corollary A.2 there is no Brauer–Manin obstruction, so Conjecture 3.2 gives the Hasse principle, as desired. If $d = 2$, then the Hasse principle holds unconditionally.

It remains to consider the case of cubic surfaces ($d = n = 3$). Here the Hasse principle does not always hold. But by **[SD93]** there is no Brauer–Manin obstruction whenever the action of $\mathrm{Gal}(\overline{\mathbf{Q}}/\mathbf{Q})$ on the 27 lines is as large as possible (namely, the Weyl group $W(E_6)$). The Galois action on the 27 lines on the *generic* cubic surface over the purely transcendental field $\mathbf{C}(a_1, \ldots, a_{20})$

is $W(E_6)$ (this follows from [**Tod35**]), so the same is true for the generic cubic surface over $\mathbf{Q}(a_1, \dots, a_{20})$, and it then follows by Hilbert irreducibility (see §9.2 and §13 of [**Ser97**]) that the same holds for a density 1 set of cubic surfaces over $\mathbf{Q}$. Such cubic surfaces, under Conjecture 3.2, satisfy the Hasse principle as desired. □

Remark 3.5. For n large compared to d, the conclusion of Proposition 3.4 can be proved unconditionally by using the circle method.

Part (ii) of Conjecture 2.2 would follow from the conclusion of Proposition 3.4 and the following result.

Theorem 3.6. *If* $n, d \geq 2$ *and* $(n,d) \neq (2,2)$, *then* $N_{\text{loc}}(H)/N_{\text{tot}}(H) \to c$ *for some* $c > 0$. *Moreover,* $c = \prod c_v$ *where* c_v *is as in Conjecture 2.2.*

Proof. By Hensel's Lemma, a hypersurface $f = 0$ will have a point in $\mathbf{Q}_p$ if its reduction modulo p has a smooth point in $\mathbf{F}_p$. If f is absolutely irreducible modulo p and p is sufficiently large (in terms of n and d), then the existence of a smooth point in $\mathbf{F}_p$ is ensured by the Lang–Weil estimate [**LW54**]. Lemmas 20 and 21 of [**PS99a**] will now imply the theorem, provided that we can show that the space of reducible polynomials is of codimension at least 2 in the space of all polynomials. The lower bound on the codimension follows from the inequalities

$$\left(\binom{n+r}{n} - 1\right) + \left(\binom{n+(d-r)}{n} - 1\right) \leq \left(\binom{n+d}{n} - 1\right) - 2$$

for $0 < r < d$, which hold provided that $n, d \geq 2$ and $(n,d) \neq (2,2)$. (Here r and $d-r$ represent degrees in a potential factorization.) See also [**PS99b**] for an exposition of the application of the density lemmas from [**PS99a**]. □

Remark 3.7. It follows from Theorem 3.6 and part (i) of Conjecture 2.2 that for each pair (d,n) with $n \geq 2$ and $d > n+1$, there are hypersurfaces of degree d in $\mathbf{P}^n$ for which the Hasse principle fails. There does not seem to be an unconditional proof of this statement yet, for any such (d,n) with $n \geq 3$. For results conditional on various conjectures see [**SW95**] and [**Poo01**].

4. Generalization to number fields

Let k be a number field. A hypersurface can be described by its vector of coefficients, viewed as a point in $\mathbf{P}^{m-1}(k)$ where $m = \binom{n+d}{d}$ as before. Instead of counting polynomials with bounded coefficients, we let $N_{\text{tot}}(H)$ be the number of hypersurfaces whose corresponding point in $\mathbf{P}^{m-1}(k)$ has (exponential)

Weil height $\leq H$. There is no longer an exact formula for $N_{\mathrm{tot}}(H)$, but its asymptotics are given by Schanuel's Theorem (see [**Ser97**, §2.5] for an exposition). We define $N(H)$ and $N_{\mathrm{loc}}(H)$ in a similar way. In the special case $k = \mathbf{Q}$, these definitions do not agree with the earlier ones (since we are now counting hypersurfaces instead of polynomials), but the ratios of interest have the same limit, by Remark 2.1(2).

The statement of Conjecture 2.2 remains unchanged, except that the c_v will now be defined by counting hypersurfaces, and the constant c will depend on k as well as d and n. The statement and proof of Proposition 3.4 remain valid over number fields. Finally, the statement and proof of Theorem 3.6 also generalize to number fields in a straightforward way, although the proof is somewhat tedious, since it requires generalizing the statements and proofs Lemmas 20 and 21 of [**PS99a**].

Appendix A
The Brauer–Manin obstruction for complete intersections of dimension ≥ 3
(by Jean-Louis Colliot-Thélène)

It seems that a full proof of the following proposition has never before appeared in print, though a sketch can be found in §2 of [**SW95**]. Let H^i below denote étale cohomology (or profinite group cohomology) unless otherwise specified, and let $\operatorname{Br} X$ denote the cohomological Brauer group $\mathrm{H}^2(X, \mathbf{G}_m)$ of a scheme X.

Proposition A.1. *Let k be a field of characteristic* 0. *Let X be a smooth complete intersection in $\mathbf{P}^n_k$ satisfying* $\dim X \geq 3$. *Then the natural map* $\operatorname{Br} k \to \operatorname{Br} X$ *is an isomorphism.*

Proof. Let $\overline{k}$ denote an algebraic closure of k, let $G = \operatorname{Gal}(\overline{k}/k)$, and let $\overline{X} = X \times_k \overline{k}$. Let $p : X \to \operatorname{Spec} k$ denote the structure map. In the Leray spectral sequence

$$E_2^{p,q} := \mathrm{H}^p(k, \mathrm{R}^q p_* \mathbf{G}_m)) \implies E^{p+q} := \mathrm{H}^{p+q}(X, \mathbf{G}_m),$$

the étale sheaf $\mathrm{R}^q p_* \mathbf{G}_m$ on $\operatorname{Spec} k$ corresponds to the G-module $\mathrm{H}^q(\overline{X}, \mathbf{G}_m)$. A smooth complete intersection of positive dimension is geometrically connected [**Har77**, Exercise III.5.5(b)], so $\mathrm{H}^0(\overline{X}, \mathbf{G}_m) = \overline{k}^\times$, and [**BLR90**, p. 203] shows that $\mathrm{H}^1(\overline{X}, \mathbf{G}_m) = \mathrm{H}^1_{\text{Zariski}}(\overline{X}, \mathbf{G}_m) =: \operatorname{Pic} \overline{X}$. Thus the exact sequence

$$E_2^{1,0} \to E^1 \to E_2^{0,1} \to E_2^{2,0} \to \ker\left(E^2 \to E_2^{0,2}\right) \to E_2^{1,1}$$

from the spectral sequence becomes

$$0 \to \operatorname{Pic} X \to \left(\operatorname{Pic} \overline{X}\right)^G \to \operatorname{Br} k \to \ker\left(\operatorname{Br} X \to \operatorname{Br} \overline{X}\right) \to \mathrm{H}^1(k, \operatorname{Pic} \overline{X}).$$

It remains to prove that $\operatorname{Pic} X \to \left(\operatorname{Pic} \overline{X}\right)^G$ is an isomorphism, $\mathrm{H}^1(k, \operatorname{Pic} \overline{X}) = 0$, and $\operatorname{Br} \overline{X} = 0$.

For smooth complete intersections of dimension ≥ 3 in $\mathbf{P}^n$, M. Noether proved that the restriction map $\operatorname{Pic} \overline{\mathbf{P}^n} \to \operatorname{Pic} \overline{X}$ is an isomorphism (see Corollary 3.3 on p. 180 of **[Har70]** for a modern proof). The commutative square

$$\begin{array}{ccc} \operatorname{Pic} \mathbf{P}^n & \longrightarrow & \operatorname{Pic} X \\ \big\downarrow \wr & & \big\downarrow \\ \operatorname{Pic} \overline{\mathbf{P}^n} & \xrightarrow{\sim} & \operatorname{Pic} \overline{X} \end{array}$$

shows that the injections

$$\operatorname{Pic} X \hookrightarrow \left(\operatorname{Pic} \overline{X}\right)^G \hookrightarrow \operatorname{Pic} \overline{X}$$

are isomorphisms, and that $\mathrm{H}^1(k, \operatorname{Pic} \overline{X}) = \mathrm{H}^1(G, \mathbf{Z}) = \operatorname{Hom}_{\text{conts}}(G, \mathbf{Z}) = 0$.

Finally we need to show that if Y is a complete intersection of dimension ≥ 3 in $\mathbf{P}^n$ over an algebraically closed field L of characteristic 0, then $\operatorname{Br} Y = 0$. For each prime ℓ, the Kummer sequence yields the exact rows of the diagram

$$\begin{array}{ccccccccc} 0 & \longrightarrow & \operatorname{Pic}(\mathbf{P}^n)/\ell & \longrightarrow & \mathrm{H}^2(\mathbf{P}^n, \mathbf{Z}/\ell\mathbf{Z}) & & & & \\ & & \big\downarrow & & \big\downarrow & & & & \\ 0 & \longrightarrow & \operatorname{Pic}(Y)/\ell & \longrightarrow & \mathrm{H}^2(Y, \mathbf{Z}/\ell\mathbf{Z}) & \longrightarrow & (\operatorname{Br} Y)[\ell] & \longrightarrow & 0, \end{array}$$

where for any abelian group A, the notation A/ℓ denotes $A/\ell A$, and $A[\ell]$ is the kernel of multiplication-by-ℓ on A. The top horizontal injection $\operatorname{Pic}(\mathbf{P}^n)/\ell \to \mathrm{H}^2(\mathbf{P}^n, \mathbf{Z}/\ell\mathbf{Z})$ is an isomorphism since both groups are of rank 1 over $\mathbf{Z}/\ell\mathbf{Z}$. The right vertical map $\mathrm{H}^2(\mathbf{P}^n, \mathbf{Z}/\ell\mathbf{Z}) \to \mathrm{H}^2(Y, \mathbf{Z}/\ell\mathbf{Z})$ is an isomorphism by a version of the Weak Lefschetz Theorem: see Corollary B.6 in Appendix B of this paper. The diagram then implies that $(\operatorname{Br} Y)[\ell] = 0$. This holds for all ℓ, and $\operatorname{Br} Y$ is torsion **[Gro68**, Proposition 1.4], so $\operatorname{Br} Y = 0$. □

Corollary A.2. *If in addition, k is a number field, then the Brauer–Manin obstruction for X is vacuous.*

Proof. This follows from Proposition A.1, since the elements of $\operatorname{Br} X$ coming from $\operatorname{Br} k$ do not give any obstruction to rational points. □

Appendix B
Applications of the Weak Lefschetz Theorem
(by Nicholas M. Katz)

We work over an algebraically closed field k. Take as ambient space any separated V/k of finite type which is smooth, and everywhere of dimension N (i.e., each connected component of V has the same dimension N). In V, we are given a certain number $r \geq 1$ of closed subschemes H_i, each of which has the property that its complement $V - H_i$ is *affine*. Define the closed subscheme X of V to be the intersection of the H_i. Its complement $V - X$ is covered by r affine open sets, each of dimension (at most) N, namely the $V - H_i$.

Lemma B.1. *The scheme $V - X$ has cohomological dimension at most $N + r - 1$, i.e., for any constructible torsion sheaf $\mathscr{F}$ on $V - X$, we have $\mathrm{H}^i(V - X, \mathscr{F}) = 0$ for $i \geq N + r$.*

This is a special case of

Lemma B.2. *If a separated k-scheme W/k of finite type is the union of r affine opens U_i, each of dimension at most N, then W has cohomological dimension at most $N + r - 1$.*

Proof. For $r = 1$, this is just the Lefschetz affine theorem [**sga73**, Exposé XIV, Corollaire 3.2]. For general r, one proceeds by induction on r, writing W as the union of the two open sets $A := U_r$ and $B := \bigcup_{i<r} U_i$. Then $A \cap B$ is the union of $r - 1$ affines, and we use the long exact sequence

$$\ldots \to \mathrm{H}^i(A \cup B, \mathscr{F}) \to \mathrm{H}^i(A, \mathscr{F}) \oplus \mathrm{H}^i(B, \mathscr{F}) \to \mathrm{H}^i(A \cap B, \mathscr{F}) \to \ldots$$

to get the assertion. □

Suppose now that ℓ is a prime number invertible in the field k, and that $\mathscr{F}$ is a $\mathbf{Z}/\ell\mathbf{Z}$ sheaf on V whose restriction to $V - X$ is lisse, for instance $\mathbf{Z}/\ell\mathbf{Z}$ itself. Because $V - X$ is smooth, everywhere of dimension N, the Poincaré dual of Lemma B.1 is the vanishing of compact cohomology up through dimension $N - r$:

Lemma B.3. *For any integer $i \leq N - r$, we have $\mathrm{H}^i_c(V - X, \mathscr{F}) = 0$.*

Now use the excision sequence (this is why we need $\mathscr{F}$ to be a sheaf on V) in compact cohomology

$$\ldots \to \mathrm{H}^i_c(V - X, \mathscr{F}) \to \mathrm{H}^i_c(V, \mathscr{F}) \to \mathrm{H}^i_c(X, \mathscr{F}) \to \ldots$$

to conclude

Theorem B.4. *For any integer $i < N - r$, the restriction map*

$$\mathrm{H}^i_c(V, \mathscr{F}) \to \mathrm{H}^i_c(X, \mathscr{F})$$

is an isomorphism. For $i = N - r$, this map is injective.

Corollary B.5. *If V/k is in addition assumed proper, then we have the same result for non-compact cohomology: For any integer $i < N - r$, the restriction map*

$$\mathrm{H}^i(V, \mathscr{F}) \to \mathrm{H}^i(X, \mathscr{F})$$

is an isomorphism. For $i = N - r$, this map is injective.

As a special case of Corollary B.5, we get

Corollary B.6. *Suppose X is a closed subscheme of projective space $\mathbf{P}^N$ which is defined by the vanishing of $N - d$ homogeneous forms. Then for $i < d$, the restriction map*

$$\mathrm{H}^i(\mathbf{P}^N, \mathbf{Z}/\ell\mathbf{Z}) \to \mathrm{H}^i(X, \mathbf{Z}/\ell\mathbf{Z})$$

is an isomorphism. For i=d, this map is injective.

Remark B.7. In Corollary B.6, it is enough if X is defined set-theoretically by the vanishing of $N - d$ homogeneous forms, since X and X^{red} have the same étale cohomology.

References

[BLR90] S. Bosch, W. Lütkebohmert & M. Raynaud – *Néron Models*, Springer-Verlag, Berlin, 1990.

[CT03] J.-L. Colliot-Thélène – Points rationnels sur les fibrations, *Higher dimensional varieties and rational points, Proceedings of the 2001 Budapest conference* (K. Böröczky, J. Kollár & T. Szamuely, eds.), Springer-Verlag, 2003, Bolyai Society Colloquium Publications.

[CTCS80] J.-L. Colliot-Thélène, D. Coray & J.-J. Sansuc – Descente et principe de Hasse pour certaines variétés rationnelles, *J. Reine Angew. Math.* **320** (1980), 150–191.

[CTS80] J.-L. Colliot-Thélène & J.-J. Sansuc – La descente sur les variétés rationnelles, Journées de Géometrie Algébrique d'Angers, Juillet 1979/Algebraic Geometry, Angers, 1979, Sijthoff & Noordhoff, Alphen aan den Rijn, 1980, 223–237.

[CTS82] ———, Sur le principe de Hasse et l'approximation faible, et sur une hypothèse de Schinzel, *Acta Arith.* **41** (1982), no. 1, 33–53.

[CTS87] ———, La descente sur les variétés rationnelles. II, *Duke Math. J.* **54** (1987), no. 2, 375–492.

[CTSD94] J.-L. Colliot-Thélène & P. Swinnerton-Dyer – Hasse principle and weak approximation for pencils of Severi-Brauer and similar varieties, *J. Reine Angew. Math.* **453** (1994), 49–112.

[Gro68] A. Grothendieck – Le groupe de Brauer. II. Théorie cohomologique, Dix Exposés sur la Cohomologie des Schémas, North-Holland, Amsterdam, 1968, 67–87.

[Har70] R. Hartshorne – *Ample subvarieties of algebraic varieties*, Notes written in collaboration with C. Musili. Lecture Notes in Mathematics, Vol. 156, Springer-Verlag, Berlin, 1970.

[Har77] ———, *Algebraic geometry*, Springer-Verlag, New York, 1977, Graduate Texts in Mathematics, No. 52.

[Kol96] J. Kollár – *Rational Curves on Algebraic Varieties*, Ergebnisse der Mathematik und ihrer Grenzgebiete. 3. Folge. A Series of Modern Surveys in Mathematics [Results in Mathematics and Related Areas. 3rd Series. A Series of Modern Surveys in Mathematics], vol. 32, Springer-Verlag, Berlin, 1996.

[LW54] S. Lang & A. Weil – Number of points of varieties in finite fields, *Amer. J. Math.* **76** (1954), 819–827.

[Poo01] B. Poonen – The Hasse principle for complete intersections in projective space, Rational points on algebraic varieties, Progr. Math., vol. 199, Birkhäuser, Basel, 2001, 307–311.

[PS99a] B. Poonen & M. Stoll – The Cassels-Tate pairing on polarized abelian varieties, *Ann. of Math. (2)* **150** (1999), no. 3, 1109–1149.

[PS99b] ———, A local-global principle for densities, Topics in number theory (University Park, PA, 1997), Math. Appl., vol. 467, Kluwer Acad. Publ., Dordrecht, 1999, 241–244.

[SD93] P. Swinnerton-Dyer – The Brauer group of cubic surfaces, *Math. Proc. Cambridge Philos. Soc.* **113** (1993), no. 3, 449–460.

[Ser97] J.-P. Serre – *Lectures on the Mordell–Weil theorem*, third ed., Friedr. Vieweg & Sohn, Braunschweig, 1997.

[sga73] *Théorie des topos et cohomologie étale des schémas. Tome 3* – Springer-Verlag, Berlin, 1973, Séminaire de Géométrie Algébrique du Bois-Marie 1963–1964 (SGA 4), Dirigé par M. Artin, A. Grothendieck et J. L. Verdier. Avec la collaboration de P. Deligne et B. Saint-Donat, Lecture Notes in Mathematics, Vol. 305.

[SW95] P. Sarnak & L. Wang – Some hypersurfaces in $\mathbf{P}^4$ and the Hasse principle, *C. R. Acad. Sci. Paris Sér. I Math.* **321** (1995), no. 3, 319–322.

[Tod35] J. A. Todd – On the topology of certain threefold loci, *Proc. Edinburgh Math. Soc. (2)* (1935), no. 4, 175–184.

Arithmetic of Higher-dimensional Algebraic Varieties
(B. Poonen, Yu. Tschinkel, eds.), p. 185–204
Progress in Mathematics, Vol. 226, © 2004 Birkhäuser Boston, Cambridge, MA

DESCENT ON SIMPLY CONNECTED SURFACES OVER ALGEBRAIC NUMBER FIELDS

Wayne Raskind
Department of Mathematics, University of Southern California, Los Angeles, CA 90089-1113, USA • *E-mail :* raskind@usc.edu

Victor Scharaschkin
Department of Mathematics, The University of Queensland, Brisbane, Qld 4072, Australia • *E-mail :* victors@maths.uq.edu.au

Abstract. We study descent and obstructions to the Hasse principle on simply connected surfaces over number fields.

1. Introduction

Let X be a smooth projective surface over an algebraic number field k, and denote by $X(k)$ its set of k-rational points. Let $\overline{k}$ be an algebraic closure of k and $\overline{X} = X \times_k \overline{k}$. If $\mathrm{Pic}(\overline{X})$ is finitely generated and torsion free, Colliot-Thélène and Sansuc **[CTS80]**, **[CTS87]** have established a theory of descent, which associates to X a finite set of auxiliary objects, *universal torsors.* These are simpler arithmetically than X, in that there is no Brauer–Manin obstruction to the Hasse principle for the algebraic part of the Brauer group of a smooth compactification of such a universal torsor (see below for more details

Key words and phrases. Descent, torsors, obstructions.

The first author was partially supported by NSF grant 0070850, SFB 476 (Münster), CNRS France, and sabbatical leave from the University of Southern California. The second author was partially supported by NSF-grant 0070850.

and definitions). The images of the rational points of the universal torsors via structure morphisms to X give a finite partition of $X(k)$. This works well in a large number of cases for surfaces of geometric genus zero. For example, if X is a surface that becomes birational to the projective plane over $\overline{k}$, the theory has led to a much greater understanding of the arithmetic of X, and in some examples, an almost complete solution to most diophantine problems (see e.g. [**CTSSD87a**], [**CTSSD87b**] and [**Sko01**], §7).

On the other hand, we have little general understanding of the arithmetic of surfaces of positive geometric genus. In particular, we do not know very well the role played by the "transcendental" part of the Brauer group in the existence and distribution of rational points. Is the Brauer–Manin obstruction to the Hasse principle the only one for any geometrically simply connected surface? If a K3 surface X has a k-rational point, does it have infinitely many, and is $X(k)$ Zariski dense in X? For a surface X of general type, are almost all of the rational points in the images of the k-rational points of finitely many rational maps to X from abelian and rational varieties? If U denotes the complement of the images of these maps, is $U(L)$ finite for all number fields L (Bombieri–Lang conjecture)? See e.g., ([**HS00**], Part F.5, Conjectures F.5.2.1 and F.5.2.2) for more precise statements of the conjectures. For K3 surfaces, see [**BM90**] for conjectures on the growth of rational points of bounded height, and [**BT99**], [**BT00**] for some results on the density of rational points.

The purpose of this paper is to establish a formal framework that we hope will lead to a better understanding of the arithmetic of a (geometrically) simply connected surface X of nonzero geometric genus. Our theory associates to X a conjecturally finite set of auxiliary objects that are not schemes, in general. Rather, they are *gerbes* which are bound by the second étale cohomology group of $\overline{X}$ with $\mathbf{Z}/n\mathbf{Z}(2)$-coefficients. The images of the rational points of these objects via the structure morphisms to X partition $X(k)$, and we expect, but cannot show at the moment, that they are "simpler" than X, in terms of the Brauer–Manin obstruction to the Hasse principle. We have also not made much progress in describing these gerbes in terms of explicit "equations," so that they may be computationally useful. Nonetheless, we hope that this note might inspire others to go further, and we believe that K3 surfaces of geometric Picard number 20 provide a promising class of varieties for computation. We develop the theory in some detail in this case below. For surfaces of general type, we hope that our theory might eventually yield some insight into the Bombieri–Lang conjecture, but this is probably a long way off.

The exposition here is a bit uneven, in that some proofs are given in quite a bit of detail, while others are only sketched or omitted. Complete details as well as more examples will be given elsewhere, but we hope what is written here will give the reader a good idea of how the theory should work.

The first author would like to thank J.-L. Colliot-Thélène for introducing him to the theory of descent on rational surfaces some years ago, and for several helpful discussions about this work. Since that time, there have been major advances in the theory of algebraic cycles (especially torsion 0-cycles on surfaces of nonzero geometric genus), and this paper may be regarded as an attempt to apply these to make a more general theory. The idea that gerbes could be useful in this theory and are reasonable algebro-geometric objects was inspired by the work of Harari and Skorobogatov (see **[HS02]** and **[Sko01]**). Harari **[Har96]** was also the first person to consider obstructions to the Hasse principle for the transcendental part of the Brauer group. We thank E. Peyre and N. Yui for helpful discussions. This paper was completed while the first author enjoyed the hospitality of Université de Paris-Sud, Institut Fourier (Grenoble) and Université Denis Diderot (Paris 7). The audience in the first author's course on this material in Grenoble listened patiently as he explained some rather raw material. Finally, we thank B. Poonen and Y. Tschinkel for organizing and giving us the opportunity to present this work at this excellent conference, and the American Institute of Mathematics for providing a very stimulating atmosphere.

2. Notation and preliminaries

Let k be a field and X a smooth, projective, geometrically connected variety over k. We denote by $\overline{k}$ a separable closure of k, by G the absolute Galois group of k and $\overline{X} = X \times_k \overline{k}$. We will say that X is *geometrically simply connected* if $\pi_1^{alg}(\overline{X}) = \{1\}$. When we assume this condition, we really only need that the abelianized fundamental group of $\overline{X}$ is trivial. Let ℓ be a prime number different from the characteristic of k. We denote by $\mathbf{Z}/\ell^m\mathbf{Z}(r)$ the étale sheaf $\mathbf{Z}/\ell^m\mathbf{Z}$ Tate-twisted r-times. The notation $\mathrm{Br}(X)$ will be used for the cohomological Brauer group, $H^2_{et}(X, \mathbf{G}_m)$. The Hochschild–Serre spectral sequence:

$$E_2^{r,s} = H^r(k, H^s(\overline{X}, \mathbf{G}_m)) \Longrightarrow H^{r+s}(X, \mathbf{G}_m)$$

gives an exact sequence:

$$\mathrm{Br}(k) \to \ker[\mathrm{Br}(X) \to \mathrm{Br}(\overline{X})] \xrightarrow{f} H^1(k, \mathrm{Pic}(\overline{X})) \to H^3(k, \mathbf{G}_m). \tag{1}$$

The group $\ker[\mathrm{Br}(X) \to \mathrm{Br}(\overline{X})]$ is called the *algebraic part* of the Brauer group. If k is a number field, then $H^3(k, \mathbf{G}_m) = 0$, and so the map f is surjective. The image of $\mathrm{Br}(X)$ in $\mathrm{Br}(\overline{X})$ will be called the *transcendental part* of the Brauer group. Of course, this is a quotient of $\mathrm{Br}(X)$.

For A an abelian group and ℓ a prime number, we denote by $A\{\ell\}$ the ℓ-primary part of A and $A[\ell^n]$ the subgroup of A of elements killed by ℓ^n. *The ℓ-Tate module* of A is $\varprojlim_n A[\ell^n]$ and will be denoted $T_\ell(A)$. If A is either a finitely generated $\mathbf{Z}_\ell$-module or an ℓ-primary torsion module of finite cotype, we denote by A^* the Pontryagin dual of A.

If X is smooth and projective over an algebraically closed field, we denote by $\mathrm{NS}(X)$ the group of divisors modulo algebraic equivalence (the Néron–Severi group). This is a finitely generated abelian group. If X is simply connected, then this group is equal to the Picard group, $\mathrm{Pic}(X)$. By the *geometric Picard number* of X over an arbitrary field k, we shall mean the rank of the Néron–Severi group of $\overline{X}$. By the *geometric genus* of a smooth projective surface X over k, we mean the dimension of the k-vector space of global holomorphic 2-forms on X.

If k is an algebraic number field of finite degree over $\mathbf{Q}$, we denote places of k by v, the completion of k at such a place by k_v and $\mathbf{A}_k$ the ring of adeles of k. If S is a finite set of places of k, we write G_{S} for the Galois group of a maximal extension of k that is unramified outside S and put $\prod_{\mathsf{S}} := \prod_{v \in \mathsf{S}}$. If X is a variety over k and we are speaking of the ℓ-adic étale cohomology of $\overline{X}$, then S will always contain the archimedean places, the places above ℓ and the places of bad reduction of X.

We shall denote by $CH^i(X)$ the group of codimension i-cycles modulo rational equivalence; when $i = \dim(X)$, we denote this group by $CH_0(X)$ and by $A_0(X)$ the subgroup of zero-cycles of degree zero.

It will be helpful for the reader to have some familiarity with the theory of descent of Colliot-Thélène–Sansuc (see **[CTS80]**, **[CTS87]** and **[Sko01]**), but in the next section, we will quickly review their theory.

3. Rapid review of the theory of Colliot-Thélène–Sansuc

The basic references for this section are **[CTS80]**, **[CTS87]** and **[Sko01]**. Let X a smooth projective surface such that $\mathrm{Pic}(\overline{X})$ is torsion free, let S be the algebraic torus whose group of characters is $\mathrm{Pic}(\overline{X})$, and put $S_X = S \times_k X$. Recall the fundamental exact sequence of descent (see Théorème 1.5.1 of **[CTS87]**):

$$0 \to H^1(k,S) \to H^1(X,S) \xrightarrow{\chi} H^1(\overline{X},S)^G \xrightarrow{\partial} H^2(k,S) \to H^2(X,S) \to \cdots .$$

As shown by Harari–Skorobogatov (**[HS02]**, Proposition in Appendix B and **[Sko01]**, Proposition 2.3.1), this can be obtained from the exact sequence of terms of low degree for the Hochschild–Serre spectral sequence:

$$H^r(k, H^s(\overline{X},S)) \Longrightarrow H^{r+s}(X,S).$$

We have that $H^1(\overline{X},S)^G \cong \mathrm{End}_G(\mathrm{Pic}(\overline{X}))$. A *universal torsor* is an element of $H^1(X,S)$ whose image via χ is the identity map of $\mathrm{End}_G(\mathrm{Pic}(\overline{X}))$. The image of the identity map in $H^2(k,S)$ via ∂ in the exact sequence above is called the *elementary obstruction*. It is proved in (**[CTS87]**, Théorème 2.1.2(a)) that the Picard group of a smooth compactification $\overline{\mathcal{T}^c}$ of a universal torsor is a permutation module. Hence, by the exact sequence (1) of Section 2, the algebraic part of the Brauer group is reduced to the image of the Brauer group of k. Thus there is no Brauer–Manin obstruction to the Hasse principle for the algebraic part of the Brauer group on $\overline{\mathcal{T}^c}$. Colliot-Thélène–Sansuc also derive explicit equations for the restriction of the universal torsors to suitable open subsets of X (**[CTS87]**, Théorème 2.3.1 and **[Sko01]**, §4.2, p. 71 and §4.3), and they interpret the Brauer Manin obstruction to the Hasse principle in terms of the universal torsors (**[CTS87]**, Théorème 3.5.1; see also **[Sko01]**, §6.1).

If the geometric genus of X is zero, and X is geometrically simply connected, then we have that

$$\mathrm{Pic}(\overline{X})/n \cong H^2(\overline{X}, \mathbf{Z}/n\mathbf{Z}(1)) \quad \text{and} \quad S[n] = H^2(\overline{X}, \mathbf{Z}/n\mathbf{Z}(2)).$$

We in effect take $S[n]$ instead of S as the point of departure for our theory for surfaces of nonzero geometric genus, because in this case there is no group-scheme S such that $S[n] = H^2(\overline{X}, \mathbf{Z}/n\mathbf{Z}(2))$ for all n. *This is the fundamental reason why gerbes are required instead of torsors.*

It is somewhat tedious but straightforward to verify that we can recover the theory of **[CTS80]**, **[CTS87]** for surfaces of geometric genus zero using our theory below, although for the applications in those papers our theory is much clumsier.

4. The homological algebra of descent

In this section, we outline the homological algebra needed to set up the general theory of descent. Let

$$p: X \to \operatorname{Spec} k$$

be the structure morphism. If Y is another geometrically integral k-variety, let:

$$q: X \times_k Y \to X$$

be projection onto the first factor. Consider the Leray spectral sequence:

$$E_2^{r,s} = H^r(X, R^s q_* \mathbf{Z}/n\mathbf{Z}(2)) \Longrightarrow H^{r+s}(X \times_k Y, \mathbf{Z}/n\mathbf{Z}(2)).$$

Assume now that Y is proper over k. Then by the proper base change theorem, the stalk of $R^s q_* \mathbf{Z}/n\mathbf{Z}(2)$ at a geometric point $\overline{x} \in X$ is equal to the étale cohomology group

$$H^s(Y \times_X \overline{x}, \mathbf{Z}/n\mathbf{Z}(2)).$$

By the smooth and proper base change theorems, this sheaf is locally constant, represented by $H^s(\overline{Y}, \mathbf{Z}/n\mathbf{Z}(2))$.

Let

$$H^4(X \times_k Y, \mathbf{Z}/n\mathbf{Z}(2))^0 = \ker[H^4(X \times_k Y, \mathbf{Z}/n\mathbf{Z}(2)) \to H^0(X, R^4 q_* \mathbf{Z}/n\mathbf{Z}(2))].$$

Now assume that Y is a smooth, proper, geometrically connected and simply connected surface over k. Then from the spectral sequence, we get a map:

$$H^4(X \times_k Y, \mathbf{Z}/n\mathbf{Z}(2))^0 \to H^2(X, R^2 q_* \mathbf{Z}/n\mathbf{Z}(2)).$$

Consider the cycle map:

$$CH^2(X \times_k Y)/n \to H^4(X \times_k Y, \mathbf{Z}/n\mathbf{Z}(2)).$$

Let $CH^2(X \times_k Y)^0$ denote the inverse image of $H^4(X \times_k Y, \mathbf{Z}/n\mathbf{Z}(2))^0$. We denote by $c_{2,2}$ the composite map

$$c_{2,2}: CH^2(X \times_k Y)^0/n \to H^4(X \times_k Y, \mathbf{Z}/n\mathbf{Z}(2))^0 \to H^2(X, R^2 q_* \mathbf{Z}/n\mathbf{Z}(2))$$

Now take $Y = X$ in this discussion. Then from the Künneth formula (see e.g., **[Mil80]**, Ch. VI, Lemma 8.7), we get:

Lemma 4.1.

$$H^2(\overline{X}, R^2 q_* \mathbf{Z}/n\mathbf{Z}(2)) \cong \operatorname{End}(H^2(\overline{X}, \mathbf{Z}/n\mathbf{Z}(1))).$$

Definition 4.2. A universal n-gerbe is an element of $H^2(X, R^2q_*\mathbf{Z}/n\mathbf{Z}(2))$, which is in the subgroup generated by the image of $c_{2,2}$ and the image of $H^2(k, R^2q_*\mathbf{Z}/n\mathbf{Z}(2))$, and whose image in $H^2(\overline{X}, R^2q_*\mathbf{Z}/n\mathbf{Z}(2))$ via the natural map is the identity endomorphism. A universal ℓ-adic gerbe is an inverse system of universal ℓ^m-gerbes $\mathscr{G}_{\ell^m}$.

For some basic notions about gerbes, see the Appendix to this paper, and for more details, see ([**Mil80**], Ch. IV, §2 and [**LMB00**], §3). Note that since X and Y are geometrically simply connected, the kernel of the map:

$$H^2(X, R^2q_*\mathbf{Z}/n\mathbf{Z}(2)) \to H^2(\overline{X}, R^2q_*\mathbf{Z}/n\mathbf{Z}(2)),$$

is the image of $H^2(k, H^2(\overline{X}, \mathbf{Z}/n\mathbf{Z}(2)))$. Thus the set of universal n-gerbes is either empty or a principal homogeneous space under the image of

$$H^2(k, H^2(\overline{X}, \mathbf{Z}/n\mathbf{Z}(2))) \quad \text{in} \quad H^2(X, R^2q_*\mathbf{Z}/n\mathbf{Z}(2)).$$

Let $\operatorname{Ger}(X, k, n)$ be the set of universal n-gerbes. If this is nonempty, then we have a pairing:

$$X(k) \times \operatorname{Ger}(X, k, n) \to H^2(k, H^2(\overline{X}, \mathbf{Z}/n\mathbf{Z}(2)))$$
$$(P, \mathscr{G}) \mapsto \mathscr{G}_P,$$

where $\mathscr{G}_P$ denotes the pullback of $\mathscr{G}$ via the inclusion $P \to X$. We have that $\mathscr{G}_P = 0$ if and only if $P \in p_{\mathscr{G}}\mathscr{G}(k)$, where $p_{\mathscr{G}} : \mathscr{G} \to X$ is the structure morphism. Assume there is a universal n-gerbe, $\mathscr{G}$. For $\alpha \in H^2(k, H^2(\overline{X}, \mathbf{Z}/n\mathbf{Z}(2)))$, let $\mathscr{G}_\alpha$ be the universal n-gerbe whose class in $H^2(X, R^2q_*\mathbf{Z}/n\mathbf{Z}(2))$ is given by $[\mathscr{G}] - \alpha$. Then we have

$$X(k) = \bigsqcup_{\alpha \in H^2(k, H^2(\overline{X}, \mathbf{Z}/n\mathbf{Z}(2)))} p_{\mathscr{G}_\alpha}\mathscr{G}_\alpha(k).$$

If k is a number field, then this disjoint union should be finite and bounded independently of n, and we will show this below for K3 surfaces of geometric Picard number 20 (see Theorem 7.3).

5. The elementary obstruction

An obstruction to X having a rational point is that it not have a universal n-gerbe for all n. If X has a k-rational point, then the class of the diagonal in $CH^2(X \times_k X)$ can be modified by subtracting the class of $X \times P$, where $P \in X(k)$, to give a class in $H^2(X, R^2q_*\mathbf{Z}/n\mathbf{Z}(2))$ that gives the identity map in

$$H^2(\overline{X}, R^2q_*\mathbf{Z}/n\mathbf{Z}(2)) \cong \operatorname{End}(H^2(\overline{X}, \mathbf{Z}/n\mathbf{Z}(1))).$$

The same can be done if X has a zero-cycle of degree one. But it is not true, in general, that X has a universal n-gerbe, and we call this *the elementary obstruction.*

Lemma 5.1. *Every section*

$$j : \mathrm{Br}(X)[n] \to \mathrm{Br}(k)[n]$$

to the natural map:

$$\mathrm{Br}(k)[n] \to \mathrm{Br}(X)[n]$$

determines a unique universal n-gerbe $\mathscr{G}$ with the property that the image of $\mathscr{G}$ under the map

$$H^2(X, R^2q_*\mathbf{Z}/n\mathbf{Z}(2)) \to H^2(k, R^2q_*\mathbf{Z}/n\mathbf{Z}(2))$$

is trivial. Conversely, a universal n-gerbe determines such a section.

Unfortunately, we have not been able as of yet to compute the elementary obstruction very effectively. We believe strongly that it should be described by the G-extension class of the exact sequence obtained from the Gersten–Quillen complex:

$$H^1(\overline{X}, \mathscr{K}_2) \hookrightarrow \bigoplus_{x\in\overline{X}^1} \overline{k}(x)^*/[K_2\overline{k}(X)/H^0(\overline{X}, \mathscr{K}_2)] \to \bigoplus_{x\in\overline{X}^2} \mathbf{Z} \to CH^2(\overline{X}) \to 0.$$

While it is expected that for k a number field and X geometrically simply connected over k, $CH^2(\overline{X}) = \mathbf{Z}$, this is not known for one single surface of positive geometric genus. Nonetheless, since $A_0(\overline{X})$ is uniquely divisible in this case (Rojtman theorem [**Roj80**]), we have

$$\mathrm{Ext}^2_G(CH^2(\overline{X}), H^1(\overline{X}, \mathscr{K}_2)) \cong \mathrm{Ext}^2_G(\mathbf{Z}, H^1(\overline{X}, \mathscr{K}_2)),$$

and so the elementary obstruction should be an element of $H^2(k, H^1(\overline{X}, \mathscr{K}_2))$, which can be regarded as an element of $H^2(k, H^2(\overline{X}, \mathbf{Q}/\mathbf{Z}(2))$, since

$$H^1(\overline{X}, \mathscr{K}_2)_{tors} \cong H^2(\overline{X}, \mathbf{Q}/\mathbf{Z}(2)),$$

and the torsion free quotient of $H^1(\overline{X}, \mathscr{K}_2)$ is uniquely divisible (see [**CTR85**], Theorems 2.1 and 2.2 for basic results about the structure of $\mathscr{K}_2$-cohomology). Note that when X has geometric genus zero and is geometrically connected, it follows from ([**CTR85**], Theorem 2.12) that the map

$$\mathrm{NS}(\overline{X}) \otimes_{\mathbf{Z}} \overline{k}^* \to H^1(\overline{X}, \mathscr{K}_2)$$

is injective with uniquely divisible cokernel, and $S(k) \cong \mathrm{NS}(\overline{X}) \otimes \overline{k}^*$. Thus

$$H^2(k, S) \cong H^2(k, H^1(\overline{X}, \mathscr{K}_2)),$$

and our proposal for the elementary obstruction would generalize that of Colliot-Thélène and Sansuc.

Now suppose the elementary obstruction is trivial everywhere locally. Then a diagram chase, using the fact that the map

$$H^2(k, \mathrm{Br}(\overline{X})(1)) \to \bigoplus_v H^2(k_v, \mathrm{Br}(\overline{X})(1))$$

is injective ([**Jan89**], Theorem 3d)), would then show:

Proposition 5.2. *Assume that the elementary obstruction lies in*

$$H^2(k, H^2(\overline{X}, \mathbf{Q}/\mathbf{Z}(2))),$$

as explained above, and that it vanishes on X_v for all v. Then the elementary obstruction for X lies in the group:

$$\mathrm{Im}\, f \cap \mathrm{Im}\, \partial,$$

where

$$H^2(k, \mathrm{Pic}(\overline{X}) \otimes \mathbf{Q}/\mathbf{Z}(1)) \xrightarrow{f} \bigoplus_v H^2(k_v, \mathrm{Pic}(\overline{X}) \otimes \mathbf{Q}/\mathbf{Z}(1))$$

is the localization map, and

$$\bigoplus_v H^1(k_v, \mathrm{Br}(\overline{X})(1)) \xrightarrow{\partial} \bigoplus_v H^2(k_v, \mathrm{Pic}(\overline{X}) \otimes \mathbf{Q}/\mathbf{Z}(1))$$

is the boundary map in the Galois cohomology sequence for the exact sequence of $Gal(\overline{k}_v/k_v)$-modules:

$$(**)\ 0 \to \mathrm{Pic}(\overline{X}) \otimes \mathbf{Q}/\mathbf{Z}(1)) \to H^2(\overline{X}, \mathbf{Q}/\mathbf{Z}(2)) \to \mathrm{Br}(\overline{X})(1) \to 0.$$

Remark 5.3. It should be possible to reformulate the theory using the group $H^1_{et}(X, R^1q_*\mathcal{K}_2)$. We should have an injection:

$$H^1_{et}(X, R^1q_*\mathcal{K}_2)/n \to H^2(X, R^2q_*\mathbf{Z}/n\mathbf{Z}(2)),$$

but we have been unable to prove this.

6. The Tate conjecture and the Brauer group

Let X be smooth and projective over a field k that is finitely generated over its prime subfield. We take a finitely generated $\mathbf{Z}$-algebra A with fraction field k and a smooth proper model of X over A. Denote by (T_ℓ) the statement that the divisor class map

$$\mathrm{Pic}(X)\otimes_{\mathbf{Z}}\mathbf{Q}_\ell \to H^2(\overline{X},\mathbf{Q}_\ell(1))^G$$

is surjective (the Tate conjecture for divisors). Note that this has been proved for abelian varieties by Faltings [**Fal83**], and together with the Kuga–Satake construction ([**KS67**]; see also [**Del72**]), this implies that (T_ℓ) is true for any ℓ for $K3$ surfaces over a field of characteristic 0. Note also that the statement (T_ℓ) is trivial for varieties X for which $\mathrm{Br}(\overline{X})$ is finite, which is the case for surfaces of geometric genus zero. This is why (T_ℓ) does not figure into the theory of Colliot-Thélène–Sansuc.

Proposition 6.1. *Let X be a geometrically simply connected surface over k as above, and assume statement (T_ℓ) is true. Then $\mathrm{Br}(X)\{\ell\}/\mathrm{Br}(k)\{\ell\}$ is a finite group. If (T_ℓ) is true for every ℓ and there exists a smooth specialization Y of X modulo a maximal ideal of A with the same geometric Picard number as X and for which (T_ℓ) is true for some ℓ for Y, then the whole group $\mathrm{Br}(X)/\mathrm{Br}(k)$ is finite.*

Sketch of Proof. This follows easily from the following facts:

(i) The simple connectivity of X and the Hochschild–Serre spectral sequence give an isomorphism

$$H^2(X,\mathbf{Q}_\ell/\mathbf{Z}_\ell(1))/H^2(k,\mathbf{Q}_\ell/\mathbf{Z}_\ell(1)) \cong H^2(\overline{X},\mathbf{Q}_\ell/\mathbf{Z}_\ell(1))^G.$$

(ii) If (T_ℓ) is true, then the group $\mathrm{Br}(X)\{\ell\}/\mathrm{Br}(k)\{\ell\}$ is the quotient of $H^2(\overline{X},\mathbf{Q}_\ell/\mathbf{Z}_\ell(1))^G$ by its maximal ℓ-divisible subgroup, which is given by the image of $\mathrm{NS}(X)\otimes\mathbf{Q}_\ell/\mathbf{Z}_\ell$.

(iii) The cohomology sequence of the Kummer sequence for the étale topology identifies the quotient

$$[H^2(X,\mathbf{Q}_\ell/\mathbf{Z}_\ell(1))/H^2(k,\mathbf{Q}_\ell/\mathbf{Z}_\ell(1))]/\mathrm{NS}(X)\otimes\mathbf{Q}_\ell/\mathbf{Z}_\ell$$

with $\mathrm{Br}(X)\{\ell\}/\mathrm{Br}(k)\{\ell\}$.

(iv) If (T_ℓ) is true for all ℓ and there is a smooth specialization Y of X modulo a maximal ideal v of A with the same Picard number for which (T_ℓ) is true, then Tate has shown ([**Tat95**], Theorem 5.2) that (T_ℓ) is true for every $\ell \neq \mathrm{char}\, F_v$, and the whole prime-to-char F_v-part of the Brauer group of Y is finite. Here F_v is the (finite) residue field of v. Comparing the fixed modules of $\mathrm{Br}(\overline{X})^G$ with $\mathrm{Br}(\overline{Y})^{Gal(\overline{F}_v/F_v)}$ using the Kummer sequence and the smooth and proper base change theorems and the fact that the (co)-specialization map:

$$\operatorname{Br}(\overline{X}) \to \operatorname{Br}(\overline{Y})$$

is an isogeny on prime to char F_v-parts because the Picard numbers are the same, we get the finiteness of the prime to $p = \operatorname{char} F_v$-part of $\operatorname{Br}(X)/\operatorname{Br}(k)$. Since (T_p) is true, we get finiteness of $\operatorname{Br}(X)\{p\}/\operatorname{Br}(k)\{p\}$. This completes the proof of the proposition.

□

***Proposition 6.2* (Shioda–Inose [SI77]).** *Let X be a K3 surface over* $\mathbf{C}$ *of Picard number 20. Then X is defined over a number field, and there exist isogenous elliptic curves with complex multiplication E, E', such that X may be realized as a double cover of the Kummer surface associated to the abelian surface $E \times E'$.*

Theorem 6.3. *Let X be a K3 surface of geometric Picard number 20 over a field k that is finitely generated over the prime subfield. Then the group* $\operatorname{Br}(X)/\operatorname{Br}(k)$ *is finite.*

Proof. By the Shioda–Inose theorem, there are isogenous CM elliptic curves E and E' such that X is a double cover of the Kummer surface Y of $E \times E'$. Since the Picard numbers and second Betti numbers of Y and X are the same, the natural map:

$$\operatorname{Br}(X) \to \operatorname{Br}(Y)$$

is an isogeny, with kernel killed by 2. Since (T_2) is true for X, the group $\operatorname{Br}(X)\{2\}/\operatorname{Br}(k)\{2\}$ is finite, and it will then suffice to prove that the prime-to-2 part of $\operatorname{Br}(Y)/\operatorname{Br}(k)$ is finite. Let K be the CM field of E and let $\wp$ be a prime ideal of k with residue field $\mathbf{F}_p$, for p an odd prime number. Then E and E' have good ordinary reductions $E_\wp$ and $E'_\wp$ modulo $\wp$. The geometric Picard number of the Kummer surface $Y_\wp$ associated to $E_\wp \times E'_\wp$ is 20, and since this is ordinary, the Tate conjecture is known (see **[Nyg83]** or **[NO85]**). By Proposition 1, we then get finiteness of $\operatorname{Br}(Y)/\operatorname{Br}(k)$. This completes the proof. □

Remark 6.4. Using results of Morrison **[Mor84]**, some of the results of this section can be extended to some K3 surfaces of geometric Picard number 19.

7. Universal gerbes and the higher Abel–Jacobi mapping

Assume that X has a universal n-gerbe, $\mathscr{G}$. Then we can define a map

$$\theta_{\mathscr{G}} : CH_0(X) \to H^2(k, H^2(\overline{X}, \mathbf{Z}/n\mathbf{Z}(2)))$$

by sending a rational point $P \in X(L)$ to $cor_{L/k}\mathscr{G}_P$, and extending by linearity. Here

$$cor_{L/k} : H^2(L, H^2(\overline{X}, \mathbf{Z}/n\mathbf{Z}(2))) \to H^2(k, H^2(\overline{X}, \mathbf{Z}/n\mathbf{Z}(2)))$$

is the corestriction map. If we restrict this map to $A_0(X)$, then it is independent of the choice of $\mathscr{G}$, since if we chose another one, say $\mathscr{G} - \alpha$, the α will cancel out after taking the difference between two cycles of the same degree.

Recall the higher Abel–Jacobi mapping [**Ras95**]: the Hochschild–Serre spectral sequence

$$H^r(k, H^s(\overline{X}, \mathbf{Z}/n\mathbf{Z}(2))) \Longrightarrow H^{r+s}(X, \mathbf{Z}/n\mathbf{Z}(2))$$

and the geometric simple connectivity of X allow us to define a map

$$d_{2,n} : A_0(X) \to H^2(k, H^2(\overline{X}, \mathbf{Z}/n\mathbf{Z}(2))).$$

The proof of the following proposition is hypertechnical, and we will give it elsewhere. It uses the explicit description of $d_{2,n}$ in terms of certain 2-extensions obtained from the long exact sequence of cohomology of support in a codimension 2 cycle (see [**Jan00**]).

Proposition 7.1. *If X is geometrically simply connected, the map $\theta_{\mathscr{G}}$, restricted to $A_0(X)$, is the same (up to sign) as the higher Abel–Jacobi mapping $d_{2,n}$.*

Now take $n = \ell^m$, a power of a prime number ℓ. We can then consider the higher Abel–Jacobi mapping in continuous ℓ-adic étale cohomology (see [**Ras95**] for this mapping with $\mathbf{Q}_\ell$-coefficients; the geometric simple connectivity of X allows us to define it with $\mathbf{Z}_\ell$-coefficients using the same arguments as used above for $\mathbf{Z}/n\mathbf{Z}$-coefficients):

$$d_{2,\ell} : A_0(X) \to H^2(k, H^2(\overline{X}, \mathbf{Z}_\ell(2))).$$

Proposition 7.2. *The image of $d_{2,\ell}$ is a finitely generated $\mathbf{Z}_\ell$-module (which is conjecturally torsion).*

Proof. This follows easily from the fact that for a suitable finite set of places S of k, we can factor $d_{2,\ell}$ through $H^2(G_S, H^2(\overline{X}, \mathbf{Z}_\ell(2)))$ (see Section 2 for notation), and this group is a finitely generated $\mathbf{Z}_\ell$-module. □

Theorem 7.3. *If X is a K3 surface of geometric Picard number 20 over $\mathbf{Q}$ or K (the CM-field of the elliptic curves in the Shioda–Inose theorem), then the image of the higher ℓ-adic Abel–Jacobi mapping $d_{2,\ell}$ is finite for all ℓ and zero for almost all ℓ.*

Sketch of Proof. As in the proof of the last result, we are reduced to the case of a Kummer surface X associated to the product A of two isogenous CM elliptic curves. Now the map

$$\mathrm{Br}(\overline{X}) \to \mathrm{Br}(\overline{A})$$

is an isogeny, with kernel killed by 2, as one can see by comparing the Picard numbers (4 for A, 20 for X) with the second Betti numbers (6 for A, 22 for X). Thus it suffices to prove the result for A. In this case, the statement follows from work of Wiles **[Wil95]** and Dee **[Dee99]** on the finiteness of the Selmer group of the symmetric square of a CM elliptic curve. This completes the proof of the theorem. □

8. Geometric interpretation of the Brauer–Manin obstruction

Let X be a geometrically integral variety over a number field k, and assume that for every place v of k, X has rational points in k_v. If $\mathscr{A} \in \mathrm{Br}(X)$ and $P_v \in X(k_v)$, let $\mathscr{A}_{P_v}$ denote the pullback of $\mathscr{A}$ via the morphism

$$P_v : \operatorname{Spec} k_v \to X$$

determined by the point P_v. Let $inv_v : \mathrm{Br}(k_v) \to \mathbf{Q}/\mathbf{Z}$ denote the homomorphism giving the invariant of a central simple algebra, which is an isomorphism for v non-archimedean and an injection onto the subgroup of elements of order 2 in $\mathbf{Q}/\mathbf{Z}$ for v real. Let T be a subgroup of $\mathrm{Br}(X)$ containing $\mathrm{Br}(k)$, and let

$$X(\mathbf{A}_k)^T = \{(P_v) \in \prod_v X(k_v) : \forall \mathscr{A} \in T, \sum_v inv_v \mathscr{A}_{P_v} = 0\}.$$

The Brauer–Hasse–Noether theorem implies that

$$X(k) \subseteq X(\mathbf{A}_k)^T.$$

If $T = \mathrm{Br}(X)$, we will denote this set by $X(\mathbf{A}_k)^{\mathrm{Br}}$.

We will say that X *is a counterexample to the Hasse principle* if $X(\mathbf{A}_k) \neq \emptyset$, but $X(k) = \emptyset$; that *there is no Brauer Manin obstruction for T to the Hasse principle for X* if $X(\mathbf{A}_k)^T \neq \emptyset$; and that X *is a counterexample to the Hasse principle explained by the Brauer–Manin obstruction* if $X(\mathbf{A}_k)^{\mathrm{Br}} = \emptyset$.

Theorem 8.1. *Let X be a smooth projective geometrically simply connected surface over an algebraic number field k. Let ℓ be a prime number, and assume*

(i) $X(\mathbf{A}_k) \neq \emptyset$;
(ii) *there is no elementary obstruction to the existence of an ℓ-adic gerbe;*
(iii) *the Tate conjecture for divisors (T_ℓ) is true (see Section 6).*

Then there is no Brauer–Manin obstruction for $\mathrm{Br}(X)\{\ell\}$ *if and only if there exists a universal ℓ-adic gerbe with rational points in every k_v.*

Proof. This proof is similar in outline to the one in ([**CTS87**], Théorème 3.5.1), except we replace Tate–Nakayama duality with Poitou–Tate duality. By Poincaré duality, we have a nondegenerate pairing:

$$H^2(\overline{X}, \mathbf{Z}_\ell(2)) \times H^2(\overline{X}, \mathbf{Q}_\ell/\mathbf{Z}_\ell(1)) \to H^4(\overline{X}, \mathbf{Q}_\ell/\mathbf{Z}_\ell(3)) \cong \mathbf{Q}_\ell/\mathbf{Z}_\ell(1).$$

By Lemma 5.1, the vanishing of the elementary obstruction gives us a section of the natural map $\mathrm{Br}(k) \to \mathrm{Br}(X)$ and a universal gerbe, $\mathscr{G}$, that is trivial on that section.

Let S be a finite set of places of k including the archimedean places, the places above ℓ and the bad reduction places of X. By the Poitou–Tate global duality theorem, we have an exact sequence

$$H^2(G_{\mathsf{S}}, H^2(\mathbf{Z}_\ell(2))) \to \prod_{v \in \mathsf{S}} H^2(G_v, H^2(\mathbf{Z}_\ell(2))) \stackrel{\rho}{\twoheadrightarrow} H^0(G_{\mathsf{S}}, H^2(\mathbf{Q}_\ell/\mathbf{Z}_\ell(1)))^*,$$

where $H^2(-) := H^2(\overline{X}, -)$. The map ρ is derived from the perfect pairings

$$H^2(k_v, H^2(\overline{X}, \mathbf{Z}_\ell(2))) \times H^0(k_v, H^2(\overline{X}, \mathbf{Q}_\ell/\mathbf{Z}_\ell(1))) \to H^2(k_v, \mathbf{Q}_\ell/\mathbf{Z}_\ell(1))$$

by taking the sum of the invariants. Consider the following diagram

$$\begin{array}{ccccc} \prod_{\mathsf{S}} X(k_v) & \times & \prod_{\mathsf{S}} \mathrm{Br}(X_v) & \to & \mathrm{Br}(k_v) \\ \downarrow & & \uparrow & & \\ \prod_{\mathsf{S}} H^2(k_v, H^2(\overline{X}, \mathbf{Z}_\ell(2))) & \times & \prod_{\mathsf{S}} H^0(k_v, H^2(\overline{X}, \mathbf{Q}_\ell/\mathbf{Z}_\ell(1))) & \to & \mathrm{Br}(k_v). \end{array}$$

The map on the left is given by $\theta_{\mathscr{G}_v}$ (see Section 7), where $\mathscr{G}$ was fixed at the beginning of the proof here. The map on the right is the surjection

$$\begin{array}{ccc} H^2(\overline{X}, \mathbf{Q}_\ell/\mathbf{Z}_\ell(1))^{G_v} & \stackrel{\sim}{\leftarrow} & H^2(X_v, \mathbf{Q}_\ell/\mathbf{Z}_\ell(1))/H^2(k_v, \mathbf{Q}_\ell/\mathbf{Z}_\ell(1)) \\ & & \downarrow \\ & & \mathrm{Br}(X_v)\{\ell\}/\,\mathrm{Br}(k_v)\{\ell\} \end{array}$$

explained in the proof of Proposition 5.2, followed by the map

$$\mathrm{Br}(X_v)\{\ell\}/\,\mathrm{Br}(k_v)\{\ell\} \to \mathrm{Br}(X_v)\{\ell\}$$

that we get from the universal ℓ-adic gerbe, $\mathscr{G}_v$ and Lemma 5.1. This diagram is commutative.

Now suppose there is no Brauer–Manin obstruction for $\mathrm{Br}(X)\{\ell\}$, and let (P_v) be a family of points such that

$$\sum_v inv_v \mathscr{A}_{P_v} = 0$$

for all $\mathscr{A} \in \mathrm{Br}(X)\{\ell\}$. Since $\mathrm{Br}(X)\{\ell\}/\mathrm{Br}(k)\{\ell\}$ is finite, by enlarging S, if necessary, we may assume that there is a smooth proper model $\mathscr{X}_\mathsf{S}$ of X over the ring of S-integers of k and a surjection $\mathrm{Br}(\mathscr{X}_\mathsf{S})\{\ell\} \to \mathrm{Br}(X)\{\ell\}/\mathrm{Br}(k)\{\ell\}$. Then the Poitou–Tate exact sequence above and commutativity of the diagram show that the family $(\theta_{\mathscr{G}_v}(P_v)) \in \prod_{v \in \mathsf{S}} H^2(G_v, H^2(\overline{X}, \mathbf{Z}_\ell(2)))$ comes from an element α of $H^2(G_\mathsf{S}, H^2(\overline{X}, \mathbf{Z}_\ell(2)))$. Let $\mathscr{G}$ be the universal ℓ-adic gerbe that was fixed at the beginning of the proof and let α' be the image of α in $H^2(k, H^2(\overline{X}, \mathbf{Z}_\ell(2)))$ Then the universal gerbe $\mathscr{G}_{\alpha'}$ with class $\mathscr{G} - \alpha'$ has the property that $(\mathscr{G}_{\alpha'}, P_v) = 0$ for all v, so that $\mathscr{G}_{\alpha'}$ has points everywhere locally, as desired.

For the other direction, if there is a universal gerbe $\mathscr{G} \xrightarrow{q} X$ with points everywhere locally, then choosing a family $(P_v) \in \mathscr{G}_v(k_v)$, we have $\theta_\mathscr{G}(q(P_v)) = 0$, so this element pairs to zero with any $\mathscr{A} \in \mathrm{Br}(X)$. Thus $(P_v) \in X(\mathbf{A}_k)^{\mathrm{Br}(X)\{\ell\}}$. This completes the proof of the theorem. □

Remark 8.2.

(i) Theorem 8.1 is a much weaker analogue of Proposition 3.3.2 and Théorème 3.5.1 of **[CTS87]**, which prove the result without assuming the existence of a universal ℓ-adic gerbe (universal torsor in their situation). We hope to be able to remove this assumption, but we have had trouble expressing the elementary obstruction (see Section 5 above) in terms of other computable cohomological invariants.

(ii) We hope that this theorem may lead to other obstructions to the Hasse principle. For the theorem effectively allows one to replace the Brauer–Manin obstruction by the existence of an auxiliary algebro-geometric object that must have rational points everywhere locally. Using analogues of the homological algebra developed in Section 4 above, one can define other auxiliary objects associated to higher cohomology groups of varieties of higher dimension. For example, let X be a threefold in $\mathbf{P}^4$. Then since $\mathrm{Br}(X)$ is the image of $\mathrm{Br}(k)$, there is no Brauer–Manin obstruction to the Hasse principle. However, there is still the interesting cohomology group $H^3(\overline{X}, \mathbf{Z}/n\mathbf{Z}(3))$. One can define the notion of "universal 3-gerbe" associated to this group using similar formalism as used above, and this can be used to formulate an obstruction to the Hasse principle, that there should be one such with points everywhere locally. This sounds rather abstract, but maybe it could be made more concrete in some cases.

9. Descent on curves

If X is a smooth projective curve over a number field, we can develop a similar theory using $H^1(\overline{X}, \mathbf{Z}/n(1))$. In this case, we are dealing with principal homogeneous spaces over X under the group-scheme $J[n]$, where J is the Jacobian of X, and these are parametrized by $J(k)/n$. As pointed out to us by Skorobogatov, these are related to the heterogeneous spaces of Coombes–Grant **[CG89]**. The elementary obstruction is an element of $H^2(k, H^1(\overline{X}, \mathbf{Z}/n(1)))$ given by the class of the 2-extension

$$0 \to H^1(\overline{X}, \mathbf{Z}/n(1)) \to \overline{k}(X)^*/n \to Div(\overline{X})/n \xrightarrow{deg} \mathbf{Z}/n \to 0$$

that is obtained by reducing the exact sequence

$$0 \to \overline{k}(X)^*/\overline{k}^* \to Div(\overline{X}) \to \mathrm{Pic}(\overline{X}) \to 0$$

modulo n, using the fact that $J(\overline{k})$ is divisible. If the genus of X is greater than one, the universal torsors will be of higher genus than X, and seemingly more complicated. It should be the case that the Brauer–Manin obstruction to the Hasse principle is simpler on these spaces than on X, but we cannot prove this. See **[Sko01]** §6.2 for more on descent on curves and abelian varieties.

10. Concluding remarks

Our theory above suffers from two major shortcomings:

(i) We are not able to show that the Brauer–Manin obstruction on a universal n-gerbe is "simpler" than on X, and we cannot even say at the moment what "simpler" should mean. Even in the case of curves of genus at least two, we face a similar problem.

(ii) We have also not been able to describe the universal n-gerbes in terms of explicit local "equations." In this case, we have some idea of what form this might take, but have not been able to describe these explicitly in any concrete examples. Briefly, our idea is to replace a suitable Zariski open set U in the theory of Colliot-Thélène–Sansuc (with $\mathrm{Pic}(\overline{U}) = 0$) with a quasi-finite étale morphism $U \to X$ which splits the n-torsion of the Brauer group of X. We can write a diagram similar to the one in ([**CTS87**], 1.6.10), but we have not been able to find explicit "equations" to describe the universal n-gerbes. If the Galois group acts trivially on $H^2(\overline{X}, \mathbf{Z}/n(2))$, we can describe them as follows: the universal torsor associated to $S = \mathrm{Hom}(\mathrm{Pic}(\overline{X}), \mathbf{G}_m)$ may be described by taking line bundles $\mathscr{L}_i$ that form a basis of $\mathrm{Pic}(\overline{X})$, removing their zero sections and taking the product. Thus we need to describe the universal gerbe bound by $\mathrm{Br}(\overline{X})[n](1)$. This may be done as in ([**Mil80**], Ch. IV, §2) by taking a spanning set of elements of $\mathrm{Br}(\overline{X})[n]$, describing them as gerbes as in Appendix B below, removing their zero sections and taking the product. It is a great challenge to be able to describe them in a more arithmetic situation, as is needed here.

Appendix A
Stacks and gerbes

In this appendix, we briefly recall the definition of stacks and gerbes. We pick and choose material from the books of Milne ([**Mil80**], Ch. IV, §2, p. 144-45) and Laumon–Moret-Bailly ([**LMB00**], §§1-3). Let $\varphi : F \to (\mathscr{C}/X)_E$ be a functor from a category to the underlying category of a site. For our purposes, this site will almost always be the big étale site. Given an object U of $(\mathscr{C}/X)_E$, we denote by $F(U)$ the category consisting of objects u of F such that $\varphi(u) = U$ and morphisms f between such objects that cover the identity morphism of U. Given a covering $U_i \xrightarrow{g_i} U$ and an element of $F(U)$, we get via g_i^* elements of $F(U_i)$ which agree on $F(U_i \times U_j)$ and satisfy the cocycle condition on three-fold fibre products. If any family $F(U_i)$ satisfying these condition arises from such an element of $F(U)$, and if for any $u_1, u_2 \in F(U)$, the functor

$$(V \xrightarrow{g} U) \mapsto \mathrm{Hom}_{F(V)}(g^* u_1, g^* u_2)$$

is a sheaf, then φ is a *stack* (champ). It is a *gerbe* if it is a stack of groupoids, there is a covering U_i of U such that each $F(U_i)$ is nonempty, and if any two objects of $F(U)$ are locally isomorphic. A gerbe is bound by an abelian sheaf $\mathscr{F}$ if for any object U of $\mathscr{C}/X$ and any $u \in F(U)$, we have

$$\mathscr{F}(U) \cong Aut_{F(U)}(u).$$

A basic theorem of Giraud is that the set of gerbes bound by an abelian sheaf $\mathscr{F}$ up to equivalence is isomorphic to the second cohomology group $H^2(X_E, \mathscr{F})$.

In ([**LMB00**], Définition 3.1.5), a gerbe is defined to be a stack of groupoids $\mathscr{G}$ with a structure morphism $\mathscr{G} \xrightarrow{A} X$ such that both A and the diagonal morphism:

$$\mathscr{G} \times_{A,X,A} \mathscr{G}$$

are epimorphisms.

Any scheme X is a stack and a gerbe for the Zariski topology via the sheaf $\mathrm{Hom}(-, X)$ it represents. Thus a stack for the étale topology may be vaguely (but incorrectly) regarded as a "scheme for the étale topology".

Appendix B
Description of the Brauer group in terms of gerbes

The group $H^2(X, \mathbf{G}_m)$ has a nice description in terms of the gerbes it classifies (up to equivalence). This may be done as follows (see [**Mil80**], Ch. IV, §2, p. 145): let $\mathscr{A}$ be an Azumaya algebra on X. For U étale over X, let $F(U)$ be the set of pairs (E, α) where E is a locally free sheaf of $\mathscr{O}_U$-modules and $\alpha : \mathscr{A}(U) \to \mathrm{End}_U(E)$ is an isomorphism. Then descent theory shows that this is a stack, and the definition of an Azumaya algebra (see e.g., ibid. Ch. IV, §2, 2.1) shows that it is a gerbe. It is bound by $\mathbf{G}_m$ since the map

$$\mathbf{G}_m(U) \to Aut_U(E, \alpha)$$

that sends an element a of $\mathbf{G}_m(U)$ to multplication by a is an isomorphism.

References

[BM90] V. V. Batyrev & Y. I. Manin – Sur le nombre des points rationnels de hauteur borné des variétés algébriques, *Math. Ann.* **286** (1990), no. 1-3, 27–43.

[BT99] F. A. Bogomolov & Y. Tschinkel – On the density of rational points on elliptic fibrations, *J. Reine Angew. Math.* **511** (1999), 87–93.

[BT00] ———, Density of rational points on elliptic $K3$ surfaces, *Asian J. Math.* **4** (2000), no. 2, 351–368.

[CG89] K. R. Coombes & D. R. Grant – On heterogeneous spaces, *J. London Math. Soc. (2)* **40** (1989), no. 3, 385–397.

[CTR85] J.-L. COLLIOT-THÉLÈNE & W. RASKIND – K_2-cohomology and the second Chow group, *Math. Ann.* **270** (1985), no. 2, 165–199.

[CTS80] J.-L. COLLIOT-THÉLÈNE & J.-J. SANSUC – La descente sur les variétés rationnelles, Journées de Géometrie Algébrique d'Angers, Juillet 1979/Algebraic Geometry, Angers, 1979, Sijthoff & Noordhoff, Alphen aan den Rijn, 1980, 223–237.

[CTS87] ______, La descente sur les variétés rationnelles. II, *Duke Math. J.* **54** (1987), no. 2, 375–492.

[CTSSD87a] J.-L. COLLIOT-THÉLÈNE, J.-J. SANSUC & P. SWINNERTON-DYER – Intersections of two quadrics and Châtelet surfaces. I, *J. Reine Angew. Math.* **373** (1987), 37–107.

[CTSSD87b] ______, Intersections of two quadrics and Châtelet surfaces. II, *J. Reine Angew. Math.* **374** (1987), 72–168.

[Dee99] J. A. T. DEE – 1999, Thesis.

[Del72] P. DELIGNE – La conjecture de Weil pour les surfaces $K3$, *Invent. Math.* **15** (1972), 206–226.

[Fal83] G. FALTINGS – Endlichkeitssätze für abelsche Varietäten über Zahlkörpern, *Invent. Math.* **73** (1983), no. 3, 349–366.

[Har96] D. HARARI – Obstructions de Manin transcendantes, Number theory (Paris, 1993–1994), London Math. Soc. Lecture Note Ser., vol. 235, Cambridge Univ. Press, Cambridge, 1996, 75–87.

[HS00] M. HINDRY & J. H. SILVERMAN – *Diophantine Geometry*, Graduate Texts in Mathematics, vol. 201, Springer-Verlag, New York, 2000.

[HS02] D. HARARI & A. N. SKOROBOGATOV – Non-abelian cohomology and rational points, *Compositio Math.* **130** (2002), no. 3, 241–273.

[Jan89] U. JANNSEN – On the l-adic cohomology of varieties over number fields and its Galois cohomology, Galois groups over **Q** (Berkeley, CA, 1987), Math. Sci. Res. Inst. Publ., vol. 16, Springer, New York, 1989, 315–360.

[Jan00] ______, Letter from Jannsen to Gross on higher Abel-Jacobi maps, The arithmetic and geometry of algebraic cycles (Banff, AB, 1998), NATO Sci. Ser. C Math. Phys. Sci., vol. 548, Kluwer Acad. Publ., Dordrecht, 2000, 261–275.

[KS67] M. KUGA & I. SATAKE – Abelian varieties attached to polarized K_3-surfaces, *Math. Ann.* **169** (1967), 239–242.

[LMB00] G. LAUMON & L. MORET-BAILLY – *Champs Algébriques*, Ergebnisse der Mathematik und ihrer Grenzgebiete. 3. Folge. A Series of Modern Surveys in Mathematics [Results in Mathematics and Related Areas. 3rd Series. A Series of Modern Surveys in Mathematics], vol. 39, Springer-Verlag, Berlin, 2000.

[Mil80] J. S. MILNE – *Étale Cohomology*, Princeton Mathematical Series, vol. 33, Princeton University Press, Princeton, N.J., 1980.

[Mor84] D. R. Morrison – On $K3$ surfaces with large Picard number, *Invent. Math.* **75** (1984), no. 1, 105–121.

[NO85] N. Nygaard & A. Ogus – Tate's conjecture for $K3$ surfaces of finite height, *Ann. of Math. (2)* **122** (1985), no. 3, 461–507.

[Nyg83] N. O. Nygaard – The Tate conjecture for ordinary $K3$ surfaces over finite fields, *Invent. Math.* **74** (1983), no. 2, 213–237.

[Ras95] W. Raskind – Higher l-adic Abel-Jacobi mappings and filtrations on Chow groups, *Duke Math. J.* **78** (1995), no. 1, 33–57.

[Roj80] A. A. Rojtman – The torsion of the group of 0-cycles modulo rational equivalence, *Ann. of Math. (2)* **111** (1980), no. 3, 553–569.

[SI77] T. Shioda & H. Inose – On singular $K3$ surfaces, Complex Analysis and Algebraic Geometry, Iwanami Shoten, Tokyo, 1977, 119–136.

[Sko01] A. N. Skorobogatov – *Torsors and Rational Points*, Cambridge Tracts in Mathematics, vol. 144, Cambridge University Press, Cambridge, 2001.

[Tat95] J. Tate – On the conjectures of Birch and Swinnerton-Dyer and a geometric analog, Séminaire Bourbaki, Vol. 9, Soc. Math. France, Paris, 1995, Années 1964/65-1965/66, Exp. No. 306, 415–440.

[Wil95] A. Wiles – Modular elliptic curves and Fermat's last theorem, *Ann. of Math. (2)* **141** (1995), no. 3, 443–551.

Arithmetic of Higher-dimensional Algebraic Varieties
(B. Poonen, Yu. Tschinkel, eds.), p. 205–233
Progress in Mathematics, Vol. 226, © 2004 Birkhäuser Boston, Cambridge, MA

RATIONAL POINTS ON COMPACTIFICATIONS OF SEMI-SIMPLE GROUPS OF RANK 1

Joseph Shalika

Department of Mathematics, Johns Hopkins University, 3400 N. Charles St., Baltimore, MD 21218-2686, USA • *E-mail :* shalika@math.jhu.edu

Ramin Takloo-Bighash

Department of Mathematics, Princeton University, Fine Hall, Washington Road, Princeton, NJ 08544-1000, USA • *E-mail :* rtakloo@math.princeton.edu

Yuri Tschinkel

Department of Mathematics, Princeton University, Fine Hall, Washington Road, Princeton, NJ 08544-1000, USA • *E-mail :* ytschink@math.princeton.edu

Abstract. We explain our approach to the problem of counting rational points of bounded height on equivariant compactifications of semi-simple groups.

1. Introduction

Let G be a linear algebraic group, $S \subset G$ a subgroup and $S\backslash G$ the corresponding homogeneous space, all assumed to be defined over a number field F. We consider G-equivariant embeddings

$$S\backslash G \hookrightarrow \mathbb{P}^n.$$

Such embeddings arise from a choice of an F-rational projective representation $\varrho : G \to \mathrm{PGL}_{n+1}$, together with a point $p_0 \in \mathbb{P}^n(F)$ with stabilizer S.

Key words and phrases. Rational points, height zeta functions.

The second author was partially supported by the Clay Mathematics Institute.
The third author was partially supported by NSF grant 0100277.

The standard height on F-rational points of $\mathbb{P}^n$ is defined by

$$\begin{array}{rclcl} H : \mathbb{P}^n(F) & \rightarrow & \mathbb{R}_{>0}, & & \\ x = (x_0, ..., x_n) & \mapsto & H(x) & := & \prod_v H_v(x), \\ & & H_v(x) & := & \max_j(|x_j|_v), \end{array}$$

the product over all valuations v of F. More generally, one considers heights whose local factors coincide with the above at almost all v and differ from the above by a globally bounded function at the remaining v. For example, one could choose $H_v(x) = (\sum_j |x_j|_v^2)^{1/2}$ at a real v.

We are interested in the asymptotic distribution of the number $N(\varrho, B)$ of F-rational points on $S\backslash G$ of height $\leq B$, as $B \rightarrow \infty$. In many cases, one finds

$$N(\varrho, B) \sim \mathsf{c} \cdot B^{\mathsf{a}} \log(B)^{\mathsf{b}-1}, \tag{1.1}$$

with $\mathsf{a} \in \mathbb{Q}$, $\mathsf{b} \in \mathbb{N}$, and $\mathsf{c} \in \mathbb{R}_{>0}$ (see [**7**], [**2**], [**18**], [**4**]). There is a conceptual interpretation of these constants in terms of global geometric and arithmetic invariants of the associated algebraic variety X, the closure of the G-orbit through p_0 (see [**7**], [**1**], [**14**] and [**3**]).

The proofs of the above asymptotics rely on harmonic analysis on corresponding adelic groups. Note that results of type (1.1) are quite nontrivial even when unitary representations of the adeles $G(\mathbb{A})$ are well understood. For example, we still don't know how to treat general equivariant compactifications of the Heisenberg group (the case of bi-equivariant compactifications is considered in [**17**]). For semi-simple groups we need to appeal to rather nontrivial results from the theory of automorphic forms: multiplicity one, uniform bounds of matrix coefficients, Eisenstein series, spectral theory, etc.

We now give an outline of our approach in a special case. Let G be a split semi-simple group of adjoint type over $\mathbb{Q}$. Consider the Cartan decomposition $G(\mathbb{A}) = KA^+K$, where $K = \prod_v K_v$ is a maximal compact subgroup and $A^+ = \prod_v A_v^+$ (and the product is over all valuations v of $\mathbb{Q}$). Here A_p^+ (resp. A_∞^+) can be identified (via the logarithmic map) with the monoid $\mathfrak{a}^+$ (resp. cone $\mathfrak{a}_\infty^+$) in the Lie algebra $\mathfrak{a}$ (resp. $\mathfrak{a}_\infty := \mathfrak{a} \otimes \mathbb{R}$), dual to the monoid (resp. cone) spanned by simple roots of the corresponding maximal torus A. Fix a triangulation Σ of $\mathfrak{a}^+$ (resp. $\mathfrak{a}_\infty^+$ for archimedean v) into simplicial subcones σ and let λ be a continuous $\mathbb{R}$-valued function on $\mathfrak{a}$ (resp. $\mathfrak{a}_\infty$) which is *linear* on every cone $\sigma \in \Sigma$. For example, we may take

$$\lambda(a_v) = \langle \mathbf{s}, \bar{a}_v \rangle, \tag{1.2}$$

where $\mathrm{s} \in \mathfrak{a}_v^* \otimes \mathbb{R}$ and $\overline{a}_v = \log(a_v) \in \mathfrak{a}_v$, $a_v \in A(\mathbb{Q}_v)$. Put $q_v = p$ for $v = p$, $q_v = e$ for $v = \infty$ and define

$$H_v(\lambda, g_v) = q_v^{\lambda(a_v)} \quad \text{and} \quad H(\lambda, g) := \prod_v H_v(\lambda, g_v),$$

where $g = (g_v) \in G(\mathbb{A})$, $g_v = k_v a_v k_v'$ with $k_v, k_v' \in K_v$, $a_v \in A_v^+$.

Problem 1.1. Study the analytic properties of the *zeta function*

$$\mathscr{Z}(\lambda, s, g) := \sum_{\gamma \in G(F)} H(\lambda, \gamma g)^{-s}. \tag{1.3}$$

For λ chosen as in (1.2), the zeta function (1.3) encodes information about the distribution of rational points of bounded height on "wonderful" compactifications of G studied by de Concini and Procesi in [**5**]. The study of arbitrary bi-equivariant compactifications of G can be reduced to other λ.

The main goal of this paper is to explain in detail how our approach works in the simplest case: $\mathbb{P}^3$ considered as the wonderful compactification of PGL_2 over $\mathbb{Q}$. The counting problem itself is trivial, but it allows us to focus on the method, which covers (verbatim) wonderful compactifications of rank one semi-simple groups of adjoint type and highlights the technical difficulties one faces for groups of higher rank. Compactifications of anisotropic forms of semi-simple groups of adjoint type are treated in [**16**].

2. Basic definitions and results

Throughout we will use the following notations:

- $\mathscr{V} = \mathscr{V}_{\mathbb{Q}} := \{2, 3, 5, \ldots, p, \ldots, \infty\}$ - valuations of $\mathbb{Q}$;
- $G = PGL_2$, A the (diagonal) torus, N upper triangular unipotent matrices, $P = NA$ the Borel subgroup;
  $\mathfrak{a} \simeq \mathbb{Z}$ - Lie algebra of A, $\mathfrak{a}^*$ its character lattice;
- $K_p = G(\mathbb{Z}_p)$, $K_f = \prod_p K_p$, $K_\infty = SO(2)$ and $K = K_\infty \times K_f$;
- for $v \in \mathscr{V}$, let $G(\mathbb{Q}_v) = N_v A_v K_v$ and $K_v A_v^+ K_v$ be the Iwasawa, resp. Cartan decompositions;
-
  $$\mathfrak{S} = \mathfrak{S}_{\mathsf{c}} := \left\{ \begin{pmatrix} a & \\ & 1 \end{pmatrix} \middle| a \geqslant \mathsf{c} > 0 \right\} \subset G(\mathbb{R}) \hookrightarrow G(\mathbb{A})$$
  a Siegel domain, $G(\mathbb{A}) = G(\mathbb{Q}) \cdot \mathfrak{S} \cdot \Omega$, for some compact $\Omega \subset G(\mathbb{A})$;
- $dg = \prod_v dg_v = dn\, da\, dk$ normalized Haar measure,
  $$\int_{K_v} dk_v = 1, \quad \int_{N(\mathbb{Q}) \backslash N(\mathbb{A})} dn = 1;$$

- $\mathrm{vol}_p(\ell) := \mathrm{vol}\left\{ g_p = k_p a_p^\ell k_p' \mid a_p^\ell = \begin{pmatrix} a_p & 0 \\ 0 & 1 \end{pmatrix} \in G(\mathbb{Q}_p) \text{ with } |a_p|_p = p^\ell \right\}$,
- $\mathfrak{U} = \mathfrak{U}(\mathfrak{g})$ universal enveloping algebra of $\mathfrak{g} = \mathrm{Lie}(G(\mathbb{R}))$;
- $\Delta \in \mathfrak{U}$ - the standard Casimir element (Laplacian);
- $\mathsf{T}_\delta = \{s \in \mathbb{C} \mid \Re(s) > 4 + \delta\}$;
- $\mathsf{L}^2 := \mathsf{L}^2(G(\mathbb{Q})\backslash G(\mathbb{A}))$, unless noted otherwise, $\|\cdot\|_2$ the L^2-norm.

Define two local height functions:

$$H_v : g_v \mapsto |\mathrm{Tr}(g_v g_v^*)|_v^{1/2}, \quad \text{and} \quad \chi_{v,P} : g_v = n_v a_v k_v \mapsto |a_v|_v. \tag{2.1}$$

Define the global heights by

$$H := \prod_v H_v \quad \text{and} \quad \chi_P := \prod_v \chi_{v,P} \tag{2.2}$$

Remark 2.1. For $\gamma \in P(\mathbb{Q})$ we have $\chi_P(\gamma g) = \chi_P(g)$ (by the product formula) and χ_P^{-1} is a height on $\mathbb{P}^1(\mathbb{Q}) = P(\mathbb{Q})\backslash G(\mathbb{Q})$.

Remark 2.2. The group G has a (canonical) compactification

$$G = \left\{ \begin{pmatrix} a & b \\ c & d \end{pmatrix} \right\} \hookrightarrow \mathbb{P}(\mathrm{End}(V)) = \mathbb{P}^3 = \{(a:b:c:d)\}$$

which is equivariant for the action of G on both sides. A standard height on $\mathbb{P}^3(\mathbb{Q})$ is

$$\sqrt{a^2+b^2+c^2+d^2} \cdot \prod_p \max(|a|_p, |b|_p, |c|_p, |d|_p).$$

Its local factors are identical with (2.1).

The main object of interest is the *height zeta function*

$$\mathscr{Z}(s,g) := \sum_{\gamma \in G(\mathbb{Q})} H(\gamma g)^{-s}. \tag{2.3}$$

The convergence of the series in the domain $\Re(s) \gg 0$ (for fixed g) is a special case of a general fact: let X be any projective algebraic variety over a number field F, and H any height induced from a projective embedding of X. Then the height zeta function

$$\mathscr{Z}(H,s) = \sum_{x \in X(F)} H(x)^{-s}$$

converges absolutely and uniformly on compacts in the domain $\Re(s) \gg 0$.

Proposition 2.3. *There exists a* $\sigma > 0$ *such that the series*

$$\sum_{\gamma \in G(\mathbb{Q})} H(\gamma g)^{-s}$$

converges absolutely and uniformly on compacts in the domain $\mathsf{T}_\sigma \times G(\mathbb{A})$ *to a function* $\mathscr{Z}(s, g)$ *which*

(1) *is continuous and bounded on* $G(\mathbb{Q})\backslash G(\mathbb{A})$*;*
(2) *has bounded* Δ*-derivatives.*

In particular, in this domain, $\mathscr{Z}(s, g)$ *and all its* Δ*-derivatives are in* L^2*.*

Proof. By reduction theory, $G(\mathbb{A}) = G(\mathbb{Q}) \cdot \mathfrak{S} \cdot \Omega$, where $\Omega \subset G(\mathbb{A})$ is some compact and $\mathfrak{S}$ a Siegel domain. Now we use the following easy property of the height: there exist constants $\mathsf{c}, \mathsf{r} > 0$ such that

– $H(\gamma a \omega) \geq \mathsf{c} H(\gamma)^{\mathsf{r}}$ for all $a \in \mathfrak{S}, \omega \in \Omega$ and $\gamma \in G(\mathbb{Q})$.

In particular, there exists an $\mathsf{r} > 0$ such that for all $g \in G(\mathbb{A})$ one has

$$\mathscr{Z}(\Re(s), g) \leq \mathscr{Z}(\mathsf{r}\Re(s), e) < \infty \quad \text{for} \quad \Re(s) \gg 0.$$

Since Δ commutes with the K-action, it suffices to prove (2) on matrices in A_∞^+. Explicit formulas for the height and for Δ give the result. □

A solution of Problem 1.1 in our special case is given by

Theorem 2.4. *There exists an* $\varepsilon > 0$ *such that* $\mathscr{Z}(s, e)$ *admits a meromorphic continuation to* $\mathsf{T}_{-\varepsilon}$ *with a unique simple pole at* $s = 4$*.*

In the analysis of $\mathscr{Z}(s, g)$ we use the Eisenstein series

$$E(s, g) := \sum_{\gamma \in P(\mathbb{Q})\backslash G(\mathbb{Q})} \chi(s, \gamma g), \tag{2.4}$$

where $\chi(s, g) := \chi_P(g)^{s+1/2}$. The idea of the proof of Theorem 2.4 is to first establish an identity of continuous L^2-functions on $G(\mathbb{Q})\backslash G(\mathbb{A})$

$$\mathscr{Z}(s, g) = \mathscr{Z}^{\mathrm{res}}(s) + \mathscr{Z}^{\mathrm{cusp}}(s, g) + \mathscr{Z}^{\mathrm{eis}}(s, g), \quad \text{for} \quad \Re(s) \gg 0, \tag{2.5}$$

where

$$\mathscr{Z}^{\mathrm{res}}(s) := \int_{G(\mathbb{Q})\backslash G(\mathbb{A})} \mathscr{Z}(s, g)\, dg = \int_{G(\mathbb{A})} H(g)^{-s}\, dg \tag{2.6}$$

is the contribution from the trivial representation, $\mathscr{Z}^{\mathrm{cusp}}(s,g)$ is the projection of $\mathscr{Z}(s,g)$ onto the cuspidal spectrum, and

$$\mathscr{Z}^{\mathrm{eis}}(s,g) := \frac{1}{2\pi}\int_{\mathbb{R}}\left(\int_{G(\mathbb{Q})\backslash G(\mathbb{A})}\mathscr{Z}(s,g)\overline{E(it,g)}dg\right)E(it,g)dt \tag{2.7}$$

is the projection onto the continuous spectrum. This is accomplished in Propositions 5.1 and 7.1. Then we use (2.5) to meromorphically continue $\mathscr{Z}(s,e)$ (see Propositions 3.4, 7.6 and 5.1). Finally, a Tauberian theorem implies

Corollary 2.5. *There is a constant* $\mathsf{c}>0$*, such that*

$$\#\{\gamma\in G(\mathbb{Q})\mid H(\gamma)\leqslant B\} = \mathsf{c}B^4(1+o(1)), \quad \textit{as } B\to\infty.$$

3. Heights and height integrals

Let $\varphi_p=\varphi_p(\chi_p)$ be a bi-K_p-invariant function on $G(\mathbb{Q}_p)$ such that

$$\varphi_p(a_p^\ell) = \frac{p^{-\frac{\ell}{2}}}{1+p^{-1}}\left(\frac{1-\bar{\chi}_p(p)^2/p}{1-\bar{\chi}_p(p)^2}\chi_p(p^\ell)+\frac{1-\chi_p(p)^2/p}{1-\chi_p(p)^2}\bar{\chi}_p(p^\ell)\right), \tag{3.1}$$

for $\ell\geq 1$, where χ_p is a nontrivial unramified quasi-character of $\mathbb{Q}_p^*$,

$$\chi_p(p)=\bar{\chi}_p(p)^{-1}=p^{\alpha_p}, \text{ with parameter } \alpha_p\in\mathbb{C}^*. \tag{3.2}$$

We write also $\varphi_p=\varphi_p(s,\cdot)$, if $\alpha_p=s$ for all p. Define

$$I_v(s) := \int_{G(\mathbb{Q}_v)} H_v(g_v)^{-s}\,dg_v, \quad I_f(s) := \prod_p I_p(s) \tag{3.3}$$

and, for $\varphi_p=\varphi_p(\chi_p)$ and $\varphi=\prod_p\varphi_p$,

$$I_p(s,\varphi_p(\chi_p)) := \int_{G(\mathbb{Q}_v)}\varphi_p(g_p)H_p(g_p)^{-s}\,dg_p, \quad I_f(s,\varphi) := \prod_p I_p(s,\varphi).$$

Lemma 3.1. *The functions* $I_p(s)$ *are holomorphic in* T_{-2}*. Moreover,* $I_f(s)$ *is holomorphic in* T_0 *and admits a meromorphic continuation to* T_{-2} *with an isolated simple pole at* $s=4$*.*

Proof. We have

$$\mathrm{vol}_p(\ell) = \begin{cases} p^\ell(1+p^{-1}) & \text{if } \ell>0,\\ 1 & \text{if } \ell=0\end{cases} \tag{3.4}$$

so that $I_p(s)$ is given by

$$1+(1+p^{-1})\sum_{\ell\geq 1} p^{-\frac{\ell s}{2}}p^{\ell} = (1-p^{-(\frac{s}{2}-1)})^{-1}(1+p^{-\frac{s}{2}}) = \zeta_p(\frac{s}{2}-1)(1+p^{-\frac{s}{2}}),$$

where ζ_p is the local factor of the Riemann zeta function ζ (the sum converges absolutely and uniformly on compacts in T_{-2}). The Euler product $\prod_p(1+p^{-\frac{s}{2}})$ converges (uniformly on compacts in T_{-2}) to a holomorphic function. It suffices to recall the analytic properties of ζ. □

Lemma 3.2. *Assume that for all p, $|\Re(\alpha_p)| < \mathsf{r}$. Then $I_p(s,\varphi_p)$ is holomorphic for $\Re(s) > 2\mathsf{r}+1$. Moreover, for all $\varepsilon > 0$ there exists a constant $\mathsf{c} = \mathsf{c}(\varepsilon)$ such that $I_f(s,\varphi)$ is holomorphic and*

$$|I_f(s,\varphi)| \leq \mathsf{c}, \quad \textit{for} \quad \Re(s) > 2\mathsf{r}+3+\varepsilon.$$

Proof. Combining (3.1) with (3.4) we have

$$I_p(s,\varphi_p) = 1 + \frac{1-\bar{\chi}_p^2(p)/p}{1-\bar{\chi}_p^2(p)}\sum_{\ell>0} p^{-\ell\frac{s-1}{2}}\chi_p^{\ell}(p) + \frac{1-\chi_p^2(p)/p}{1-\chi_p^2(p)}\sum_{\ell>0} p^{-\ell\frac{s-1}{2}}\bar{\chi}_p^{\ell}(p).$$

For $\Re(s) > 2\mathsf{r}+1$ the series converges absolutely (and uniformly) to

$$I_p(s,\varphi_p) = \frac{(1-p^{-\frac{s-1}{2}}\chi_p(p))^{-1}(1-p^{-\frac{s-1}{2}}\bar{\chi}_p(p))^{-1}}{\zeta_p(s)}. \tag{3.5}$$

The corresponding Euler product is holomorphic and bounded by some constant $\mathsf{c}(\varepsilon)$, provided $\Re(s) > 2\mathsf{r}+3+\varepsilon$. □

Lemma 3.3. *For all $\varepsilon > 0$ and $\mathsf{n} \in \mathbb{N}$ there is a $\mathsf{c} = \mathsf{c}(\varepsilon,\mathsf{n})$ such that*

$$I_\infty(s) \quad \textit{and} \quad I_{\mathsf{n},\infty}(s) := \int_{G(\mathbb{R})} \Delta^{\mathsf{n}}\cdot H_\infty(g_\infty)^{-s}\, dg_\infty$$

are holomorphic for all $s \in \mathsf{T}_{-2+\varepsilon}$ with absolute value bounded by c.

Proof. Write

$$g = k\begin{pmatrix}1 & \\ & a\end{pmatrix}k', \quad \text{with} \quad k,k' \in SO(2),\ 0 < a \leqslant 1. \tag{3.6}$$

By a standard integration formula (cf. p. 142 of [**11**])

$$I_\infty(\sigma) \ll \int_0^1 a^{\frac{\sigma}{2}}(a^{-1}-a)\,da^* < \int_0^1 a^{\frac{\sigma}{2}-1}\,da^* < \infty$$

(for $\sigma > 2$). A similar explicit computation proves the second claim. □

Proposition 3.4. *The function $\mathscr{Z}^{\mathrm{res}}(s)$ is holomorphic in T_0 and admits a meromorphic continuation to T_{-2} with an isolated pole at $s = 4$.*

Proof. By definition,

$$\mathcal{Z}^{\mathrm{res}}(s)=\int_{G(\mathbb{Q})\backslash G(\mathbb{A})}\sum_{\gamma\in G(\mathbb{Q})}H(\gamma g)^{-s}dg=\int_{G(\mathbb{A})}H(g)^{-s}\,dg=I_f(s)\cdot I_\infty(s).$$

It suffices to apply Lemma 3.1 and 3.3. □

4. Eisenstein series

We will use the following notations:

- $\int_{\mathbb{R}}:=\int_{-\infty}^{+\infty}$;
- $d\mu_{\mathsf{n}}(t):=(1+t^2)^{-\mathsf{n}}dt$ - a measure on $\mathbb{R}$;
- $c(s):=\prod_v c_v(s)$, with $c_p(s)=\frac{\zeta_p(2s)}{\zeta_p(2s+1)}$ and $c_\infty(s)=\pi\frac{\Gamma(s)}{\Gamma(s+1/2)}$;
- $W(t):=1-\sum_\eta\frac{2\Re(\eta)}{\Re(\eta)^2+(t-\Im(\eta))^2}$, sum over all poles, with multiplicity, of $c(s)$ with $\Re(\eta)<0$;
- $E(s,g)$ - Eisenstein series (2.4), $E_s:=E(s,e)$ and

$$E_P(s,g):=\int_{N(\mathbb{Q})\backslash N(\mathbb{A})}E(s,ng)dn;$$

- truncations:

$$\wedge_T E(s,g):=\sum\nolimits_T E_P(s,\gamma g)\quad\text{and}\quad\wedge^T E(s,g):=E(s,g)-\wedge_T E(s,g),$$

  the sum over $\gamma\in P(\mathbb{Q})\backslash G(\mathbb{Q})$, with $\chi_P(\gamma g)\geq T>0$.

We recall some basic facts from the theory of Eisenstein series. We follow closely the exposition in [**8**], [**9**].

Theorem 4.1. (1) *the poles of $E(s,g)$ coincide with the poles of $c(s)$;*
(2) *away from the poles, $E(s,g)=c(s)E(-s,g)$;*
(3) *away from the poles,*

$$\int_K E(s,kg)\,dk=E(s,e)\varphi(s,g),$$

where $\varphi(s,\cdot)=\prod_v\varphi_v(s,\cdot)$ is the spherical function of the corresponding principal series representation;
(4) $\int_{G(\mathbb{Q})\backslash G(\mathbb{A})}E(it,g)dg=0$.

Proof. Property (4) follows easily from the fact that the p-adic principal series representations corresponding to a purely imaginary parameter do not have the trivial representation as a quotient. $\square$

The following key facts about Eisenstein series will be crucial for the analysis of the height zeta function. In the special case at hand, they could be deduced from standard facts about the Riemann zeta function. However, we give proofs applicable in more general situations (see also [**13**]).

Theorem 4.2. *For* $\mathsf{n} \gg 0$, *one has*

(1) $\int_{\mathbb{R}} \| \wedge^T E(it,\cdot)\|_2^2 d\mu_{\mathsf{n}}(t) < \infty$;

(2) $\int_{\mathbb{R}} \|E(it,\cdot)\|_{2,\Omega}^2 d\mu_{\mathsf{n}}(t) < \infty$, *where* $\Omega \subset G(\mathbb{A})$ *is a compact subset;*

(3) *for all* $g \in G(\mathbb{A})$

$$\int_{\mathbb{R}} | \wedge^T E(it,g)| d\mu_{\mathsf{n}}(t) < \infty \quad \text{and} \quad \int_{\mathbb{R}} |E(it,g)|^2 d\mu_{\mathsf{n}}(t) < \infty;$$

(4) *the function* $g \mapsto \int_{\mathbb{R}} |E(it,g)| d\mu_{\mathsf{n}}(t)$ *is continuous.*

We will need the following.

Lemma 4.3. *We have*

$$\| \wedge^T E(it,\cdot)\|_2^2 - 2T - \frac{c'(it)}{c(it)} + \Psi(it,T), \tag{4.1}$$

where $\Psi(it,T)$ *is a bounded function.*

Proof. Following [**9**], p. 102, we get, away from the poles

$$\| \wedge^T E(s,\cdot)\|_2^2 = (s+\bar{s})^{-1}\left(e^{T(s+\bar{s})} - |c(s)|^2 e^{-T(s+\bar{s})}\right) + \Psi(s-\bar{s},T), \tag{4.2}$$

where

$$\Psi(s,T) := \frac{1}{s-\bar{s}}\left(\overline{c(s)}e^{(s-\bar{s})T} - c(s)e^{-(s-\bar{s})T}\right).$$

The singularities from $(s+\bar{s})^{-1}$ and $(s-\bar{s})^{-1}$ are removable ([**8**], p. 231). For $s = \sigma + it$, and $\sigma \to 0$ the right side of (4.2) equals

$$\lim_{\sigma\to 0} \frac{1}{2\sigma}\left\{e^{2T\sigma} - |c(\sigma+it)|^2 e^{-2T\sigma}\right\} + \Psi(it,T) = 2T - \frac{c'(it)}{c(it)} + \Psi(it,T).$$

Since $|c(it)| = 1$, it suffices to show that Ψ is bounded near $t = 0$. Write

$$\Psi(it,T) = \overline{c(it)}\frac{e^{2itT}-1}{2it} - c(it)\frac{e^{-2itT}-1}{2it} + \frac{c(it)-\overline{c(it)}}{2it}.$$

Since $c(0) \in \mathbb{R}$ and $c(it)$ is differentiable in t, $\lim_{t\to 0} \Psi(it, T)$ exists. □

Corollary 4.4. *The function $s \mapsto c(s)$ is bounded in any region $0 \leq \Re(s) \leq \varepsilon$, $\Im(s) \geq 1$.*

Proof. Follows from the positivity of $\| \wedge^T E(s, \cdot)\|^2$ and (4.2). □

Lemma 4.5. *For $\mathsf{n} \gg 0$,*

$$\int_{\mathbb{R}} \left| \frac{c'(it)}{c(it)} \right| d\mu_{\mathsf{n}}(t) < \infty.$$

Proof. The only pole of $c(s)$ in the right half-plane is at $s = 1/2$. By Theorem 6.9 in [**12**], there is a $\mathsf{q} > 0$, such that

$$\frac{c'(it)}{c(it)} = \log(\mathsf{q}) + \frac{1}{t^2 + \frac{1}{4}} + (1 - W(t)).$$

By Theorem 7.1 of [**12**], there is a polynomial Q with positive coefficients such that

$$\left| \int_{-T}^{+T} \frac{c'(it)}{c(it)}\, dt \right| \leqslant Q(T),$$

for all $T \in \mathbb{R}_{>0}$. By definition, for all t, $W(t) > 1$, so that

$$\left| \int_{-T}^{+T} \frac{c'(it)}{c(it)}\, dt \right| \geqslant \int_{-T}^{+T} (W(t) - 1) dt - \int_{-T}^{+T} \left(\log(\mathsf{q}) + \frac{1}{t^2 + \frac{1}{4}} \right) dt,$$

and

$$\int_{-T}^{+T} W(t) dt \leqslant 2T + Q(T) + \int_{-T}^{+T} \left(\log(\mathsf{q}) + \frac{1}{t^2 + \frac{1}{4}} \right) dt.$$

In particular, there is a polynomial R with positive coefficients such that for all $T \in \mathbb{R}_{>0}$ one has

$$\int_{-T}^{+T} W(t)\, dt \leqslant R(T).$$

Now we use the following easy statement: if a function $f : \mathbb{R} \to \mathbb{R}_{>0}$ and a polynomial R with positive coefficients are such that

$$\int_{-T}^{+T} f(t)\, dt \leqslant R(T) \text{ then } \int_{\mathbb{R}} f(t)\, d\mu_{\mathsf{n}}(t) < \infty, \quad \text{for } \mathsf{n} \gg 0.$$

If follows that, for $\mathsf{n} \gg 0$,

$$\int_{\mathbb{R}} W(t) d\mu_{\mathsf{n}}(t) < \infty \quad \text{and} \quad \int_{\mathbb{R}} \left| \frac{c'(it)}{c(it)} \right| d\mu_{\mathsf{n}}(t) < \infty.$$

□

Combining Lemma 4.3 with Lemma 4.5 we see that for all polynomials R there exists an $\mathsf{n}' > 0$ such that for all $\mathsf{n} > \mathsf{n}'$ one has

$$\mathscr{J}^T := \int_{\mathbb{R}} |R(it)| \cdot \| \wedge^T E(it, \cdot)\|_2^2 d\mu_{\mathsf{n}}(t) < \infty. \tag{4.3}$$

This proves (1) of Theorem 4.2.

Next, fix a compact $\Omega \subset G(\mathbb{A})$, $g \in \Omega$, and let $\| \cdot \|_{\infty,\Omega}$, resp. $\| \cdot \|_{2,\Omega}$ be the $\mathsf{L}^\infty(\Omega)$, resp. $\mathsf{L}^2(\Omega)$, norms. We bound $\|E(it,\cdot)\|_{2,\Omega}$ by

$$\| \wedge_T E(it, \cdot)\|_{2,\Omega} + \| \wedge^T E(it, \cdot)\|_{2,\Omega} \leqslant \mathsf{c}\| \wedge_T E(it, \cdot)\|_{\infty,\Omega} + \| \wedge^T E(it, \cdot)\|_2,$$

for some $\mathsf{c} > 0$. We proceed to estimate

$$\mathscr{J}_T := \int_{\mathbb{R}} |R(it)| \cdot \| \wedge_T E(it, \cdot)\|_{\infty,\Omega}^2 \, d\mu_{\mathsf{n}}(t). \tag{4.4}$$

Observe that

$$\wedge_T E(it, g) = \sum_{\gamma \in \mathscr{S}} E_P(it, \gamma g) \mathscr{X}_T(\chi_P(\gamma g)),$$

where $\mathscr{S}$ is a finite subset of $P(\mathbb{Q})\backslash G(\mathbb{Q})$, depending only on $\mathfrak{G}$ and T, and $\mathscr{X}_T$ is the characteristic function of $[T, \infty)$. Thus

$$| \wedge_T E(it, g)| \leq \sum_{\gamma \in \mathscr{S}} |\chi(it, \gamma g)| + |\chi(-it, g)| \leqslant 2 \sum_{\gamma \in \mathscr{S}} \|\chi(0, \gamma \cdot)\|_{\infty,\Omega},$$

and the sum on the right side is bounded. It follows that the integral $\mathscr{J}_T$ converges, for $\mathsf{n} \gg 0$. Combined with (4.3) this proves (2) of Theorem 4.2.

An argument based on Sobolev's lemma (see the appendix or the proof of Lemma 3.4.7 [**15**]), shows that there exists a polynomial R such that

$$|E(it, g)| \leqslant |R(it)| \cdot \|E(it, \cdot)\|_{2,\Omega}.$$

Hence the integral in (3), Theorem 4.2, is bounded by $\mathscr{J}^T + \mathscr{J}_T$, from (4.3) and (4.4)). This proves (3) and (4).

5. Eisenstein integrals

For $g \in G(\mathbb{A})$, and $s, w \in \mathbb{C}$, we put (formally)

$$\hat{\mathscr{Z}}(s, w) := \int_{G(\mathbb{Q})\backslash G(\mathbb{A})} \mathscr{Z}(s, g) E(-w, g) \, dg. \tag{5.1}$$

The main technical result of this section is the following.

Proposition 5.1. *For $s \in \mathsf{T}_{-1}$, the integral*

$$\mathscr{Z}^{\mathrm{eis}}(s, g) := \frac{1}{2\pi} \int_{\mathbb{R}} \hat{\mathscr{Z}}(s, it) E(it, g) \, dt \tag{5.2}$$

(1) *is absolutely and uniformly convergent to a holomorphic in s and continuous in g function;*
(2) $\mathscr{Z}^{\mathrm{eis}}(s,g) \in \mathsf{L}^2$.

Proof. Since $\int_K E(s,gk)dk = E(s,e)\varphi(s,g)$, (formally) $\hat{Z}(s,w)$ equals

$$\int_{G(\mathbb{A})} H(g)^{-s} E(-w,g)\, dg = \left(\int_{G(\mathbb{A})} H(g)^{-s}\varphi(-w,g)\, dg\right)\cdot E(-w,e).$$

The local integrals

$$I_v(s,w) := \int_{G(\mathbb{Q}_v)} H_v(g)^{-s}\varphi_v(-w,g)dg_v$$

have been computed for $v = p$ in (3.5) (the character corresponding to the spherical function $\varphi_v(w,\cdot)$ is $|\cdot|_p^w$, see (3.1) and (3.2)). We get

$$I_f(s,w) = \prod_p I_p(s,w) = \frac{\zeta(\frac{s-1}{2}-w)\zeta(\frac{s-1}{2}+w)}{\zeta(s)}. \tag{5.3}$$

In particular, for $\varepsilon \in (0,1)$ and all $w = \sigma + it$ with $0 \leqslant \sigma \leqslant \varepsilon$, the map

$$s \mapsto I(s,w) = I_f(s,w)\cdot I_\infty(s,w)$$

is holomorphic for $s \in \mathsf{T}_{-1+\varepsilon}$. Furthermore, for s and w as above,

$$|I(s,\sigma+it)| \ll_{\mathsf{n},\varepsilon} (1+t^2)^{-\mathsf{n}}. \tag{5.4}$$

Indeed, the function $\varphi_\infty(it,\cdot)$ is bounded. Moreover, by Lemma 4.1, it is an eigenfunction for Δ. We now apply repeatedly integration by parts (with respect to Δ) combined with Lemma 3.3 to $I_\infty(s,w)$ and use the standard bounds for ζ.

We have

$$\int_{\mathbb{R}} |\hat{\mathscr{Z}}(s,it)|\cdot|E(it,g)|\, dt \leqslant \int_{\mathbb{R}} |I(s,it)|\cdot|E(it,e)|\cdot|E(it,g)|\, dt.$$

Let $\mathsf{K} \subset \mathsf{T}_{-1}$ and $\Omega \subset G(\mathbb{A})$ be compact sets. By the estimates above, for every $\mathsf{n} \in \mathbb{N}$, there is a constant $\mathsf{c} = \mathsf{c}(\mathsf{n},\mathsf{K})$ such that

$$\int_{\mathbb{R}} \sup_{s\in\mathsf{K}} |I(s,it)|\cdot|E(it,e)|^2\, dt \le \mathsf{c}\int_{\mathbb{R}} |E(it,e)|^2\, d\mu_{\mathsf{n}}(t). \tag{5.5}$$

To show the absolute and uniform convergence of the integral (5.2) for $(s,g) \in \mathsf{K}\times\Omega$ we need to check that, for $\mathsf{n} \gg 0$,

$$\int_{\mathbb{R}} |E(it,e)|\cdot\|E(it,\cdot)\|_{\infty,\Omega}\, d\mu_{\mathsf{n}}(t) < \infty,$$

or, by Cauchy–Schwartz, that

$$\int_{\mathbb{R}} |E(it,e)|^2 \, d\mu_{\mathsf{n}}(t) < \infty \quad \text{and} \quad \int_{\mathbb{R}} \|E(it,\cdot)\|_{\infty,\Omega} \, d\mu_{\mathsf{n}}(t) < \infty,$$

which follows from Theorem 4.2.

To prove that $\mathscr{Z}^{\mathrm{eis}}(s,\cdot) \in \mathsf{L}^2$, for $s \in \mathsf{T}_{-1}$, write $E_s := E(s,e)$,

$$2\pi \cdot \mathscr{Z}^{\mathrm{eis}}(s,g) = \mathscr{J}^T(s,g) + \mathscr{J}_T(s,g),$$

$$\mathscr{J}^T(s,g) := \int_{\mathbb{R}} I(s,it) E_{-it} \wedge^T E(it,g) \, dt,$$

similarly $\mathscr{J}_T(s,g)$, with $\wedge^T$ replaced by $\wedge_T$. We bound $\|\mathscr{J}^T(s,\cdot)\|_2^2$ as

$$\begin{aligned}
&\int_{G(\mathbb{Q})\backslash G(\mathbb{A})} \left| \int_{\mathbb{R}} I(s,it) E_{-it} \wedge^T E(it,g) dt \right| \cdot \left| \int_{\mathbb{R}} \overline{I(s,i\tau)} E_{i\tau} \overline{\wedge^T (i\tau,g)} d\tau \right| dg \\
&\leqslant \int_{\mathbb{R}^2} |I(s,it)I(s,i\tau)E_{it}E_{i\tau}| \int_{G(\mathbb{Q})\backslash G(\mathbb{A})} | \wedge^T E(it,g) \wedge^T E(i\tau,g)| dg dt d\tau \\
&\leqslant \int_{\mathbb{R}^2} |I(s,it)I(s,i\tau)E_{it}E_{i\tau}| \cdot \| \wedge^T E(it,\cdot)\|_2 \cdot \| \wedge^T E(i\tau,\cdot)\|_2 dt d\tau \\
&= \left(\int_{\mathbb{R}} |I(s,it)E_{it}| \cdot \| \wedge^T E(it,\cdot)\|_2 dt \right)^2
\end{aligned}$$

by repeatedly applying Fubini's theorem, and then Cauchy–Schwartz to the inner integral. Thus it suffices to bound the integrals

$$\int_{\mathbb{R}} |E(it,e)| d\mu_{\mathsf{n}}(t) \text{ and } \int_{\mathbb{R}} \| \wedge^T E(it,\cdot)\|_2^2 \, d\mu_{\mathsf{n}}(t),$$

which has been accomplished in Theorem 4.2.

We turn to $\mathscr{J}_T(s,\cdot)$. It suffices to consider the integral over $\mathfrak{S}_T$ (since $\mathfrak{S}^T$ is compact). For $T > 1$ we have, for all $g \in G(\mathbb{A})$,

$$\wedge_T E(it,g) = E_P(it,g) = \chi(it,g) + c(it)\chi(-it,g).$$

Substituting this into the definition of $\mathscr{J}_T$ (and rewriting the integral $\int_{\mathbb{R}}$) we see that we need to bound the $\mathsf{L}^2(\mathfrak{S}_T)$-norms of

$$\int_{\Re(w)=0} I(s,w)E_{-w}\chi(w,\cdot) dw \quad \text{and} \quad \int_{\Re(w)=0} I(s,w)E_{-w}c(w)\chi(-w,\cdot) \, dw.$$

By Corollary 4.4, $c(s)$ is bounded in the region $\Re(s) \in [0,\varepsilon]$, $\Im(s) \geq 1$ and we can shift the contour of integration of the first integral slightly to the left and that of the second to the right. This gives

$$\chi(-\varepsilon,\cdot)\int_{\Re(w)=-\varepsilon} I(s,w)E_{-w}dw \quad\text{and}\quad \chi(-\varepsilon,\cdot)\int_{\Re(w)=\varepsilon} I(s,w)E_{-w}dw.$$

Now it suffices to observe that, for $\varepsilon > 0$,

$$\|\chi(-\varepsilon,\cdot)\|_{2,\mathfrak{S}_T} = \int_T^\infty a^{2(-\varepsilon+\frac{1}{2})}a^{-1}da^* < \infty,$$

and to use the estimate (5.4). This completes the proof of Proposition 5.1 □

6. P-series

Definition 6.1. Let $\varphi : G(\mathbb{A}) \to \mathbb{C}$ satisfy

- $\varphi(ngk) = \varphi(g)$, for $n \in N(\mathbb{A}), k \in K$ and $g \in G(\mathbb{A})$;
- $\varphi(ag) = \varphi(g)$, for $a \in A(\mathbb{Q})$, $g \in G(\mathbb{A})$;
- $\varphi : A(\mathbb{Q})\backslash A(\mathbb{A}) \to \mathbb{C}$ is smooth on compact support.

For $g \in G(\mathbb{A})$, set

$$\theta_\varphi(g) := \sum_{\gamma\in P(\mathbb{Q})\backslash G(\mathbb{Q})} \varphi(\gamma g) \quad\text{and}\quad \hat{\theta}_\varphi(s) := \int_{G(\mathbb{Q})\backslash G(\mathbb{A})} \theta_\varphi(g)E(s,g)dg.$$

Since φ is left $P(\mathbb{Q})$-invariant and compactly supported modulo $P(\mathbb{Q})$ the sum defining θ_φ is finite so that $\theta_\varphi \in \mathsf{C}_c^\infty(G(\mathbb{Q})\backslash G(\mathbb{A}))$ and $\hat{\theta}_\varphi$ is meromorphic, analytic in $\Re(s) \geq 0$, except possibly at $s = 1/2$.

In the following we identify $\mathbb{A}^*$ with $A(\mathbb{A})$:

$$a := \begin{pmatrix} a & 0 \\ 0 & 1 \end{pmatrix},$$

write da^*, resp. da, for the corresponding Haar measure and regard φ as being (simultaneously) in $\mathsf{C}_c^\infty(\mathbb{Q}^* \cdot \hat{\mathbb{Z}}^*\backslash\mathbb{A}^*) = \mathsf{C}_c^\infty(\mathbb{R}_{>0}^*)$.

For $\Re(s) \gg 0$ we may use Fubini to derive

$$\begin{aligned}
\hat{\theta}_\varphi(s) &= \int_{P(\mathbb{Q})\backslash G(\mathbb{A})} \theta_\varphi(g)\chi(s,g)dg \\
&= \int_{N(\mathbb{A})A(\mathbb{A})\backslash G(\mathbb{A})} \theta_{\varphi,P}(g)\chi(s,g)dg \\
&= \int_{A(\mathbb{Q})\backslash A(\mathbb{A})} \theta_{\varphi,P}(a)\chi(s,a)|\delta(a)|_{\mathbb{A}}^{-1}da \\
&= \int_{\mathbb{Q}^*\backslash\mathbb{A}^*} \theta_{\varphi,P}(a)|a|_{\mathbb{A}}^{s-1/2}da^*,
\end{aligned}$$

where $\theta_{\varphi,P}$ is the constant term of θ_φ. By Bruhat's lemma,

$$\theta_\varphi(g) = \varphi(g) + \sum_{\gamma \in N(\mathbb{Q})} \varphi(\mathsf{w}\gamma g) \quad \text{and} \quad \theta_{\varphi,P}(g) = \varphi(g) + \int_{N(\mathbb{A})} \varphi(\mathsf{w}ng)dn,$$

(with w the nontrivial element of the Weyl group), so that

$$\hat{\theta}_\varphi(s) = \int_{\mathbb{Q}^*\backslash\mathbb{A}^*} \varphi(a)|a|_{\mathbb{A}}^{s-1/2} da^* + \int_{\mathbb{Q}^*\backslash\mathbb{A}^*} \int_{N(\mathbb{A})} \varphi(\mathsf{w}na)dn \cdot |a|_{\mathbb{A}}^{s-1/2} da^*.$$

To justify the above equation note that the double integral on the right is absolutely convergent. Indeed, for $g \in G(\mathbb{A})$ and φ as above, set

$$f_\varphi(s,g) = \int_{\mathbb{Q}^*\backslash\mathbb{A}^*} \varphi(ag)|a|_{\mathbb{A}}^{-1/2-s} da^*.$$

Then

$$f_\varphi(s,nag) = |a|_{\mathbb{A}}^{1/2+s} f_\varphi(s,g)$$

and the double integral may be written as

$$\int_{N(\mathbb{A})} f_\varphi(s,\mathsf{w}n)dn = c(s) \cdot f_\varphi(-s,e). \tag{6.1}$$

The Euler product defining $c(s)$ and hence the integral (6.1) converge for $\Re(s) \gg 0$. We have, for general φ, and, at first for $\Re(s) \gg 0$ and then, by analytic continuation, for all s,

$$\hat{\theta}_\varphi(s) - f_\varphi(-s,e) + c(s)f_\varphi(s,e). \tag{6.2}$$

Notice that $\hat{\theta}_\varphi(it)$ is rapidly decreasing ($f_\varphi(it,e)$ is essentially the Fourier transform of a function in $\mathsf{C}_c^\infty(\mathbb{R})$) and define, for $g \in G(\mathbb{A})$,

$$\theta_\varphi^c(g) := \frac{1}{2\pi} \int_{\mathbb{R}} \hat{\theta}_\varphi(it)E(it,g)dt. \tag{6.3}$$

Proposition 6.2. *If $\varphi \in \mathsf{C}_c^\infty(\mathbb{R}_{>0}^*)$, then θ_φ^c is continuous and in L^2.*

Proof. From (6.3) and (6.6) we have $|\theta_\varphi^c(g)| \le \mathscr{J}^T(g) + \mathscr{J}_T(g)$, with

$$\mathscr{J}^T(g) := \int_{\mathbb{R}} |\wedge^T E(it,g)| d\mu_{\mathsf{n}}(t) \quad \text{and} \quad \mathscr{J}_T(g) := \int_{\mathbb{R}} |\wedge_T E(it,g)| d\mu_{\mathsf{n}}(t).$$

Both integrals are finite by Theorem 4.2, (3). To prove continuity, let Ω be a pre-compact neighborhood of g and recall that (by the same theorem)

$$\int_{\mathbb{R}} \sup_{g\in\Omega} |E(it,g)| d\mu_{\mathsf{n}}(t) < \infty \quad \text{for} \quad \mathsf{n} \gg 0.$$

∎

Theorem 6.3. *For all* $\varphi \in \mathsf{C}_c^\infty(\mathbb{Q}^* \cdot \hat{\mathbb{Z}}^* \backslash \mathbb{A}^*)$ *one has*

$$\theta_\varphi^c = \theta_\varphi - \langle \theta_\varphi, \mathbf{1} \rangle \mathbf{1}.$$

Proof. We first prove the identity for constant terms:

$$\theta_{\varphi,P}^c = \theta_{\varphi,P} - \langle \theta_\varphi, \mathbf{1} \rangle \mathbf{1}.$$

By Theorem 4.2, $g \mapsto \int_{\mathbb{R}} |E(-it, g)| d\mu_{\mathsf{n}}(t)$ is continuous so that

$$\int_{N(\mathbb{Q})\backslash N(\mathbb{A})} \int_{\mathbb{R}} |\hat{\theta}_\varphi(it) E(-it, ng)| dt dn < \infty.$$

It follows that, for $g \in G(\mathbb{A})$,

$$\theta_{\varphi,P}^c(g) = \int_{\mathbb{R}} \hat{\theta}_\varphi(it) E_P(-it, g) dt = \int_{\mathbb{R}} \hat{\theta}_\varphi(it) \left(\chi(it, g) + c(-it)\chi(-it, g)\right) dt.$$

Since

$$c(-s)\hat{\theta}_\varphi(s) = f_\varphi(s, e) + c(-s) f_\varphi(-s, e) = \hat{\theta}_\varphi(-s)$$

we obtain that

$$\theta_{\varphi,P}^c(g) = \frac{1}{\pi} \int_{\mathbb{R}} \hat{\theta}_\varphi(it) \chi(it, g) dt. \tag{6.4}$$

We now compare $\theta_{\varphi,P}^c$ and $\theta_{\varphi,P}$. We have already seen that

$$\hat{\theta}_\varphi(s) = \int_{\mathbb{Q}^*\backslash\mathbb{A}^*} \theta_{\varphi,P}(a) |a|_{\mathbb{A}}^{s-1/2} da^*, \quad \text{for} \quad \Re(s) \gg 0. \tag{6.5}$$

For $\Re(s) \geq 1$, the function

$$a \mapsto \theta_{\varphi,P}(a) |a|_{\mathbb{A}}^{-1/2+s}$$

is left-invariant under $\hat{\mathbb{Z}}^*$ and in $\mathsf{L}^1(\mathbb{Q}^* \cdot \hat{\mathbb{Z}}^* \backslash \mathbb{A}^*) = \mathsf{L}_1(\mathbb{R}_{>0}^*)$. On the other hand, $c(s)$ is bounded in the region $0 \leq \Re(s) \leq \sigma_0$ (for any $\sigma_0 > 0$), $|\Im(s)| \geq 1$ (by Corollary 4.4). It follows from (6.2) that in this domain

$$\hat{\theta}_\varphi(\sigma + it) = \int_{\mathbb{Q}^*\backslash\mathbb{A}^*} \theta_{\varphi,P}(a) |a|_{\mathbb{A}}^{-1/2+\sigma} |a|_{\mathbb{A}}^{it} da^*$$

is rapidly decreasing in t and we may apply Fourier inversion so that

$$\theta_{\varphi,P}(a) = |a|_{\mathbb{A}}^{-\sigma+1/2} \int_{\mathbb{R}} \hat{\theta}_\varphi(\sigma + it) |a|_{\mathbb{A}}^{-it} dt. \tag{6.6}$$

But we also have

$$\theta_{\varphi,P}^c(g) = \frac{1}{2\pi i} \int_{\Re(s)=0} \hat{\theta}_\varphi(s) |a|_{\mathbb{A}}^{1/2-s} ds. \tag{6.7}$$

We shift the contour to $\Re(s) = \sigma$, taking $\sigma > 1/2$. The shift is justified by using (6.6) and the fact that $f_\varphi(\sigma + it, e)$ is rapidly decreasing in t, uniformly in σ, for $|\sigma| \leq \sigma_0$. It follows that

$$\theta^c_{\varphi,P}(a) = \frac{1}{2\pi i}\int_{\Re(s)=\sigma} \hat{\theta}_\varphi(s)|a|_{\mathbb{A}}^{1/2-s} ds - \mathrm{res}_{s=1/2}\hat{\theta}_\varphi(s)|a|_{\mathbb{A}}^{1/2-s}.$$

Now it suffices to compute the residue of $c(s)$ at $s = 1/2$ and to see that

$$f_\varphi(1/2, e) = \int_{\mathbb{Q}^*\backslash\mathbb{A}^*} \varphi(a)|a|_{\mathbb{A}}^{-1} da^* = \int_{G(\mathbb{Q})\backslash G(\mathbb{A})} \theta_\varphi(g)dg,$$

(for appropriately normalized measures).

Let us now set

$$\theta_\varphi^{\mathrm{cusp}} := \theta_\varphi - \langle\theta_\varphi, \mathbf{1}\rangle \cdot \mathbf{1} - \theta^c_\varphi.$$

We have proved that $\theta_\varphi^{\mathrm{cusp}}$ is continuous, in L_2 and has a vanishing constant term. Thus $\theta_\varphi^{\mathrm{cusp}}$ is a cusp form, i.e., orthogonal to all the P-series θ_φ. To complete the proof of Theorem 6.3 it suffices to prove that θ^c_φ is orthogonal to all cusp forms. Since $\theta^c_\varphi \in \mathsf{L}^2$, it will suffice to prove that

$$\langle\theta^c_\varphi, \psi * \alpha\rangle = 0$$

for all cusp forms ψ and all $\alpha \in \mathsf{C}_c^\infty(G(\mathbb{A}))$. Replacing ψ by $\psi * \alpha$ we may assume that ψ is rapidly decreasing (and continuous). Recall that

$$\theta^c_\varphi(g) = \int_{\mathbb{R}} \hat{\theta}_\varphi(t)E(-it, g)dt.$$

Since $\langle E(-it, \cdot), \psi\rangle = 0$, for all $t \in \mathbb{R}$, it suffices to prove that

$$\int_{G(\mathbb{Q})\backslash G(\mathbb{A})}\int_{\mathbb{R}} |E(it, g)\psi(g)|d\mu_{\mathsf{n}}(t)dg < \infty \quad \text{for } \mathsf{n} \gg 0.$$

As before, we consider the integral

$$\mathscr{J}^T := \int_{\mathfrak{S}}\int_{\mathbb{R}} |\wedge^T E(it, g)| \cdot |\psi(g)|d\mu_{\mathsf{n}}(t)dg$$

and a similar integral $\mathscr{J}_T$ with $\wedge^T$ replaced by $\wedge_T$. Using Theorem 4.2,

$$\mathscr{J}^T \ll \|\psi\|_2 \cdot \int_{\mathbb{R}} \| \wedge^T E(it, \cdot)\|_2 d\mu_{\mathsf{n}}(t) < \infty.$$

To treat the integral $\mathscr{J}_T$ we decompose $\mathfrak{S} = \mathfrak{S}^T \cup \mathfrak{S}_T$. Since $\mathfrak{S}_T$ is compact, ψ continuous and $|\wedge_T E(it, g)|$ bounded in this domain, the double integral is absolutely convergent. It remains to estimate

$$\int_{\mathfrak{S}^T}\int_{\mathbb{R}} |\wedge_T E(it, g)| \cdot |\psi(g)|d\mu_{\mathsf{n}}(t)dg.$$

To complete the proof of Theorem 6.3 observe that for $T > 1$,

$$| \wedge_T E(it, g)| = |E_P(it, g)| \leq 2\chi(0, g)$$

and, since ψ is rapidly decreasing,

$$\int_{\mathfrak{S}^T} |\psi(g)| \cdot \chi(0, g) dg < \infty.$$

□

Proposition 6.4. *For $\Re(s) \gg 0$, the function $\mathscr{Z}^{\mathrm{eis}}(s, \cdot)$ is orthogonal to all cusp forms and to the constant function.*

Proof. Recall that

$$\mathscr{Z}^{\mathrm{eis}}(s, g) = \frac{1}{2\pi i} \int_{\mathbb{R}} \hat{\mathscr{Z}}(s, it) E(-it, g) dt,$$

where

$$\hat{\mathscr{Z}}(s, it) = I(s, it) E(it, e) \quad \text{with} \quad I(s, w) = \int_{G(\mathbb{A})} H(g)^{-s} \varphi_w(g) dg.$$

As above, we show first that the double integral

$$\int_{G(\mathbb{Q})\backslash G(\mathbb{A})} \int_{\mathbb{R}} \hat{\mathscr{Z}}(s, it) E(-it, g) \psi(g) dt dg,$$

is absolutely convergent, for ψ continuous and bounded, and that it converges to zero. The last integral is majorized by

$$\int_{\mathbb{R}} \int_{G(\mathbb{Q})\backslash G(\mathbb{A})} |E(it, e)| \cdot |E(-it, g)| \cdot |\psi(g)| d\mu_{\mathsf{n}}(t) dg.$$

We set

$$\mathscr{J}^T := \int_{\mathbb{R}} \int_{G(\mathbb{Q})\backslash G(\mathbb{A})} |E(-it, e)| \cdot | \wedge^T E(it, g)| \cdot |\psi(g)| d\mu_{\mathsf{n}}(t) dg.$$

and a similar integral $\mathscr{J}_T$ (with $\wedge^T$ replaced by $\wedge_T$). By Cauchy–Schwartz, to treat $\mathscr{J}^T$ it suffices to consider

$$\int_{\mathbb{R}} |E(-it, e)| \cdot \| \wedge^T E(it, \cdot)\|_2 d\mu_{\mathsf{n}}(t),$$

and again by Cauchy–Schwartz, the two integrals

$$\int_{\mathbb{R}} |E(-it, e)|^2 d\mu_{\mathsf{n}}(t) \quad \text{and} \quad \int_{\mathbb{R}} \| \wedge^T E(it, \cdot)\|_2^2 d\mu_{\mathsf{n}}(t),$$

which are finite by Theorem 4.2. To bound $\mathscr{J}_T$ we decompose $\mathfrak{S} = \mathfrak{S}_T \cup \mathfrak{S}^T$. The contribution from (the compact) $\mathfrak{S}_T$ is estimated using the boundedness of

$\wedge_T E(it,\cdot)\cdot\psi(\cdot)$. To treat $\mathfrak{S}^T$ recall that in this domain, $\wedge_T E(it,\cdot) = E_P(it,\cdot)$ and

$$|E_P(it,g)\psi(g)| \leq \mathsf{c}\cdot\chi(0,g).$$

Using the fact that the right side is integrable on $\mathfrak{S}$ and, once again, Theorem 4.2, we see that $\mathscr{J}^T$ is also finite. Since

$$\int_{\mathbb{R}} |E(-it,e)|d\mu_{\mathsf{n}}(t) \tag{6.8}$$

is finite for $\mathsf{n} \gg 0$, the integral J^T is also finite.

As we have remarked, $\langle E(-it,\cdot),\psi(\cdot)\rangle = 0$, for $t\in\mathbb{R}$, at least if ψ is rapidly decreasing. It remains to recall Proposition 4.1 (4):

$$\int_{G(\mathbb{Q})\backslash G(\mathbb{A})} E(it,g)dg = 0.$$

This completes the proof. □

We will need the following

Lemma 6.5. *For all $\varphi\in\mathsf{C}_c^\infty(\mathbb{R}^*_{>0})$, we have*

$$\langle \mathscr{Z}^{\mathrm{eis}}(s,\cdot),\theta_\varphi\rangle = \langle \mathscr{Z}(s,\cdot),\theta_\varphi^c\rangle.$$

Proof. First observe that

$$\langle \mathscr{Z}^{\mathrm{eis}},\theta_\varphi\rangle = \langle \hat{\mathscr{Z}},\hat{\theta}_\varphi\rangle_{\mathsf{L}^2(\mathbb{R})} = \frac{1}{2\pi}\int_{\mathbb{R}} \hat{\mathscr{Z}}(s,it)\hat{\theta}_\varphi(t)dt. \tag{6.9}$$

For this it suffices to prove that

$$\int_{G(\mathbb{Q})\backslash G(\mathbb{A})}\int_{\mathbb{R}} |\hat{\mathscr{Z}}(s,it)E(it,g)\theta_\varphi(g)|dtdg < \infty.$$

We have $\hat{\mathscr{Z}}(s,it) = I(s,it)E(it,e)$, where, by (5.4), $I(s,it)$ is rapidly decreasing in t, and

$$\int_{G(\mathbb{A})\backslash G(\mathbb{A})}\int_{\mathbb{R}} |E(-it,e)E(it,g)\theta_\varphi(g)|d\mu_{\mathsf{n}}(t)dg < \infty,$$

for $\mathsf{n}\gg 0$, by Theorem 4.2. To show that

$$\langle \hat{\mathscr{Z}},\hat{\theta}_\varphi\rangle_{\mathsf{L}^2(\mathbb{R})} = \langle \mathscr{Z},\theta_\varphi^c\rangle$$

it suffices to check that, for $\Re(s)\gg 0$,

$$\int_{G(\mathbb{A})\backslash G(\mathbb{A})}\int_{\mathbb{R}} |\mathscr{Z}(s,g)E(it,g)\hat{\theta}_\varphi(it)|dtdg < \infty.$$

Since $\mathscr{X}(s,g)$ is bounded in g, for $\Re(s) \gg 0$, and $\hat{\theta}_\varphi(it)$ is rapidly decreasing, it suffices to recall that

$$\int_{\mathbb{R}} |E(it,\cdot)| d\mu_{\mathsf{n}}(t) \in \mathsf{L}^2, \quad \text{for} \quad \mathsf{n} \gg 0.$$

□

7. The cuspidal spectrum

Write $G(\mathbb{A}) = G(\mathbb{Q}) \cdot G(\mathbb{R}) \cdot K_f$ and let $\Gamma = \mathrm{pr}_\infty(G(\mathbb{Z}))$. The map

$$\begin{array}{rccc} j : & \mathsf{L}^2(G(\mathbb{Q})\backslash G(\mathbb{A})) & \to & \mathsf{L}^2(\Gamma\backslash G(\mathbb{R})) \\ & \varphi & \mapsto & \varphi|_{G(\mathbb{R})} \end{array}$$

is an isometry. We consider the right regular representation of $G(\mathbb{R})$ on

$$\mathsf{H} := \mathsf{L}^2(G(\mathbb{Q})\backslash G(\mathbb{A}))^{K_f} = \mathsf{L}^2(\Gamma\backslash G(\mathbb{R}))$$

and denote by $\mathsf{H}^\infty \subset \mathsf{H}$ the subspace of smooth vectors. If $M = \Gamma\backslash G(\mathbb{R})$ and ω is the gauge form on M whose accociated measure is a fixed right-invariant measure on $\Gamma\backslash G(\mathbb{R})$, then $\mathsf{H}^\infty = \mathsf{H}^\infty(M)$ as defined in the appendix. Write $\mathsf{L}_0^2 := \mathsf{L}_0^2(G(\mathbb{Q})\backslash G(\mathbb{A}))$ for the space of cusp forms, $\mathsf{H}_0 = \mathsf{L}_0^2(G(\mathbb{Q})\backslash G(\mathbb{A}))^{K_f}$ for the closed $G(\mathbb{R})$-stable subspace of H and let

$$\mathscr{P} : \mathsf{H} \to \mathsf{H}_0$$

be the orthogonal projection, mapping H^∞ to H_0^∞, the subspace of smooth vectors in H_0. We have decompositions

$$\mathsf{L}_0^2 = \oplus_\pi \mathsf{H}_\pi \subset \mathsf{L}^2(G(\mathbb{Q})\backslash G(\mathbb{A})), \quad \text{and} \quad \mathsf{H}_0 = \oplus \mathsf{H}_\pi^{K_f}, \tag{7.1}$$

as a $G(\mathbb{A})$-modules, into a countable direct sum of closed irreducible subspaces (each occurring with finite multiplicity). Let $\mathscr{A}_0$ be the set of all π occurring in H_0. Note that each $\mathsf{H}_\pi^{K_f}$ contains an essentially unique K_∞-fixed vector φ_π normalized so that $\|\varphi_\pi\|_2 = 1$. Let $\mathsf{H}_0(M) \subset \mathsf{L}^2(M,\omega)$ be the Hilbert subspace spanned by the φ_π's. We remark that the φ_π are necessarily in H^∞ and are eigenfunctions of Δ, with eigenvalue, say, λ_π. By Proposition 8.8, for $\varphi \in \mathsf{H}_0 \cap \mathsf{H}^\infty$, the Fourier series

$$\sum_{\pi \in \mathscr{A}_0} \langle \varphi, \varphi_\pi \rangle \varphi_\pi \tag{7.2}$$

converges to φ, uniformly of compact sets.

We apply the above arguments to the height zeta function as follows. For $\Re(s) \gg 0$, put $\mathscr{Z}^{\mathrm{res}} = \langle \mathscr{Z}, \mathbf{1}\rangle \mathbf{1}$ and define $\mathscr{Z}^{\mathrm{cusp}}$ so that

$$\mathscr{Z}(s, \cdot) = \mathscr{Z}^{\mathrm{res}}(s) + \mathscr{Z}^{\mathrm{cusp}}(s, \cdot) + \mathscr{Z}^{\mathrm{eis}}(s, \cdot).$$

Proposition 7.1. *For* $\Re(s) \gg 0$,

$$\mathscr{Z}^{\mathrm{cusp}}(s, \cdot) = \mathscr{P}(\mathscr{Z}(s, \cdot)) \in \mathsf{L}^2_0.$$

Proof. By Proposition 5.1,

$$\mathscr{Z}^{\mathrm{eis}}(s, \cdot) \in \mathsf{L}^2$$

for $\Re(s) \gg 0$ and is continuous. Same holds for all its Δ-derivatives. It follows that $\mathscr{Z}^{\mathrm{cups}}(s, \cdot) \in \mathsf{L}^2$ and is also continuous.

For $\varphi \in \mathsf{C}_c^\infty(\mathbb{R}^*_{>0})$ we have

$$\langle \mathscr{Z}^{\mathrm{cusp}}(s, \cdot), \theta_\varphi\rangle = \langle \mathscr{Z}(s, \cdot), \theta_\varphi\rangle - \langle \mathscr{Z}^{\mathrm{eis}}(s, \cdot), \theta_\varphi\rangle - \langle \mathscr{Z}(s, \cdot), \mathbf{1}\rangle \mathbf{1}.$$

By Lemma 6.5, $\langle \mathscr{Z}^{\mathrm{eis}}(s, \cdot), \theta_\varphi\rangle = \langle \mathscr{Z}(s, \cdot), \theta^{\mathrm{c}}_\varphi\rangle$ and, by Theorem 6.3,

$$\theta^{\mathrm{c}}_\varphi = \theta_\varphi - \langle \theta_\varphi, \mathbf{1}\rangle \mathbf{1}.$$

We get at once

$$\langle \mathscr{Z}^{\mathrm{cusp}}(s, \cdot), \theta_\varphi\rangle = 0$$

and $\mathscr{Z}^{\mathrm{cusp}} \in \mathsf{L}^2_0$. (Here we used the right K-invariance of $\mathscr{Z}^{\mathrm{cusp}}(s, \cdot)$.)

To prove that $\mathscr{Z}^{\mathrm{cusp}} = \mathscr{P}(\mathscr{Z})$, we need to show that

$$\langle \mathscr{Z}^{\mathrm{cusp}}(s, \cdot), \psi\rangle = \langle \mathscr{Z}(s, \cdot), \psi\rangle,$$

for all $\psi \in \mathsf{L}^2_0$. This is an immediate consequence of Proposition 6.4. □

We proceed to investigate the meromorphic properties of the Fourier expansion of $\mathscr{Z}^{\mathrm{cusp}}$ as in (7.2). For $v = p$, let φ_p be the corresponding local spherical function with parameter $\alpha_p \in \mathbb{C}^*$. They are given by (3.1) and (3.2). The crucial facts in our further analysis are

Theorem 7.2. [10] *For* $\pi = \otimes_\nu \pi_\nu \in \mathscr{A}_0$ *let* φ_p *be the normalized spherical function corresponding to* π_p *and* α_p *its parameter. Then*

$$|\Re(\alpha_p)| \le 1/6.$$

Remark 7.3. Any nontrivial uniform bound towards the Ramanujan conjecture suffices for our purposes.

Theorem 7.4. *For all* $\mathsf{r} > 0$ *there is a* $\mathsf{c} > 0$ *such that*

$$\mathscr{Z}_\Delta(s) := \sum_{\pi \in \mathscr{A}_0, \lambda_\pi \neq 0} \frac{\|\varphi\|^{\mathsf{r}}_\infty}{|\lambda_\pi|^{\mathsf{n}}} < \infty, \quad \textit{for all} \quad \mathsf{n} > \mathsf{c}. \tag{7.3}$$

Proof. Estimate the $\|\cdot\|_\infty$-norm of eigenfunctions in terms of the corresponding eigenvalues as in the appendix and use the following fact: there exist constants $\mathsf{c}, \mathsf{r} > 0$ such that the number of linearly independent Δ-eigenfunctions with eigenvalue less than B is bounded by $\mathsf{c}(1 + B^{\mathsf{r}})$, for all $B \geqslant 0$ (see **[12]**, for example). □

Lemma 7.5. *For all $\varepsilon > 0$ and $\mathsf{n} \in \mathbb{N}$ there is a constant $\mathsf{c} = \mathsf{c}(\varepsilon, \mathsf{n})$ such that for all $s \in \mathsf{T}_{-2/3+\varepsilon}$ and all π the function*

$$\langle \mathscr{Z}(s,\cdot), \varphi_\pi \rangle = \int_{G(\mathbb{Q})\backslash G(\mathbb{A})} \mathscr{Z}(s,g)\overline{\varphi_\pi(g)}dg$$

is holomorphic, with absolute value bounded by $\mathsf{c}\|\varphi\|_\infty |\lambda_\pi|^{-\mathsf{n}}$ (for $\lambda_\pi \neq 0$).

Proof. For φ_π (with $\lambda_\pi \neq 0$) and s such that $\mathscr{Z}(s,\cdot) \in \mathsf{L}^2$

$$\begin{aligned}\langle \mathscr{Z}(s,\cdot), \varphi_\pi \rangle &= \int_{G(\mathbb{A})} H(g)^{-s}\overline{\varphi_\pi(g)}\, dg = \lambda_\pi^{-\mathsf{n}} \int_{G(\mathbb{A})} H(g)^{-s}\overline{\Delta^{\mathsf{n}}\varphi_\pi(g)}\, dg \\ &= \lambda_\pi^{-\mathsf{n}} \int_{G(\mathbb{A})} \Delta^{\mathsf{n}} H(g)^{-s}\overline{\varphi_\pi(g)}\, dg.\end{aligned}$$

Since the (right) actions of Δ and K commute, $\Delta^{\mathsf{n}} H(g)^{-s}$ is invariant under K (which has volume 1), so that the above expression equals

(7.4)
$$\lambda_\pi^{-\mathsf{n}} \int_K \int_{G(\mathbb{A})} \Delta^{\mathsf{n}} H(kg)^{-s}\overline{\varphi_\pi(g)}dgdk = \lambda_\pi^{-\mathsf{n}} \int_{G(\mathbb{A})} \Delta^{\mathsf{n}} H(g)^{-s} \int_K \overline{\varphi_\pi(kg)}dkdg.$$

As is well known,

$$\int_K \varphi_\pi(kg)dk = \varphi_\pi(g)\cdot\varphi_\pi(e),$$

where φ_π is the spherical function attached to π, i.e.,

$$\varphi_\pi(g) = \langle \pi(g)\varphi_\pi, \varphi_\pi\rangle, \quad g \in G(\mathbb{A}).$$

Indeed, the functional

$$\varphi \mapsto \int_K \varphi(k\cdot)dk$$

is a bounded, left K-invariant functional on H_π^∞ (and H_π). Thus it is proportional to the functional

$$\varphi \mapsto \langle \varphi, \varphi_\pi\rangle.$$

Taking $\varphi = \pi(g)\varphi_\pi$, we find that the proportionality constant is $\varphi_\pi(e)$.

Thus the integral in (7.4) is computed as

$$I_f(s,\overline{\varphi})\cdot\int_{G(\mathbb{R})}\Delta^{\mathsf{n}}H_\infty(g_\infty)^{-s}\overline{\varphi_\infty(g_\infty)}\,dg_\infty\cdot\overline{\varphi_\pi(e)} \tag{7.5}$$

Combining Lemma 3.2 with Theorem 7.2 (giving $\mathsf{r}=1/6$), we find that $I_f(s,\overline{\varphi})$ is holomorphic for $s\in\mathsf{T}_{-2/3}$ and *uniformly* bounded by a constant $\mathsf{c}(\varepsilon)$ for $s\in\mathsf{T}_{-2/3+\varepsilon}$ (and $\varepsilon>0$). Since φ is bounded, Lemma 3.3 shows that

$$\int_{G(\mathbb{R})}\Delta^{\mathsf{n}}H_\infty(g_\infty)^{-s}\overline{\varphi_\infty(g_\infty)}\,dg_\infty$$

is absolutely convergent, in T_{-2}, to a holomorphic function which, in $\mathsf{T}_{-2+\varepsilon}$, is bounded by $\mathsf{c}\|\varphi\|_\infty$ for some constant $\mathsf{c}=\mathsf{c}(\varepsilon,\mathsf{n})$. □

Proposition 7.6. *For all $\varepsilon>0$ the function*

$$\mathscr{Z}^{\mathrm{cusp}}(s,g):=\sum_{\pi\in\mathscr{A}_0}\langle\mathscr{Z}(s,\cdot),\varphi_\pi\rangle\varphi_\pi(g) \tag{7.6}$$

is holomorphic in s and continuous in $g\in G(\mathbb{A})$ for all $s\in\mathsf{T}_{-2/3+\varepsilon}$.

Proof. Combine Theorem 7.4 and Lemma 7.5. The uniform convergence on compacts $\Omega\subset G(\mathbb{A})$ follows from the estimate

$$\sum_\pi\sup_{g\in\Omega}|\langle\mathscr{Z}(s,\cdot),\varphi_\pi\rangle\varphi_\pi(g)|\ll_{\mathsf{n},\varepsilon,\Omega}\sum_{\pi\in\mathscr{A}_0}\frac{\|\varphi_\pi\|_\infty^2}{|\lambda_\pi|^{\mathsf{n}}}$$

and the convergence of the spectral zeta function (7.3) for $\mathsf{n}\gg 0$. □

8. Appendix

We will use the following notations:

- $dx=dx_1\cdots dx_n$ - Lebesgue measure on $\mathbb{R}^n$, $|x|^2=x_1^2+\cdots+x_n^2$;
- $\alpha=(\alpha_1,\ldots,\alpha_n)\in\mathbb{N}^n$, $|\alpha|=\alpha_1+\cdots+\alpha_n$ and $\partial^\alpha=(\frac{\partial}{\partial x_1})^{\alpha_1}\cdots(\frac{\partial}{\partial x_n})^{\alpha_n}$ the corresponding differential operator;
- $\mathbb{B},\mathbb{B}'\subset\mathbb{R}^n$ - open balls such that the closure $\bar{\mathbb{B}}'\subset\mathbb{B}$;
- $\mathsf{H}^\infty(\mathbb{B}):=\{u\in\mathsf{C}^\infty(\mathbb{B})\,|\,\partial^\alpha u\in\mathsf{L}^2(\mathbb{B}),\quad\forall\alpha\}$;
- Δ - a fixed second order elliptic operator on $\mathbb{B}$;
- $u\mapsto\hat{u}$ the usual Fourier transform.

For $u\in\mathsf{C}_c^\infty(\mathbb{B})\subset\mathsf{C}_c^\infty(\mathbb{R}^n)$, put

$$\|u\|_{\mathbb{B}}^2:=\|u\|_{(2,0),\mathbb{B}}^2=\int_{\mathbb{B}}|u(x)|^2dx=\int_{\mathbb{R}^n}|u(x)|^2dx$$

and, more generally,

$$\|u\|_{(2,\mathsf{r}),\mathbb{B}} := \sum_{|\alpha|\leq \mathsf{r}} \|\partial^\alpha u\|_{\mathbb{B}}, \quad \|u\|^2_{(2,-1)} := \int_{\mathbb{R}^n} (1+|x|^2)^{-1} |\hat{u}(x)|^2 dx.$$

Since ∂^α preserves $\mathsf{H}^\infty(\mathbb{B})$, we can extend the norm $\|\cdot\|_{(2,\mathsf{r}),\mathbb{B}}$ to $\mathsf{H}^\infty(\mathbb{B})$.

Proposition 8.1. *For every* $\mathsf{r}\geq 2$, *there is a* $\mathsf{c}=\mathsf{c}_{\mathsf{r},\mathbb{B}}>0$ *such that*

$$\|u\|_{(2,\mathsf{r}),\mathbb{B}'} \leq \mathsf{c}\left(\|u\|_{\mathbb{B}} + \|\Delta u\|_{(2,\mathsf{r}-2),\mathbb{B}}\right), \quad \textit{for all} \quad u\in \mathsf{H}^\infty(\mathbb{B}). \tag{8.1}$$

Proof. Induction on r. Fix $\psi \in \mathsf{C}_c^\infty(\mathbb{B})$ with $0\leq\psi\leq 1$ and $\psi=1$ on $\bar{\mathbb{B}}'$. Let $v=u\cdot\psi\in\mathsf{C}^\infty(\mathbb{B})$. By Corollary 6.27, p. 267 in [**6**],

$$\|v\|_{(2,2),\mathbb{B}} \leq \mathsf{c}\left(\|\Delta v\|_{\mathbb{B}} + \|v\|_{\mathbb{B}}\right).$$

Next we have

$$\Delta v = \Delta(\psi\cdot u) = \psi\cdot\Delta u + \sum_{j=1}^n \psi_j \frac{\partial u}{\partial x_j} + \psi_0\cdot u, \tag{8.2}$$

with ψ_j $(0\leq j\leq n)$ fixed in $\mathsf{C}_c^\infty(\mathbb{B})$. To prove the assertion for $\mathsf{r}=2$ it suffices to show that with φ fixed in $\mathsf{C}^\infty(\mathbb{B})$,

$$\|\varphi\cdot\frac{\partial u}{\partial x_j}\|_{\mathbb{B}} \leq \mathsf{c}_1\left(\|u\|_{\mathbb{B}} + \|\Delta u\|_{\mathbb{B}}\right). \tag{8.3}$$

For this we write

$$\varphi\cdot\frac{\partial u}{\partial x_i} = \frac{\partial(u\varphi)}{\partial x_i} - u\frac{\partial\varphi}{\partial x_i}$$

and note that $w:=\varphi\cdot u\in\mathsf{C}_c^\infty(\mathbb{B})$. Apply (6.25), p. 262 of [**6**] to obtain

$$\|w\|_{(2,1),\mathbb{B}} \leq \mathsf{c}\left(\|\Delta w\|_{(2,-1)} + \|w\|_{\mathbb{B}}\right).$$

The following inequality will imply (8.3):

$$\|\Delta w\|_{(2,-1)} \leq \mathsf{c}_1\left(\|\Delta u\|_{\mathbb{B}} + \|u\|_{\mathbb{B}}\right). \tag{8.4}$$

Again, for φ as above, we have

$$\Delta w = \Delta(\varphi\cdot u) = \varphi\cdot\Delta u + \sum_{j=1}^n \varphi_j\frac{\partial u}{\partial x_j} + \varphi_0\cdot u,$$

with φ_j's fixed in $\mathsf{C}_c^\infty(\mathbb{B})$. To prove (8.4) it suffices to bound $\|\varphi \cdot \frac{\partial u}{\partial x_j}\|_{(2,-1)}$ (for $\varphi \in \mathsf{C}_c^\infty(\mathbb{B})$), or equivalently, $\|\frac{\partial w}{\partial x_j}\|_{(2,-1)}$. We have in fact

$$\|\frac{\partial w}{\partial x_j}\|_{(2,-1)}^2 = \int_{\mathbb{R}^n} (1+|x|^2)^{-1} x_j^2 |\hat{w}(x)|^2 dx$$
$$\leq \int_{\mathbb{R}^n} |\hat{w}(x)|^2 dx \leq \int_{\mathbb{R}^n} |w(x)|^2 dx \leq \|u\|_{\mathbb{B}}^2$$

This completes the proof of (8.1) for $\mathsf{r} = 2$.

Suppose that $\mathsf{r} \geq 2$ and assume that the claim holds for all s with $2 \leq \mathsf{s} \leq \mathsf{r}$. Choose a ball $\mathbb{B}_1 \subset \mathbb{R}^n$ with

$$\mathbb{B}' \subset \bar{\mathbb{B}}' \subset \mathbb{B}_1 \subset \bar{\mathbb{B}}_1 \subset \mathbb{B}.$$

For $\psi \in \mathsf{C}^\infty(\mathbb{B}_1)$ and v as before, we have

$$\|u\|_{(2,\mathsf{r}),\mathbb{B}'} \leq \|v\|_{(2,\mathsf{r}),\mathbb{B}_1} \leq \mathsf{c}(\|\Delta v\|_{(2,\mathsf{r}-2),\mathbb{B}_1} + \|v\|_{\mathbb{B}_1}), \tag{8.5}$$

again, by Corollary 6.27 in **[6]**. We need only bound $\|\Delta v\|_{(2,\mathsf{r}-2),\mathbb{B}_1}$. For this we have first from (8.2)

$$\partial^\alpha(\Delta v) = \partial^\alpha(\psi \cdot \Delta u) + \partial^\alpha(\psi_0 \cdot u) + \sum_{j=1}^n \partial^\alpha(\psi_j \frac{\partial u}{\partial x_j}), \tag{8.6}$$

where now $\psi_j \in \mathsf{C}_c^\infty(\mathbb{B}_1)$ (for $0 \leq j \leq n$). It suffices to bound the L^2-norms of the terms on the right in (8.6) (for $|\alpha| \leq \mathsf{r}-2$) in terms of $\|u\|_{\mathbb{B}}$ and $\|\Delta u\|_{(2,\mathsf{r}-2),\mathbb{B}}$. By Leibniz' rule,

$$\|\partial^\alpha(\psi_0 \cdot u)\|_{\mathbb{B}_1} \leq \mathsf{c}_2 \|u\|_{(2,r-2),\mathbb{B}_1},$$

and we may use induction on r, provided $\mathsf{r} \geq 4$. Suppose then that $\mathsf{r} = 3$ and set $w = \psi_0 \cdot u \in \mathsf{C}_c^\infty(\mathbb{B}_1)$. Since here $|\alpha| \leq 1$, we have trivially,

$$\|\partial^\alpha w\|_{\mathbb{B}_1} \leq \|w\|_{(2,1),\mathbb{B}_1}.$$

Applying **[6]** once more, this time to $\mathbb{B}_1$,

$$\|w\|_{(2,1),\mathbb{B}_1} \leq \mathsf{c}_1(\|\Delta w\|_{(2,-1)} + \|w\|_{\mathbb{B}_1})$$
$$\leq \mathsf{c}_2(\|\Delta u\|_{\mathbb{B}_1} + \|u\|_{\mathbb{B}_1})$$
$$\leq \mathsf{c}_2(\|\Delta u\|_{\mathbb{B}} + \|u\|_{\mathbb{B}}),$$

(the second inequality follows from (8.4)). Next we have, with $|\alpha| \leq \mathsf{r}-2$,

$$\|\partial^\alpha(\psi \Delta u)\|_{\mathbb{B}} \leq \mathsf{c}_3 \|\Delta u\|_{(2,\mathsf{r}-2),\mathbb{B}},$$

and finally

$$\|\partial^\alpha(\psi_j \frac{\partial u}{\partial x_j})\|_{\mathbb{B}_1} \leq \mathsf{c}_4 \|u\|_{(2,\mathsf{r}-1),\mathbb{B}}$$

and we may apply the induction hypothesis to arrive at (8.1). □

Corollary 8.2. *Suppose that* $\mathsf{r} > 0$ *is even. Then there is a constant* $\mathsf{c} = \mathsf{c}_\mathsf{r} > 0$ *such that for all* $u \in \mathsf{H}^\infty(\mathbb{B})$

$$\|u\|_{(2,\mathsf{r}),\mathbb{B}'} \leq \mathsf{c}\left(\|u\|_\mathbb{B} + \|\Delta u\|_\mathbb{B} + \cdots + \|\Delta^{\mathsf{r}/2} u\|_\mathbb{B}\right).$$

Proof. We use induction on r. The case $\mathsf{r} = 2$ follows from Proposition 8.1. Suppose that $\mathsf{r} \geq 4$ and choose a ball $\mathbb{B}_1 \subset \mathbb{R}^n$ so that

$$\mathbb{B}' \subset \bar{\mathbb{B}}' \subset \mathbb{B}_1 \subset \bar{\mathbb{B}}_1 \subset \mathbb{B}.$$

By Proposition 8.1,

$$\|u\|_{(2,\mathsf{r}),\mathbb{B}'} \leq \mathsf{c}'_\mathsf{r}\left(\|u\|_{\mathbb{B}_1} + \|\Delta u\|_{(2,\mathsf{r}-2),\mathbb{B}_1}\right).$$

Further, we have by induction

$$\|\Delta u\|_{(2,\mathsf{r}-2),\mathbb{B}_1} \leq \mathsf{c}_{\mathsf{r}/2-1}\left(\sum_{j=0}^{\mathsf{r}/2-1} \|\Delta^{j+1} u\|_\mathbb{B}\right).$$

The claim follows combining these two inequalities. □

The following form of Sobolev's lemma will be useful for our purposes.

Proposition 8.3. *Let* $\mathsf{r} > n/2$ *be an integer,* $\mathbb{B} \subset \mathbb{R}^n$ *an open ball and* $u \in \mathsf{H}^\infty(\mathbb{B})$*. Then* $u \in \mathsf{L}^\infty(\mathbb{B})$ *and there exists a* $\mathsf{c} = \mathsf{c}_{\mathsf{r},\mathbb{B}}$ *such that*

$$\sup_{x\in\mathbb{B}} |u(x)| \leq \mathsf{c}\|u\|_{(2,\mathsf{r}),\mathbb{B}}.$$

Corollary 8.4. *Assume in addition that* r *is even. Then there exists a* $\mathsf{c} = \mathsf{c}_{\mathsf{r},\mathbb{B}}$*, such that*

$$\sup_{x\in\mathbb{B}'} |u(x)| \leq \mathsf{c}\left(\sum_{j=0}^{\mathsf{r}/2} \|\Delta^j u\|_\mathbb{B}\right)$$

Let M be an n-dimensional manifold, ω a gauge form on M (a nowhere vanishing exterior n-form) and $d\mu$ the associated volume form. We define

$$\|f\|^2_\mathbb{B} := \int_\mathbb{B} |f(x)|^2 d\mu(x), \quad \text{and similarly} \quad \|f\|^2_M,$$

and the corresponding space $\mathsf{L}^2(M) = \mathsf{L}^2(M,\omega)$. Here $\mathbb{B}$ is a coordinate ball in M. We also fix a ball $\mathbb{B}'$ such that $\mathbb{B}' \subset \bar{\mathbb{B}}' \subset \mathbb{B}$. Let $\mathscr{G}$ be the Lie algebra of vector fields on M (for each $x \in M$, $\mathscr{G}_x = \mathscr{T}_x$, the tangent space to M at x). Let $\mathfrak{U} = \mathfrak{U}(\mathscr{G})$ be its universal enveloping algebra, regarded as the algebra of differential operators on M. Define $\mathsf{H}^\infty(M)$ to be the space of all C^∞-functions

$f : M \to \mathbb{C}$ such that $\partial f \in \mathsf{L}^2(M)$ for all $\partial \in \mathfrak{U}$. We fix a second order elliptic operator $\Delta \in \mathfrak{U}$.

Proposition 8.5. *Let* $\mathsf{r} > n/2$ *be an even integer. Then there exists a* $\mathsf{c} = \mathsf{c}_{\mathbb{B}} > 0$ *such that for all* $f \in \mathsf{H}^\infty(M)$ *one has*

$$\sup_{x\in\mathbb{B}'} |f(x)| \le \mathsf{c}\left(\sum_{j=0}^{\mathsf{r}/2} \|\Delta^j f\|_M\right).$$

Proof. Easy consequence of Corollary 8.4. □

Corollary 8.6. *Let* K *be a compact and* $\mathsf{r} > n/2$ *an even integer. Then there exists a constant* $\mathsf{c} = \mathsf{c}_K > 0$ *such that for all* $f \in \mathsf{H}^\infty(M)$ *one has*

$$\sup_{x\in K} |f(x)| \le \mathsf{c}\left(\sum_{j=0}^{\mathsf{r}/2} \|\Delta^j f\|_M\right).$$

Corollary 8.7. *For any even integer* $\mathsf{r} > n/2$ *there exists a constant* $\mathsf{c} = \mathsf{c}_{r,\mathbb{B}} > 0$ *such that for any* Δ*-eigenfunction* $f = f_\lambda \in \mathsf{C}^\infty(M)$ *with eigenvalue* $\lambda \neq 0$ *and any* $\mathbb{B}' \subset \mathbb{B}$ *as above one has*

$$\sup_{x\in\mathbb{B}'} |f(u)| \le \mathsf{c}|\lambda|^{\mathsf{r}/2}\|f\|_{\mathbb{B}}$$

Let M, ω, Δ be as above. Assume that

$$\langle \Delta f, f'\rangle = \langle f, \Delta f'\rangle, \quad \text{for all} \quad f, f' \in \mathsf{H}^\infty(M),$$

and let $\{f_j\}_{j\ge1}$ be an orthonormal sequence of Δ-eigenfuctions in $\mathsf{H}^\infty(M)$. Let $\mathsf{H}_0(M) \subset \mathsf{L}^2(M)$ be the Hilbert subspace spanned by the f_j's and

$$\mathsf{H}_{0,\infty} := \{f \in \mathsf{H}^\infty(M) \,|\, \Delta^j f \in \mathsf{H}_0(M) \quad \text{for all} \quad j\}.$$

Proposition 8.8. *Suppose that* $f \in \mathsf{H}_{0,\infty}(M)$. *Then the series*

$$\sum_{j\ge1} \langle f, f_j\rangle f_j$$

converges uniformly on compacts, and in particular, pointwise, to f.

Proof. Write $\Delta f_j = \lambda_j f_j$ (and note that $\lambda_j \in \mathbb{R}$). For $f \in \mathsf{H}_{0,\infty}(M)$ and $\mathsf{n} \in \mathbb{N}$ we set

$$f^{[\mathsf{n}]} := \sum_{j=1}^{\mathsf{n}} \langle f, f_j\rangle f_j.$$

We have

$$\Delta^{\mathsf{k}} f^{[\mathsf{n}]} = \sum_{j=1}^{\mathsf{n}} \langle \Delta^{\mathsf{k}} f, f_j \rangle f_j = \sum_{j=1}^{\mathsf{n}} \langle f, \Delta^{\mathsf{k}} f_j \rangle f_j = \sum_{j=1}^{\mathsf{n}} \langle f, f_j \rangle \lambda_j^{\mathsf{k}} f_j = \Delta^{\mathsf{k}} f^{[\mathsf{n}]}.$$

Consequently, for given $\varepsilon > 0$, we have

$$\|\Delta^{\mathsf{k}}(f - f^{[\mathsf{n}]})\|_M \leq \varepsilon$$

for $1 \leq \mathsf{k} \leq \mathsf{r}/2$ (with r as above), provided $\mathsf{n} \gg 0$. Since $f - f^{[\mathsf{n}]}$ belongs to $\mathsf{H}^\infty(M)$, the proof follows immediately from Corollary 8.6. □

References

[1] V. V. Batyrev & Y. I. Manin – Sur le nombre des points rationnels de hauteur borné des variétés algébriques, *Math. Ann.* **286** (1990), no. 1-3, 27–43.

[2] V. V. Batyrev & Yu. Tschinkel – Manin's conjecture for toric varieties, *J. Algebraic Geom.* **7** (1998), no. 1, 15–53.

[3] V. V. Batyrev & Y. Tschinkel – Tamagawa numbers of polarized algebraic varieties, *Nombre et répartition des points de hauteur bornée*, Astérisque, no. 251, 1998, 299–340.

[4] A. Chambert-Loir & Y. Tschinkel – On the distribution of points of bounded height on equivariant compactifications of vector groups, *Invent. Math.* **148** (2002), no. 2, 421–452.

[5] C. De Concini & C. Procesi – Complete symmetric varieties, Invariant theory (Montecatini, 1982), Lecture Notes in Math., vol. 996, Springer, Berlin, 1983, 1–44.

[6] G. B. Folland – *Introduction to Partial Differential Equations*, Princeton University Press, Princeton, N.J., 1976.

[7] J. Franke, Y. I. Manin & Y. Tschinkel – Rational points of bounded height on Fano varieties, *Invent. Math.* **95** (1989), no. 2, 421–435.

[8] S. Gelbart & H. Jacquet – Forms of GL(2) from the analytic point of view, Automorphic Forms, Representations and L-functions, Part 1, Proc. Sympos. Pure Math., XXXIII, Amer. Math. Soc., Providence, R.I., 1979, 213–251.

[9] S. Gelbart & F. Shahidi – Boundedness of automorphic L-functions in vertical strips, *J. Amer. Math. Soc.* **14** (2001), no. 1, 79–107 (electronic).

[10] H. H. Kim & F. Shahidi – Functorial products for GLb2 × GLb3 and the symmetric cube for GLb2, *Ann. of Math. (2)* **155** (2002), no. 3, 837–893.

[11] A. W. Knapp – *Representation Theory of Semisimple Groups*, Princeton Landmarks in Mathematics, Princeton University Press, Princeton, NJ, 2001.

[12] W. Müller – The trace class conjecture in the theory of automorphic forms, *Ann. of Math. (2)* **130** (1989), no. 3, 473–529.

[13] ———, On the spectral side of the Arthur trace formula, 2002, preprint.

[14] E. PEYRE – Hauteurs et mesures de Tamagawa sur les variétés de Fano, *Duke Math. J.* **79** (1995), 101–218.

[15] D. RAMAKRISHNAN – Modularity of the Rankin–Selberg L-series, and multiplicity one for SL(2), *Ann. of Math. (2)* **152** (2000), no. 1, 45–111.

[16] J. SHALIKA, R. TAKLOO-BIGHASH & Y. TSCHINKEL – Rational points on compactifications of anisotropic forms of semi-simple groups, 2003.

[17] J. SHALIKA & Y. TSCHINKEL – Height zeta functions of equivariant compactifications of the Heisenberg group, `ArXiV:math.NT/0203093`, to appear, 2003.

[18] M. STRAUCH & Y. TSCHINKEL – Height zeta functions of toric bundles over flag varieties, *Selecta Math. (N.S.)* **5** (1999), no. 3, 325–396.

Arithmetic of Higher-dimensional Algebraic Varieties
(B. POONEN, YU. TSCHINKEL, eds.), p. 235–257
Progress in Mathematics, Vol. 226, © 2004 Birkhäuser Boston, Cambridge, MA

WEAK APPROXIMATION ON DEL PEZZO SURFACES OF DEGREE 4

Sir Peter Swinnerton-Dyer
DPMMS, Centre for Mathematical Sciences, University of Cambridge,
Wilberforce Road, Cambridge, CB3 0WB, UK
E-mail : H.P.F.Swinnerton-Dyer@dpmms.cam.ac.uk

Abstract. Let V be a Del Pezzo surface of degree 4 over a number field k such that $V(k) \neq \emptyset$. We prove that $V(k) = V(\mathbf{A})^{\mathrm{Br}}$.

1. Introduction

In 1991 Salberger and Skorobogatov [**2**] proved that the only obstruction to weak approximation on Del Pezzo surfaces of degree 4 (that is, smooth intersections of two quadrics in $\mathbf{P}^4$) defined over number fields is the Brauer–Manin obstruction. More precisely, their result is as follows.

Theorem 1. *Let k be an algebraic number field and V a Del Pezzo surface of degree 4 defined over k and such that $V(k)$ is not empty. Let $\mathscr{A}$ be the subset of the adelic space $V(\mathbf{A}_k)$ consisting of the points $\prod P_v$ such that*

$$\sum \mathrm{inv}_v(A(P_v)) = 0 \text{ in } \mathbf{Q}/\mathbf{Z}$$

for all A in the Brauer group $\mathrm{Br}(V)$. Then the image of $V(k)$ is dense in $\mathscr{A}$.

The proof in [**2**] has two main ingredients:

– an explicit analysis of the geometry of certain torsors;

– an approximation argument, now known as Salberger's device.

In their use of torsors, Salberger and Skorobogatov needed the whole descent theory of Colliot-Thélène and Sansuc.

The present paper has three main objectives. The overall one is to give a simpler proof of Theorem 1. In order to do this, I have to provide substitutes for the two ingredients listed above, and also for the Brauer–Manin obstruction. Two of these substitutes are of independent interest, one being the Legendre–Jacobi functions and the other a new version of Salberger's device. Those two are described in Sections 2 and 3 respectively. The geometric argument in [**2**] is replaced by studies of the geometry of V and of the cubic surface W which is birationally equivalent to V and is described later in this section. The reason for introducing W is to show that V is fibred by a pencil of curves of genus 0, and to give an explicit description of these curves. This is done in Section 4, which concludes by translating Theorem 1 into an equivalent form (Theorem 2) in which the Brauer group does not explicitly appear. The final geometric argument on V occupies Section 5.

In the notation of Theorem 1, let P be a point of $V(k)$. By blowing up P, we obtain a nonsingular cubic surface W which is birationally equivalent to V and contains a line L defined over k. By considering the residual intersections of W with a pencil of planes through L, we obtain a fibration of W (and thus also of V) by a pencil of conics. For such a pencil, the Brauer group and the associated Brauer–Manin obstructions have been known for a considerable time, through the combined work of a number of Russian geometers. In particular the Brauer group is entirely algebraic (that is, it is killed by going to the algebraic closure $\bar{k}$) and it is killed by 2. There is therefore an alternative description of the Brauer–Manin obstruction in terms of the Legendre–Jacobi functions, which are conceptually and computationally simpler, though some readers find them notationally more complicated. These were first introduced in a rather crude form in [**3**] and more correctly in [**4**]; in [**1**] there is an account of their relationship with the Brauer–Manin obstructions. In Section 2 we give a more complete account of the Legendre–Jacobi functions than has hitherto appeared in print; and in Section 4 we use this to replace Theorem 1 by the equivalent Theorem 2. Readers who are content with Theorem 2 need not trouble themselves with the Brauer–Manin conditions. Theorem 3, though it is a prerequisite for the proof of Theorem 2, is also of independent interest, since it adds an approximation property to Theorem 5.1 of [**1**] for pencils of conics. However Colliot-Thélène has pointed out to me that a similar approximation theorem is implicit in Theorem 6.2 of [**1**].

A number of papers of which I have been the author or a co-author have depended on Schinzel's Hypothesis. Though this is generally believed to be true, there seems no likelihood of it being proved in the foreseeable future. Roughly speaking, it asserts that if $P_1(X), \ldots, P_n(X)$ are monic irreducible polynomials in $\mathbf{Z}[X]$ then we can find arbitrarily large x in $\mathbf{Z}$ such that all the $P_\nu(x)$ are prime unless there is an obvious obstruction to doing so. In many contexts the use of Schinzel's Hypothesis can be replaced, at the cost of some extra complications, by what has come to be known as Salberger's device. Ideally, this would say that for the given set of $P_\nu(X)$ and a pre-assigned large enough N one can find an algebraic number field k of degree exactly N over $\mathbf{Q}$ and an integer ξ in k such that all the ideals $(P_\nu(\xi))$ are prime — subject to the same obstruction as appears in Schinzel's Hypothesis. In practice, one only knows how to prove slightly weaker results; and the exposition is further complicated by the desirability of setting the result in a weak approximation context. Previous versions of Salberger's device have depended on a deep result of Waldschmidt. In Section 3 I give a formulation and proof which avoids this. But the reader is warned that the argument, though elementary, is intricate. In particular, the construction which provides the proof contains several steps which appear mysterious because the need for them only becomes apparent much later than the steps themselves.

I am grateful to Jean-Louis Colliot-Thélène and Alexei Skorobogatov for a number of valuable comments. The research described in this paper was completed at a conference in the American Institute of Mathematics in December 2002; I am grateful to that Institute for its hospitality.

2. The Legendre–Jacobi functions

Let $F(U,V), G(U,V)$ be homogeneous coprime square-free polynomials in $k[U,V]$. Some of the more interesting results in this section only hold when $\deg F$ is even, but this condition nearly always holds in applications; it did not appear in **[4]**, but it is already needed if we are to make use of the results of **[1]**. Let $\mathscr{B}$ be a finite set of places of k containing the infinite places, the primes dividing 2, those at which any coefficient of F or G is not integral, and any other primes $\mathfrak{p}$ at which FG does not remain separable when reduced mod $\mathfrak{p}$. However, we need not assume that $\mathscr{B}$ contains a base for the ideal class group of k.

Denote by $\mathfrak{o}$ the ring of integers of k. Let $\mathscr{N}^1 = \mathscr{N}^1(k) = \mathbf{P}^1_k$ be $k \cup \{\infty\}$ and let $\mathscr{N}^2 = \mathscr{N}^2(k)$ be the set of $\alpha \times \beta$ with α, β integral and coprime outside

$\mathscr{B}$. For $\alpha \times \beta$ in $\mathbf{A}^2(k)$ with α, β not both zero, we shall write $\lambda = \alpha/\beta$ with λ in $\mathscr{N}^1(k)$. Provided $F(\alpha,\beta)$ and $G(\alpha,\beta)$ are nonzero, we define the function

$$L(\mathscr{B}; F, G; \alpha, \beta) : \alpha \times \beta \mapsto \prod_{\mathfrak{p}} (F(\alpha,\beta), G(\alpha,\beta))_{\mathfrak{p}} \tag{1}$$

on $\mathscr{N}^2$, where the outer bracket on the right is the multiplicative Hilbert symbol and the product is taken over all primes $\mathfrak{p}$ of k outside $\mathscr{B}$ which divide $G(\alpha,\beta)$. By the definition of $\mathscr{B}$, $F(\alpha,\beta)$ is a unit at any such prime. Clearly we can restrict the product in (1) to those $\mathfrak{p}$ which divide $G(\alpha,\beta)$ to an odd power; thus we can also write this product as $\prod \chi_{\mathfrak{p}}(F(\alpha,\beta))$ where $\chi_{\mathfrak{p}}$ is the quadratic character mod $\mathfrak{p}$ and the product is taken over all $\mathfrak{p}$ outside $\mathscr{B}$ which divide $G(\alpha,\beta)$ to an odd power. This relationship with the quadratic residue symbol underlies the proof of Lemma 1.

The function L does depend on $\mathscr{B}$, but the effect on the right hand side of (1) if we increase $\mathscr{B}$ is obvious. What makes the function L important is that, somewhat unexpectedly, it is continuous in the topology on $\mathscr{N}^2$ induced by $\mathscr{B}$. Although in the applications we can usually take $\deg F$ even, in the course of the proofs we need to consider functions (1) with $\deg F$ odd; and for this reason it is expedient to introduce

$$M(\mathscr{B}; F, G; \alpha, \beta) = L(\mathscr{B}; F, G; \alpha, \beta)(L(\mathscr{B}; U, V; \alpha, \beta))^{(\deg F)(\deg G)},$$

which is the same as L whenever F or G has even degree. Here of course $L(\mathscr{B}; U, V; \alpha, \beta) = \prod (\alpha, \beta)_{\mathfrak{p}}$ taken over all $\mathfrak{p}$ outside $\mathscr{B}$ which divide β.

Lemma 1. *The value of M is a continuous function of $\alpha \times \beta$ in the topology induced on $\mathscr{N}^2$ by $\mathscr{B}$. For each v in $\mathscr{B}$ there is a function $m(v; F, G; \alpha, \beta)$ with values in $\{\pm 1\}$ continuous on $\mathscr{N}^2$ in the v-adic topology and such that*

$$M(\mathscr{B}; F, G; \alpha, \beta) = \prod_{v \in \mathscr{B}} m(v; F, G; \alpha, \beta). \tag{2}$$

Proof. If $\deg F$ is even, so that $M = L$, the neatest proof of the lemma is by means of the evaluation formula in [**1**], Lemma 7.2.4. The case when $\deg G$ is even then follows from (4), and (3) gives the general case. (The proof in [**1**] is for $k = \mathbf{Q}$, but it can easily be modified to cover all k.) However, the proof which we shall give, using the ideas of [**4**], provides a more convenient method of evaluation.

For this proof we have to impose on $\mathscr{B}$ the additional condition that it contains all primes whose absolute norm does not exceed $\deg(FG)$. As the proof in [**1**] shows, this condition is not needed for the truth of Lemma 1 itself; but we use it in the proof of (8) below, and the latter is crucial to the subsequent argument. In any case, to classify all small enough primes as bad is quite usual. In the proof of the lemma, we repeatedly use the fact that $L(\mathscr{B}; F, G)$ and $M(\mathscr{B}; F, G)$ are multiplicative in both F and G; the effect of this is that we can reduce to the case when both F and G are irreducible in $\mathfrak{o}_{\mathscr{B}}[U, V]$, where $\mathfrak{o}_{\mathscr{B}}$ is the ring of elements of k integral outside $\mathscr{B}$. Introducing M and dropping the parity condition on $\deg F$ are not real generalizations since if we further increase $\mathscr{B}$ so that the leading coefficient of F is a unit outside $\mathscr{B}$ then

$$M(\mathscr{B}; F, G) = L(\mathscr{B}; F, GV^{\deg G}) \tag{3}$$

by (5), and we can apply (4) to the right hand side. By doing this, we reduce to the case when $\deg F$ is even and therefore $M = L$.

It follows from the product formula for the Hilbert symbol that

$$L(\mathscr{B}; f, g; \alpha, \beta) L(\mathscr{B}; g, f; \alpha, \beta) = \prod_{v \in \mathscr{B}} (f(\alpha, \beta), g(\alpha, \beta))_v, \tag{4}$$

subject to conditions on $\mathscr{B}$ analogous to those stated before (1). The right hand side of (4) is the product of continuous terms each of which only depends on a single v in $\mathscr{B}$. This formula enables us to interchange F and G when we want to, and in particular to require that $\deg F \geqslant \deg G$ in the reduction process which follows. We also have

$$L(\mathscr{B}; f, g; \alpha, \beta) = L(\mathscr{B}; f - gh, g; \alpha, \beta) \tag{5}$$

for any homogeneous h in $k[U, V]$ with $\deg h = \deg f - \deg g$ provided the coefficients of h are integral outside $\mathscr{B}$, because corresponding terms in the two products are equal. Both (4) and (5) also hold for M.

We deal first with two special cases:

- G is a constant. Now $M(\mathscr{B}; F, G) = 1$ because all the prime factors of G must be in $\mathscr{B}$, so that the product in the definition of $L(\mathscr{B}; F, G)$ is empty.
- $G = V$. Choose H so that $F - GH = \delta U^{\deg F}$ for some nonzero δ. Now $M(\mathscr{B}; F, G) = 1$ follows from the previous case and (5), since all the prime factors of δ must be in $\mathscr{B}$.

We now argue by induction on $\deg(FG)$. We can assume that F and G are irreducible, and by (4) we need only consider the case when

$$\deg F \geqslant \deg G > 0, \quad G = \gamma U^{\deg G} + \dots, \quad F = \delta U^{\deg F} + \dots$$

for some nonzero γ, δ. Let $\mathscr{B}_1$ be obtained by adjoining to $\mathscr{B}$ those primes of k not in $\mathscr{B}$ at which γ is not a unit. By (5) we have

$$M(\mathscr{B}_1; F, G) = M(\mathscr{B}_1; F - \gamma^{-1} \delta G U^{\deg F - \deg G}, G). \tag{6}$$

By taking a factor V out of the middle argument on the right, and using (4), the second special case above and the induction hypothesis, we see that $M(\mathscr{B}_1; F, G)$ is continuous in the topology induced by $\mathscr{B}_1$ and is a product taken over all v in $\mathscr{B}_1$ of continuous terms each one of which depends on only one of the v. Hence the same is true of $M(\mathscr{B}; F, G)$, because this differs from $M(\mathscr{B}_1; F, G)$ by finitely many continuous factors, each of which depends only on one prime in $\mathscr{B}_1 \setminus \mathscr{B}$.

But $\mathscr{B}_1 \setminus \mathscr{B}$ only contains primes whose absolute norm is greater than $\deg(FG)$. Thus by adding a suitable multiple of U to V we can arrange that $G = \gamma_1 U^{\deg G} + \dots$ and $F = \delta_1 U^{\deg F} + \dots$ where γ_1 is a unit at each prime in $\mathscr{B}_1 \setminus \mathscr{B}$. Let $\mathscr{B}_2$ be obtained from $\mathscr{B}$ by adjoining all the primes at which γ_1 is not a unit; then $M(\mathscr{B}; F, G)$ has the same properties with respect to $\mathscr{B}_2$ that we have already shown that it has with respect to $\mathscr{B}_1$. Hence $M(\mathscr{B}; F, G)$ already has these properties with respect to $\mathscr{B}$ because $\mathscr{B}_1 \cap \mathscr{B}_2 = \mathscr{B}$. □

There may be finitely many values of α/β for which the right hand side of (2) appears to be indeterminate; but by means of a preliminary linear transformation on U, V one can in fact ensure that the formula is meaningful except when $F(\alpha, \beta)$ or $G(\alpha, \beta)$ vanishes.

When $\deg F$ is even, the value of $L(\mathscr{B}; F, G; \alpha, \beta)$ is already determined by $\lambda = \alpha/\beta$ regardless of the values of α and β separately; here λ lies in $k \cup \{\infty\}$ with the roots of $F(X, 1)$ and $G(X, 1)$ deleted. We shall therefore also write $L(\mathscr{B}; F, G; \alpha, \beta)$ as $L(\mathscr{B}; F, G; \lambda)$. But even if $\mathscr{B}$ contains a base for the ideal class group of k, $L(\mathscr{B}; F, G; \lambda)$ is not necessarily a continuous function of λ; see the discussions in **[3]** and Section 9 of **[1]**, and Lemma 4 below. Moreover if $\mathscr{B}$ does not contain a base for the ideal class group of k, then not all elements of $k \cup \{\infty\}$ can be written in the form α/β with α, β integers coprime outside $\mathscr{B}$; so as yet $L(\mathscr{B}; F, G; \lambda)$ has not been defined for all λ. In the case when $\deg F$ is even, we can modify the definition (1) so that it extends to all $\alpha \times \beta$ in $k \times k$ such that $F(\alpha, \beta)$ and $G(\alpha, \beta)$ are nonzero. For any such α, β and any $\mathfrak{p}$ not in $\mathscr{B}$, choose $\alpha_\mathfrak{p}, \beta_\mathfrak{p}$ integral at $\mathfrak{p}$, not both divisible by $\mathfrak{p}$ and such that $\alpha/\beta = \alpha_\mathfrak{p}/\beta_\mathfrak{p}$. Write

$$L(\mathscr{B}; F, G; \alpha, \beta) = \prod (F(\alpha_\mathfrak{p}, \beta_\mathfrak{p}), G(\alpha_\mathfrak{p}, \beta_\mathfrak{p}))_\mathfrak{p} \tag{7}$$

where the product is taken over all $\mathfrak{p}$ not in $\mathscr{B}$ such that $\mathfrak{p} | G(\alpha_\mathfrak{p}, \beta_\mathfrak{p})$. This is a finite product whose value does not depend on the choice of the $\alpha_\mathfrak{p}$ and $\beta_\mathfrak{p}$; indeed it only depends on $\lambda = \alpha/\beta$ and when α, β are integers coprime outside $\mathscr{B}$ it is the same as the function given by (1). Thus we can again write it as $L(\mathscr{B}; F, G; \lambda)$. This generalization is not needed until we come to (11); and even there we only need it if we want $\mathscr{B}$ to be independent of K.

Its disadvantage is that L thus modified is no longer a continuous function of $\alpha \times \beta$ because we have dropped the condition that α and β be coprime outside $\mathscr{B}$. We investigate this situation in more detail after the proof of Lemma 3.

Although $L(\mathscr{B}; F, G; \alpha, \beta)$ is a continuous function of $\alpha \times \beta$ whose value only depends on $\lambda = \alpha/\beta$, it is not necessarily continuous as a function of λ. In discussing the continuity properties of L as a function of λ, we shall need the following lemma.

Lemma 2. *Let $\lambda_0 = \alpha_0/\beta_0$ with α_0, β_0 non-zero and integral outside $\mathscr{B}$; and let $\mathfrak{a}$ be an integral ideal in k not divisible by any prime in $\mathscr{B}$. Then we can find α, β in k, integral outside $\mathscr{B}$, with $(\alpha, \beta) = \mathfrak{a}(\alpha_0, \beta_0)$ and such that $\alpha \times \beta$ is arbitrarily close to $\alpha_0 \times \beta_0$ at each finite prime in $\mathscr{B}$, α/β is arbitrarily close to α_0/β_0 at each infinite place of k and α/α_0 and β/β_0 are positive at each real infinite place of k.*

Proof. Let $\mathscr{S}$ be the set of primes which divide α_0 or β_0. We can write $\mathfrak{a} = (\gamma_1, \gamma_2)$ where γ_1 and γ_2 are units at every prime in $\mathscr{B}$ and both $\gamma_1/\mathfrak{a}$ and $\gamma_2/\mathfrak{a}$ are units at every prime in $\mathscr{S}$. Let δ in $\mathfrak{o}$, a unit outside $\mathscr{B}$, be such that $\alpha_0\delta$ and $\beta_0\delta$ are in $\mathfrak{o}$. Choose positive coprime integers a, b in $\mathbf{Z}$ which are close to 1 at every finite prime in $\mathscr{B}$ and units at all the primes which divide γ_1 or γ_2; and let M, N be large positive integers. By writing $\alpha_0\delta a^M/\gamma_1$ in terms of a base for $\mathfrak{o}/\mathbf{Z}$ and changing the coefficients by elements of $\mathbf{Q}$ which are small at each finite prime in $\mathscr{B} \cup \mathscr{S}$ and $O(a)$ at the infinite place of $\mathbf{Q}$, we can obtain an integer α_1 in $\mathfrak{o}$ which is prime to a and $\gamma_2/\mathfrak{a}$ and such that $\alpha_0\delta a^M/\alpha_1\gamma_1$ is close to 1 at each place in $\mathscr{B}$ and α_0, α_1 are divisible by the same power of $\mathfrak{p}$ for each $\mathfrak{p}$ in $\mathscr{S}$. Similarly we can obtain β_1 in $\mathfrak{o}$ which is prime to b and $\gamma_1/\mathfrak{a}$ and such that $\beta_0\delta b^N/\beta_1\gamma_2$ is close to 1 at each place of $\mathscr{B}$ and β_0, β_1 are divisible by the same power of $\mathfrak{p}$ for each $\mathfrak{p}$ in $\mathscr{S}$. We can further ensure that β_1 is prime to α_1 outside $\mathscr{B} \cup \mathscr{S}$. Now $\alpha = \alpha_1 b^N \gamma_1/\delta$ and $\beta = \beta_1 a^M \gamma_2/\delta$ satisfy all the requirements in the lemma. The only difficult thing to verify is that $(\alpha, \beta) = \mathfrak{a}(\alpha_0, \beta_0)$. So far as primes in $\mathscr{B}$ are concerned, the two sides agree; and

$$(\alpha, \beta) = (\alpha_1\gamma_1, \beta_1\gamma_2) = \mathfrak{a}(\alpha_1(\gamma_1/\mathfrak{a}), \beta_1(\gamma_2/\mathfrak{a})) = \mathfrak{a}(\alpha_1, \beta_1)$$

up to such primes. □

The proof of Lemma 1 constructs an evaluation formula all of whose terms come from the right hand side of (4) for various pairs f, g. For $\alpha \times \beta$ in $\mathscr{N}^2$, the formula can therefore be described by an equation of the form

$$m(v; F, G; \alpha, \beta) = \prod_j (\varphi_j(\alpha, \beta), \psi_j(\alpha, \beta))_v. \tag{8}$$

Here the φ_j, ψ_j are homogeneous elements of $k[U, V]$ which depend only on F and G and not on v or $\mathscr{B}$. The decomposition (8) is not unique, and our next task is to display an invariant aspect of it.

Let $\theta = \gamma_1 U + \gamma_2 V$ be a linear form with γ_1, γ_2 coprime integers in k. By using $(\varphi, \psi)_v = (\varphi, \theta\psi)_v(\varphi, \theta)_v$ and $(-\theta, \theta)_v = 1$, we can ensure that all the φ_j, ψ_j in (8) have even degree except that $\psi_0 = \theta$. Denote by Θ the group of elements of k^* which are not divisible to an odd power by any prime of k outside $\mathscr{B}$, and by $\Theta_0 \subset \Theta$ the subgroup consisting of those ξ which are quadratic residues mod $\mathfrak{p}$ for all $\mathfrak{p}$ outside $\mathscr{B}$; thus we are free to multiply φ_0 by any element of Θ_0. (Actually $\Theta_0 = k^{*2}$, but we shall not use this fact.)

Lemma 3. *Suppose that* $\deg F$ *is even. With the convention for the* φ_j, ψ_j *just adopted, we can take* φ_0 *to be in* Θ.

Proof. Let γ in k^* be a unit outside $\mathscr{B}$, and apply (8) to the identity

$$L(\mathscr{B}; F, G; \gamma\alpha, \gamma\beta) = L(\mathscr{B}; F, G; \alpha, \beta),$$

where $\alpha \times \beta$ is in $\mathscr{N}^2$. On cancelling common factors, we obtain

$$\prod_{v \in \mathscr{B}} (\varphi_0(\alpha, \beta), \gamma)_v = 1. \tag{9}$$

If we can choose $\alpha \times \beta$ in $\mathscr{N}^2$ so that $\varphi_0(\alpha, \beta)$ is not in Θ, this gives a contradiction. For let δ prime to $\varphi_0(\alpha, \beta)$ be such that $\prod(\varphi_0(\alpha, \beta), \delta)_\mathfrak{p} = -1$ where the product is taken over all primes $\mathfrak{p}$ outside $\mathscr{B}$ at which $\varphi_0(\alpha, \beta)$ is not a unit. Let $\mathscr{B}_1$ be obtained by adjoining to $\mathscr{B}$ all the primes at which δ is not a unit; then $\prod(\varphi_0(\alpha, \beta), \delta)_v = -1$ by the Hilbert product formula, where the product is taken over all places v in $\mathscr{B}_1$. Recalling that φ_0 does not depend on $\mathscr{B}$ and writing $\mathscr{B}_1, \delta$ for $\mathscr{B}, \gamma$ in (9), we obtain a contradiction. It follows that $\varphi_0(\alpha, \beta)$ lies in Θ for all α, β; this can only happen if $\varphi_0(U, V)$ is itself in Θ modulo squares of homogeneous polynomials. □

Let $\mathscr{S}$ be the set of primes $\mathfrak{p}$ outside $\mathscr{B}$ for which $\mathfrak{p} | F(\alpha_\mathfrak{p}, \beta_\mathfrak{p})$ or $\mathfrak{p} | G(\alpha_\mathfrak{p}, \beta_\mathfrak{p})$ in the notation of (7). We can write $\lambda = \alpha/\beta$ where (α, β) is not divisible by any prime in $\mathscr{S}$. Let $\mathfrak{a}$ be an integral ideal in the class of (α, β) not divisible by any prime in $\mathscr{S}$, and let γ be such that $(\gamma) = \mathfrak{a}/(\alpha, \beta)$; then $\lambda = \alpha\gamma/\beta\gamma$ and $(\alpha\gamma, \beta\gamma) = \mathfrak{a}$. If $\mathscr{B}_1$ is obtained from $\mathscr{B}$ by adjoining all the primes which divide $\mathfrak{a}$, then

$$L(\mathscr{B}; F, G; \lambda) = L(\mathscr{B}; F, G; \alpha\gamma, \beta\gamma) = L(\mathscr{B}_1; F, G; \alpha\gamma, \beta\gamma),$$

where the second equality holds because the two products involved are term by term the same. By (8) the right hand side is equal to

$$\prod_{v\in\mathscr{B}_1}\prod_j(\varphi_j(\alpha\gamma,\beta\gamma),\psi_j(\alpha\gamma,\beta\gamma))_v$$
$$=\left\{\prod_{v\in\mathscr{B}_1}\prod_j(\varphi_j(\alpha,\beta),\psi_j(\alpha,\beta))_v\right\}\prod_{v\in\mathscr{B}_1}(\varphi_0(\alpha,\beta),\gamma)_v$$

because of the parity properties above. If we further require that no prime which divides $\mathfrak{a}$ divides any of the $\varphi_j(\alpha,\beta)$ or $\psi_j(\alpha,\beta)$, then each of the terms in curly brackets with v in $\mathscr{B}_1\setminus\mathscr{B}$ is trivial; so the outer product there reduces to a product over v in $\mathscr{B}$. By the Hilbert product formula the product outside the curly brackets can be replaced by a product over all v not in $\mathscr{B}_1$. In view of Lemma 3 we can reduce this to a product over those v outside $\mathscr{B}_1$ which divide (α,β). If $\chi_\mathfrak{p}$ is again the quadratic residue symbol mod $\mathfrak{p}$, we can write the result which we have just obtained in the form

$$L(\mathscr{B};F,G;\lambda)=\left\{\prod_{v\in\mathscr{B}}\prod_j(\varphi_j(\alpha,\beta),\psi_j(\alpha,\beta))_v\right\}\prod\chi_\mathfrak{p}(\varphi_0) \tag{10}$$

where the final product is taken over those $\mathfrak{p}$ outside $\mathscr{B}$ which divide (α,β) to an odd power.

Lemma 4. Suppose that $\deg F$ *is even and the conventions of Lemma 3 hold. Then* φ_0 *is uniquely determined by* F *and* G *as an element of* Θ/Θ_0*, and* φ_0 *is in* Θ_0 *if and only if* $L(\mathscr{B};F,G;\lambda)$ *is continuous in* λ *in the topology induced by* $\mathscr{B}$.

Proof. Suppose first that φ_0 is in Θ_0. Thus the final product in (10) is trivial. Now let $\lambda=\alpha/\beta$ and let λ' be close to λ in the topology induced by $\mathscr{B}$. Let γ in $\mathfrak{o}$ be such that $\lambda'\beta\gamma$ is integral. Applying (10) to the representations

$$\lambda=\alpha\gamma/\beta\gamma\quad\text{and}\quad\lambda'=\lambda'\beta\gamma/\beta\gamma$$

we deduce that $L(\mathscr{B};F,G;\lambda)=L(\mathscr{B};F,G;\lambda')$.

Conversely suppose that φ_0 is in Θ but not in Θ_0. Choose a prime $\mathfrak{p}$ outside $\mathscr{B}$ at which φ_0 is not a quadratic residue. For $\lambda_0=\alpha_0/\beta_0$ let $\lambda=\alpha/\beta$ where α,β have the properties stated in Lemma 2 with $\mathfrak{a}=\mathfrak{p}$. Arguing as in the previous paragraph, but taking account of the final product in (10), we obtain

$$L(\mathscr{B};F,G;\lambda)=L(\mathscr{B};F,G;\lambda_0)\chi_\mathfrak{p}(\varphi_0)=-L(\mathscr{B};F,G;\lambda_0).$$

So $L(\mathscr{B};F,G;\lambda)$ is not continuous at $\lambda=\lambda_0$.

Now suppose that $L(\mathscr{B}; F, G; \alpha, \beta)$ has two representations, say by the φ_i', ψ_i' and the φ_j'', ψ_j''. Taking their quotient, we obtain

$$1 = \prod_{v \in \mathscr{B}} \left\{ (\varphi_0'/\varphi_0'', \theta(\alpha,\beta))_v \prod_{i>0} (\varphi_i'(\alpha,\beta), \psi_i'(\alpha,\beta))_v \prod_{j>0} (\varphi_j''(\alpha,\beta), \psi_j''(\alpha,\beta))_v \right\}.$$

This is a representation of a function of λ which is continuous; and it is of a kind to which we can apply the results of the previous two paragraphs. Hence φ_0'/φ_0'' is in Θ_0.

It remains only to show that φ_0 is independent of the choice of θ. Using a notation like that of the previous paragraph, there is a representation of 1 in which the terms with subscript 0 produce a quotient

$$\prod_{v \in \mathscr{B}} \{(\varphi_0'/\varphi_0'', \theta')_v (\varphi_0'', \theta'\theta'')_v\};$$

and since $\deg(\theta'\theta'')$ is even it follows as there that φ_0'/φ_0'' is in Θ_0. □

In practice, what we usually need to study is the subspace of $\mathscr{N}^2$ given by n conditions $L(\mathscr{B}; F_\nu, G_\nu; \alpha, \beta) = 1$, or the subspace of $\mathscr{N}^1$ given by the conditions $L(\mathscr{B}; F_\nu, G_\nu; \lambda) = 1$, where the $\deg F_\nu$ are all even. Let Λ be the abelian group of order 2^n whose elements are the n-tuples each component of which is ± 1; then there is a natural identification, which we shall write τ, of each element of Λ with a partial product of the $L(\mathscr{B}; F_\nu, G_\nu)$. Thus each element of Λ can be interpreted as a condition, which we shall write as $\mathscr{L} = 1$. If φ_0 is as in Lemma 3, there is a homomorphism

$$\varphi_0 \circ \tau : \Lambda \to \Theta/\Theta_0;$$

let Λ_0 denote its kernel. In view of Lemma 4, the conditions which are continuous in λ are just those which come from Λ_0; and we shall call these the *continuous* conditions. The following lemma corresponds to Harari's Formal Lemma (Theorem 3.2.1 of [**1**]); it shows that for most purposes we need only consider the conditions coming from the elements of Λ_0.

Lemma 5. *Suppose that* $\deg F$ *is even and all the conditions corresponding to* Λ_0 *hold at some given* λ_0. *Then there exists* λ *arbitrarily close to* λ_0 *such that all the conditions* $L(\mathscr{B}; F_\nu, G_\nu) = 1$ *hold at* λ.

Proof. Let $\lambda_0 = \alpha_0/\beta_0$. For a suitably chosen $\mathfrak{a} = (\gamma)$ we show that we can take $\lambda = \alpha/\beta$, where $\alpha \times \beta$ is as in Lemma 2. For any c in Λ, write $\varphi_{0c} = \varphi_0 \circ \tau(c)$ for the corresponding element of Θ/Θ_0. If θ is as defined just before Lemma 3, the corresponding partial product $\mathscr{L}$ of the $L(\mathscr{B}; F_\nu, G_\nu; \lambda)$ is equal to

$$f_c(\lambda) \prod_{v \in \mathscr{B}} (\varphi_{0c}, \theta(\alpha_0, \beta_0))_v \prod_{v \in \mathscr{B}} (\varphi_{0c}, \gamma)_v$$

where f_c comes from the φ_j, ψ_j with $j > 0$ and is therefore continuous. The map $c \mapsto f_c(\lambda)$ is a homomorphism $\Lambda \to \{\pm 1\}$ for any fixed λ; moreover if two distinct c give rise to the same φ_{0c}, their quotient comes from an element of Λ_0 and therefore the quotient of the corresponding f_c takes the value 1 at λ_0. In other words, if λ is close enough to λ_0, then $f_c(\lambda)$ only depends on the class of c in Λ/Λ_0. The map $c \mapsto \varphi_{0c}$ is an embedding $\Lambda/\Lambda_0 \to \Theta/\Theta_0$, by Lemma 4. The homomorphism Image$(\Lambda/\Lambda_0) \to \{\pm 1\}$ induced by $c \mapsto f_c(\lambda)$ can be extended to a homomorphism $\Theta/\Theta_0 \to \{\pm 1\}$ because Θ/Θ_0 is killed by 2; and any such homomorphism can be written in the form

$$\theta \to \prod_{v \in \mathscr{B}} (\theta, \gamma)_v$$

for a suitably chosen γ, because the Hilbert symbol induces a nonsingular form on Θ/Θ_0. But given any such γ we can construct $\lambda = \alpha/\beta$ having the properties listed in Lemma 2 with $\mathfrak{a} = (\gamma)$. □

We shall need analogues of these last results for positive 0-cycles, and this will require more notation. We continue to assume that $\deg F$ is even. Let K be the direct product of finitely many fields k_i each of finite degree over k, and let $\mathfrak{B}$ be the set of places of K lying over some place v in $\mathscr{B}$, and $\mathfrak{B}_i$ the corresponding set of places of k_i. (The place $\prod v_i$, where v_i is a place of k_i, lies over v if each v_i does so.) For λ in $\mathbf{P}^1(K)$ write $\lambda = \prod \lambda_i$ with λ_i in $\mathbf{P}^1(k_i)$; for each place w in k_i write $\lambda_i = \alpha_{iw}/\beta_{iw}$ where α_{iw}, β_{iw} are in k_i and integral at w and at least one of them is a unit at w. For any λ in K such that each $F(\lambda_i, 1)$ and $G(\lambda_i, 1)$ is nonzero, we define the function

$$L^*(\mathscr{B}; K; F, G; \lambda) : \lambda \mapsto \prod_{\mathfrak{P}_i} (F(\alpha_{iw}, \beta_{iw}), G(\alpha_{iw}, \beta_{iw}))_{\mathfrak{P}_i} \tag{11}$$

where w is the place associated with the prime $\mathfrak{P}_i$ in k_i and the product is taken over all i and all primes $\mathfrak{P}_i$ of k_i not lying in $\mathfrak{B}_i$ and such that $G(\alpha_{iw}, \beta_{iw})$ is divisible by $\mathfrak{P}_i$. As with (1), we can restrict the product to those $\mathfrak{P}_i$ which divide $G(\alpha_{iw}, \beta_{iw})$ to an odd power. Note that the functions φ_j, ψ_j in the evaluation formula (8) are the same for $k_i \supset k$ as they are for k. Now let $\mathfrak{a}$ be a positive 0-cycle on $\mathbf{P}^1$ defined over k and let $\mathfrak{a} = \cup \mathfrak{a}_i$ be its decomposition into irreducible components. Let λ_i be a point of $\mathfrak{a}_i$ and write $k_i = k(\lambda_i)$. If

$K = \prod k_i$ and $\lambda = \prod \lambda_i$, write

$$L^*(\mathscr{B}; F, G; \mathfrak{a}) = L^*(\mathscr{B}; K; F, G; \lambda) = \prod\nolimits_i L(\mathfrak{B}_i; F, G; \lambda_i). \tag{12}$$

This is legitimate, because the right hand side does not depend on the choice of the λ_i. If $K = k$ this L^* is the same as the previous function L. Moreover $L^*(\mathfrak{a} \cup \mathfrak{b}) = L^*(\mathfrak{a})L^*(\mathfrak{b})$. We can define a topology on the set of positive 0-cycles $\mathfrak{a}$ of given degree N by means of the isomorphism between that set and the points on the N-fold symmetric power of $\mathbf{P}^1$. With this topology, it is straightforward to extend to L^* the results already obtained for L.

The product in (11) is finite; so there is a finite set $\mathscr{S}$ of primes of k, disjoint from $\mathscr{B}$ and such that every $\mathfrak{P}_i$ which appears in this product lies above a prime in $\mathscr{S}$. For each i we can write $\lambda_i = \alpha_i/\beta_i$ with α_i, β_i integers in k_i. As in the argument which follows the proof of Lemma 3, let $(\alpha_i, \beta_i) = \mathfrak{a}_i$ and choose an integral ideal $\mathfrak{b}_i$ in k_i which is prime to $\mathfrak{a}_i$, in the same ideal class as $\mathfrak{a}_i$ and such that no prime of k_i which divides $\mathfrak{b}_i$ also divides $G(\alpha_i, \beta_i)$ or any $\varphi_j(\alpha_i, \beta_i)$ or $\psi_j(\alpha_i, \beta_i)$ or lies above any prime in $\mathscr{S}$. Let γ_i be such that $(\gamma_i) = \mathfrak{b}_i/\mathfrak{a}_i$ and let $\mathscr{B}_1$ be obtained from $\mathscr{B}$ by adjoining all the primes of k which lie below any prime of k_i which divides $\mathfrak{b}_i$. For most purposes it costs us nothing to replace $\mathscr{B}$ by $\mathscr{B}_1$, and we then have

$$\lambda = \prod \lambda_i = \prod (\alpha_i\gamma_i/\beta_i\gamma_i) \text{ where } \alpha_i\gamma_i \times \beta_i\gamma_i \text{ is in } \mathscr{N}^2(k_i).$$

The following lemma is a trivial consequence of earlier results.

Lemma 6. *Suppose that* $\deg F$ *is even, and let* $\mathscr{L} = 1$ *be a continuous condition derived from the* L *and* $\mathscr{L}^* = 1$ *the corresponding condition derived from the* L^*. *For each* v *in* $\mathscr{B}$ *there is a function* $\ell^*(v; F, G; \mathfrak{a})$ *with values in* $\{\pm 1\}$ *which is a continuous function of* $\mathfrak{a}$ *in the* v*-adic topology and is such that*

$$\mathscr{L}^*(\mathscr{B}; F, G; \mathfrak{a}) = \prod_{v \in \mathscr{B}} \ell^*(v; F, G; \mathfrak{a}). \tag{13}$$

3. Salberger's device

To prove Theorem 2, we need to construct points in the image of $V(k)$ in $\mathbf{L}^1(k)$ which satisfy preassigned local conditions; to do this, we construct a sequence of positive 0-cycles of gradually decreasing degrees, each satisfying appropriate local conditions. If N is large enough, we can generate positive 0-cycles of degree N on V satisfying preassigned local conditions by means of an argument which depends on the partial fraction formula (15); its use in this context was pioneered by Salberger. Of the various versions of the consequent algorithm, Lemma 8 seems the simplest, both in its proof and in the way in which it is used; in particular, it does not involve an auxiliary set of primes and its proof does not depend on a deep result of Waldschmidt.

We need a preliminary lemma about approximation.

Lemma 7. *Let L be an algebraic number field, $\mathfrak{B}$ a finite set of places of L and $\mathfrak{S}$ a finite set of primes of L not necessarily disjoint from $\mathfrak{B}$. Let $b > 1$ be in $\mathbf{Z}$ and such that no prime of L which divides b is in $\mathfrak{B}$. Let $M > 0$ be a rational integer and for each v in $\mathfrak{B}$ let ξ_v be in L_v. Then there exists ξ in L^* as close as we like to each ξ_v and such that $\xi = \alpha\gamma^M$, where (α) is the product of a first degree prime $\mathfrak{p}$ not in $\mathfrak{B} \cup \mathfrak{S}$ and primes in $\mathfrak{B}$, and $\gamma = \gamma_1/\gamma_2$ for coprime integers γ_1, γ_2 such that the prime factorization of γ_1 does not include any prime in $\mathfrak{B} \cup \mathfrak{S} \cup \{\mathfrak{p}\}$ and the only primes which divide γ_2 also divide b.*

Proof. By Dirichlet's theorem on primes in arithmetic progression, we can choose $\mathfrak{p}$ and α as in the statement of the lemma so that α is as close as we like to ξ_v for each finite v in $\mathfrak{B}$ and $\xi_v/\alpha > 0$ for each real v in $\mathfrak{B}$. For each infinite v in $\mathfrak{B}$ we choose γ_v in L_v so that $\gamma_v^M = \xi_v/\alpha$. Using weak approximation, choose γ' in L, a unit at every finite prime in $\mathfrak{B} \cup \mathfrak{S} \cup \{\mathfrak{p}\}$, so that γ' is arbitrarily close to 1 at every finite place in $\mathfrak{B}$ and arbitrarily close to γ_v at every infinite place v in $\mathfrak{B}$. By writing $\gamma' b^N$ for large enough N in terms of a base for $\mathfrak{o}_L/\mathbf{Z}$ and changing the coefficients by elements of $\mathbf{Q}$ which are small at each finite prime in $\mathscr{B} \cup \mathscr{S}$ and bounded at every infinite place in $\mathscr{B}$, we can obtain an integer γ_1 which is prime to $\mathscr{S} \cup \{\mathfrak{p}\}$ and to b and close to $\gamma' b^N$ at every place in $\mathscr{B}$. Now take $\gamma_2 = b^N$; then $\xi = \alpha\gamma^M$ satisfies all our requirements. □

For the statement and proof of the following lemma, we shall call a place of k *bad* if it lies in $\mathscr{B}$ or divides b; and we shall call a place in $\mathbf{Q}$ or in a field containing k *bad* if it lies below or above a bad place of k. For our purposes, the most important difference between places in $\mathscr{B}$ and primes dividing b is that the latter have no approximation conditions associated with them.

Lemma 8. *Let k be an algebraic number field and $P_1(X), \dots, P_n(X)$ monic irreducible non-constant polynomials in $k[X]$; and let $N \geqslant \sum \deg(P_i)$ be a given integer. Let $\mathfrak{B}$ be a finite set of places of k which contains the infinite places, the primes which divide 2, the primes at which some coefficient of some P_i is not integral and any other primes $\mathfrak{p}$ at which $\prod P_i(X)$ does not remain separable when reduced* mod $\mathfrak{p}$. *Let b be as in Lemma 7. For each v in $\mathfrak{B}$ let U_v be a non-empty open set of separable monic polynomials of degree N in $k_v[X]$. Let $M > 0$ be a fixed rational integer. Then we can find an irreducible monic* *polynomial $G(X)$ in $k[X]$ of degree N which lies in each U_v and for which λ, the image of X in $K = k[X]/G(X)$, satisfies*

$$(P_i(\lambda)) = \mathfrak{P}_i\mathfrak{A}_i\mathfrak{C}_i^M \tag{14}$$

for each i, where the $\mathfrak{P}_i$ are distinct first degree primes in K not lying above any prime in $\mathfrak{B}$, the $\mathfrak{A}_i$ are products of bad primes in K and the $\mathfrak{C}_i$ are integral ideals in K. Moreover we can arrange that $\lambda = \alpha/\beta$ where α is integral and β is an integer all of whose prime factors are bad.

Proof. We shall need to apply Lemma 7 repeatedly with the same value of M as in Lemma 8. We can assume, after adding a constant to X if necessary, that none of the $P_i(X)$ is a multiple of X. Write $R(X) = \prod P_i(X)$ and $R_i(X) = R(X)/P_i(X)$. Any polynomial $G(X)$ in $k[X]$ can be written in just one way in the form

$$G(X) = R(X)Q(X) + \sum R_i(X)\psi_i(X) \tag{15}$$

with $\deg \psi_i < \deg P_i$; for if λ_i is a zero of $P_i(X)$ this is just the classical partial fraction formula

$$\frac{G(X)}{\prod P_i(X)} = Q(X) + \sum \frac{\psi_i(X)}{P_i(X)}$$

with $\psi_i(\lambda_i) = G(\lambda_i)/R_i(\lambda_i)$. This property determines a unique $\psi_i(X)$ in $k[X]$ of degree less than $\deg P_i$. The same result holds over any k_v. If the coefficients of G are integral at v, for some v not in $\mathfrak{B}$, then so are those of Q and each ψ_i because R and the R_i are monic and $R_i(\lambda_i)$ is a unit outside $\mathfrak{B}$. For each v in $\mathfrak{B}$ let $G_v(X)$ be a polynomial of degree N lying in U_v, and write

$$G_v(X) = R(X)Q_v(X) + \sum R_i(X)\psi_{iv}(X)$$

with $\deg \psi_{iv} < \deg P_i$. We adjoin to $\mathfrak{B}$ a further finite place w at which b is a unit, and associate with it a monic irreducible polynomial $G_w(X)$ in $k_w[X]$ with degree N; the only purpose of G_w is to ensure that the $G(X)$ which we shall construct is irreducible over k. We build $G(X)$, close to $G_v(X)$ for every $v \in \mathfrak{B}$ including w, in the following manner.

For the first step let $k_i = k[X]/P_i(X)$ and for each $v \in \mathfrak{B}$ let φ_{iv} be the class of ψ_{iv} in $k_v[X]/P_i(X) = k_i \otimes_k k_v$. Take $\mathfrak{S}$ to consist of those primes in k at which the constant terms of the $P_i(X)$ are not all units. We apply Lemma 7 to each set of φ_{iv} in turn, replacing L by k_i and $\mathfrak{B}$ and $\mathfrak{S}$ by the sets of places of k_i which lie above $\mathfrak{B}$ and $\mathfrak{S}$ respectively; let φ_i be the element of k_i thus obtained, and let $\mathfrak{P}_i$ be the associated prime in k_i. Let $\psi_i'(X)$ be the unique polynomial in $k[X]$ with $\deg \psi_i' < \deg P_i$ whose class in k_i is φ_i. Clearly $\psi_i'(X)$ is arbitrarily close to each $\psi_{iv}(X)$, and its coefficients are integers outside $\mathfrak{B}$

because $\mathfrak{B}$ contains all the primes which ramify in k_i/k. Now choose positive c, T in $\mathbf{Z}$ so that c is a unit at all bad primes, divisible by all the primes outside $\mathfrak{B} \cup \{\mathfrak{P}_i\}$ which divide the numerator of any φ_i, and close to b^T at the real place and at all the primes below primes in $\mathfrak{B}$. Let $\psi_i(X) = (c/b^T)^M \psi_i'(X)$.

We now choose $Q(X)$ to be close to $Q_v(X)$ for each v in $\mathfrak{B}$, and to be such that each coefficient other than the leading coefficient (which is 1) is integral except perhaps at bad primes and is divisible by c. We can do this by an argument like, but very much simpler than, that in the proof of Lemma 7. This construction ensures that $G(X)$ is monic and arbitrarily close to each $G_v(X)$ including $G_w(X)$. The assumptions made about $G_w(X)$ ensure that $G(X)$ is irreducible in k_w and therefore in k. Moreover, the coefficients of $Q(X)$ are integers except perhaps at bad primes; and since $G(X)$ is monic the denominator of any $P_i(\lambda)$ only contains bad primes. A consequence of the choice of $\mathfrak{S}$ is that every λ_i, and therefore every $Q(\lambda_i)$, is prime to c.

We have still to prove (14). Let $\mathfrak{p}_i$ be the prime in k below $\mathfrak{P}_i$. By computing the resultant of $P_i(X)$ and $G(X)$ in two different ways, we obtain

$$\text{Norm}_{K/k} P_i(\lambda) = \pm \text{Norm}_{k_i/k} G(\lambda_i) = \pm \text{Norm}_{k_i/k}(\varphi_i R_i(\lambda_i)) \tag{16}$$

where λ_i is a zero of $P_i(X)$. By hypothesis $R_i(\lambda_i)$ is a unit at every place of $k(\lambda_i)$ which does not lie above a place in $\mathfrak{B}$; and we have arranged that the denominator of $\text{Norm}_{k_i/k}\varphi_i$ is only divisible by bad primes, and its numerator is the product of the first degree prime $\mathfrak{p}_i$, powers of primes in $\mathfrak{B}$ and Mth powers of norms of primes which come from the $\mathfrak{C}_i$ of Lemma 7. Also λ, and therefore $P_i(\lambda)$, is integral outside bad primes in K. None of these lie above $\mathfrak{p}_i$. Hence $P_i(\lambda)$ is an integer at each prime of K lying above $\mathfrak{p}_i$. It follows that the ideal $(P_i(\lambda))$ is divisible by just one prime of K above $\mathfrak{p}_i$, and that to the first power. It only remains to show that, apart from this prime and bad primes, what we have is an Mth power.

Let L be a splitting field for all the $P_i(X)$ and let $\mathfrak{P}$ be a prime in $L(\lambda)$ which divides the numerator of $P_i(\lambda)$. By (16) and the remarks on either side of it, $\mathfrak{P}$ must divide $\text{Norm}_{k_i/k}(\varphi_i)$ and therefore must divide c. Hence

$$\tilde{G}(X) = \tilde{R}(X)\tilde{Q}(X) \tag{17}$$

where the tilde denotes reduction mod $\mathfrak{P}$ of the coefficients. But the construction of $Q(X)$ has ensured that the resultant of $Q(X)$ and $R(X)$, which is $\pm \prod_i \text{Norm}_{k_i/k}(Q(\lambda_i))$, is prime to c; hence $\tilde{R}(X)$ and $\tilde{Q}(X)$ are coprime. Moreover $\tilde{R}(X)$ is a product of distinct linear factors over the residue field of L at $\mathfrak{P}$. It follows that (17) can be lifted to a factorization of $G(X)$ in the completion of $L(\lambda)$ at $\mathfrak{P}$; and the roots of $G(X)$ in this field consist of one near each root of each $P_i(X)$ together with roots which come (after a further field extension) from the lift of $\tilde{Q}(X)$. The latter are not close to any root of any $P_i(X)$.

I now claim that the power of $\mathfrak{P}$ which divides $P_i(\lambda)$ is $\mathfrak{P}^m$ where m is a multiple of M. For if λ is not close to a root of $P_i(X)$ then $m = 0$. On the other hand, (15) can be written

$$G(X) = R_i(X)\psi_i(X) + f_i(X)P_i(X)$$

where

$$f_i(X) = R_i(X)Q(X) + \sum_{j \neq i} \psi_j(X)R_j(X)/P_i(X).$$

By construction, if λ is close to a root of $P_i(X)$ then $f_i(\lambda)$ is a unit at $\mathfrak{P}$, as is $R_i(\lambda)$. If λ_i is that root of $P_i(X)$ which is close to λ, then the standard successive approximation process shows that $\lambda - \lambda_i$ has the same valuation as $\psi_i(\lambda_i) = \varphi_i$; and by construction $\mathfrak{P}^m \| \varphi_i$ where $M|m$. It follows that $\mathfrak{P}^m \| P_i(\lambda)$ with $M|m$, as claimed, in both cases.

Now let $\mathfrak{p}$ be a prime in k which divides c, and let $\mathfrak{q}$ be any prime of $k(\lambda)$ above $\mathfrak{p}$. The factors of $P_i(\lambda)$ coming from primes of $L(\lambda)$ above $\mathfrak{q}$ have the form

$$\prod_{\mathfrak{P}|\mathfrak{q}} \mathfrak{P}^{m(\mathfrak{P})} \text{ where each } m(\mathfrak{P}) \text{ is divisible by } M. \tag{18}$$

This is equal to the corestriction of $\mathfrak{q}^n$, where $\mathfrak{q}^n$ is the exact power of $\mathfrak{q}$ which divides $P_i(\lambda)$. But the extension $L(\lambda)/k(\lambda)$ is unramified at $\mathfrak{q}$, because it is only ramified at places above places in $\mathfrak{B}$. Hence each $m(\mathfrak{P})$ in (18) equals n, and so n is divisible by M. This holds for all primes in $k(\lambda)$ which divide c. $\square$

4. Reduction to a pencil of conics

Suppose now that V is a Del Pezzo surface $V = Q_1 \cap Q_2$ where Q_1, Q_2 are quadrics in $\mathbf{P}^4$, and assume that $V(k)$ is not empty. After a change of coordinates we can suppose that $(1, 0, 0, 0, 0)$ is a point of $V(k)$ and the tangents to Q_1, Q_2 at this point are $X_1 = 0, X_2 = 0$, respectively. Thus the equations of Q_1 and Q_2 can be written

$$X_0X_1 + f_1(X_1, \ldots, X_4) = 0, \quad X_0X_2 + f_2(X_1, \ldots, X_4) = 0 \tag{19}$$

where f_1, f_2 are homogeneous quadratic. The variety (19) is birationally equivalent to the cubic surface $X_2f_1 = X_1f_2$, which is obtained by blowing up the given point of $V(k)$; and this cubic surface is birationally equivalent to the pencil of affine conics

$$Vf_1(U,V,X_3,X_4) = Uf_2(U,V,X_3,X_4), \tag{20}$$

which with some abuse of language can be parametrized by the points (U,V) of $\mathbf{P}^1$. Diagonalizing this equation and then making it homogeneous gives a pencil of projective conics of the form

$$Z_0^2 g_1(U,V) + Z_1^2 g_2(U,V)/g_1(U,V) + Z_3^2 g_5(U,V)/g_2(U,V) = 0,$$

where g_r is homogeneous of degree r. Writing

$$Z_0 = g_2 Y_0, \quad Z_1 = g_1 Y_1, \quad Z_2 = g_1 g_2 Y_2$$

and dividing by $g_1 g_2$ we obtain

$$g_2 Y_0^2 + Y_1^2 + g_1 g_5 Y_2^2 = 0. \tag{21}$$

We shall assume that the g_r are coprime in pairs in $k[U,V]$; if not, there is a further simplification of (21) and of the subsequent argument which is left to the reader. Every point of the line $X_1 = X_2 = 0$ on the cubic surface also lies on one of the conics of the pencil (20), so there are an infinity of conics which contain points defined over k.

It costs nothing to set the next part of the argument in a broader context. Denote by W the surface fibred by the pencil of conics

$$a_0(U,V)Y_0^2 + a_1(U,V)Y_1^2 + a_2(U,V)Y_2^2 = 0, \tag{22}$$

and call the pencil *reduced* if a_0, a_1, a_2 are homogeneous elements of $k[U,V]$ coprime in pairs and such that

$$\deg a_0 \equiv \deg a_1 \equiv \deg a_2 \bmod 2.$$

After a linear transformation on U,V if necessary, we can also assume that $a_0 a_1 a_2$ is not divisible by V. Clearly any pencil of conics can be put into reduced form. Suppose that (22) is reduced and everywhere locally soluble. Let $\lambda = (\alpha, \beta)$ be a point of $\mathbf{P}^1(k)$; whether (22) is soluble at $\alpha \times \beta$ depends only on λ and not on the choice of α, β. Similar statements hold for local solubility at a place v and for solubility in the adeles. Denote by $c(U,V)$ a monic irreducible factor of $a_0 a_1 a_2$ in $k[U,V]$. Let $\mathscr{B}$ be a finite set of places of k containing the infinite places, the primes dividing 2, those at which any coefficient of any c or a_r is not integral, and any other primes $\mathfrak{p}$ at which $a_0 a_1 a_2$ does not remain separable when reduced mod $\mathfrak{p}$. Thus local solubility of (22) is trivial for any prime outside $\mathscr{B}$. For convenience, we also assume that $\mathscr{B}$ contains a base for the ideal class group of k.

We need to work not on $\mathbf{P}^1$ but on the set $\mathbf{L}^1$ obtained from $\mathbf{P}^1$ by deleting the roots of $a_0a_1a_2$; thus we do not have to worry about the singular fibres. Denote by W_0 the Zariski open subset of W which is the inverse image of $\mathbf{L}^1$, and define V_0 similarly in terms of the representation (21) of V. Let $\lambda \in k\cup\{\infty\}$ be a point of $\mathbf{L}^1(k)$, and write $\lambda = \alpha/\beta$ where α, β are integers of k coprime outside $\mathscr{B}$; it will not matter which pair α, β we choose.

There is a non-empty set $\mathscr{N} \subset \mathbf{L}^1(k)$, open in the topology induced by $\mathscr{B}$, such that the conic (22) is locally soluble at every place of $\mathscr{B}$ if and only if λ lies in $\mathscr{N}$. Let $\mathfrak{p}$ be a prime of k not in $\mathscr{B}$ and consider the solubility of (22) in $k_{\mathfrak{p}}$ at the point λ. If none of the $a_r(\alpha,\beta)$ is divisible by $\mathfrak{p}$, then solubility of (22) in $k_{\mathfrak{p}}$ at the point λ is trivial. Otherwise there is just one c such that $c(\alpha,\beta)$ is divisible by $\mathfrak{p}$; to fix ideas, suppose that this c divides a_2. Then the condition for solubility in $k_{\mathfrak{p}}$ is

$$(-a_0(\alpha,\beta)a_1(\alpha,\beta), c(\alpha,\beta))_{\mathfrak{p}} = 1. \tag{23}$$

Hence necessary conditions for the local solubility of (22) at λ for all $\mathfrak{p}$ outside $\mathscr{B}$ are the conditions like

$$L(\mathscr{B}; -a_0a_1, c; \lambda) = \prod(-a_0(\alpha,\beta)a_1(\alpha,\beta), c(\alpha,\beta))_{\mathfrak{p}} = 1$$

where the product is taken over all $\mathfrak{p}$ outside $\mathscr{B}$ which divide $c(\alpha,\beta)$. There is one of these conditions for each c, and for each of them the first argument in the Hilbert symbol has even degree. As in the discussion which follows the proof of Lemma 4, these generate a group of conditions naturally isomorphic to Λ. As there, we call a condition in this group *continuous* if it comes from Λ_0. Since increasing $\mathscr{B}$ does not alter the constant denoted in Lemma 4 by φ_0, it does not alter the set of continuous conditions.

Lemma 9. *Let W_0 be everywhere locally soluble. Then the continuous conditions derived from (22) are collectively equivalent to the Brauer–Manin conditions for the existence of points of W_0 defined over k. The continuous conditions similarly derived from the $L^*(\mathfrak{a})$ are collectively equivalent to the Brauer–Manin conditions for the existence of k-rational positive 0-cycles of degree N on W_0.*

Proof. The first assertion is proved for $k = \mathbf{Q}$ in Section 8 of [**1**]; as with Lemma 1, the proof there can be extended to our more general case. The second assertion follows trivially from the first in the light of (12). □

We now translate Theorem 1 into a form which is more convenient for our approach.

Theorem 2. *Let $\mathcal{B}$ be a finite set of places of k, satisfying the conditions for* (21) *analogous to those stated above for* (22).

(i) *For each v in $\mathcal{B}$ let A_v be a point of $V_0(k_v)$, and let λ_v be its image under the projection to $\mathbf{L}^1$. Suppose that all the conditions like*

$$\prod_{v\in\mathcal{B}} m(v; -a_0a_1, c; \lambda_v) = 1 \tag{24}$$

hold. Then there is a point of $V_0(k)$ as close as we like to each A_v.

(ii) *Let λ be a point of $\mathbf{L}^1(k)$ such that all the conditions like*

$$L(\mathcal{B}; -a_0a_1, c; \lambda) = 1 \tag{25}$$

hold. Then there is a point in $V_0(k)$ whose projection on $\mathbf{L}^1(k)$ is as close as we like to λ in the topology induced by $\mathcal{B}$.

Since we can find λ arbitrarily close to each λ_v, it follows from (2) that the two parts of the theorem are equivalent. In view of Lemma 5, the conclusion of (ii) still follows if we only require the continuous conditions to hold. By the first assertion of Lemma 9 and the fact that weak approximation holds for conics, Theorem 2(ii) is equivalent to Theorem 1.

5. Proof of Theorem 2

We now apply Lemma 8 to the surface W_0 fibred by the pencil (22), and we assume that $\mathcal{B}$ satisfies the conditions listed after (22). We state the theorem below in the form which corresponds to Theorem 2(ii), leaving it to any reader who wishes to do so to formulate a version which corresponds to Theorem 2(i). At the price of some extra complications, one could replace the condition on N in Theorem 3 below by the weaker condition that N is at least equal to the number of singular fibres of the pencil (22). For (21) this would enable us to take $N = 5$, since only the roots of g_5 give singular fibres. One could then dispense with one step in the argument in the latter part of this section.

Theorem 3. *With the notation above, let $N \geqslant \deg(a_0a_1a_2)$ be a fixed integer. Let $\mathfrak{a}$ be a positive 0-cycle of degree N on $\mathbf{L}^1$ defined over k and for each place v of k suppose that W_0 contains a positive 0-cycle $\mathfrak{b}_v$ of degree N defined over k_v; for v in $\mathcal{B}$ suppose further that $\mathfrak{b}_v$ is so chosen that its projection on $\mathbf{L}^1$ is $\mathfrak{a}$. If all the continuous conditions derived from the conditions* (25) *hold, then there is a positive 0-cycle of degree N on W_0 defined over k whose projection is arbitrarily near to $\mathfrak{a}$ in the topology induced by $\mathcal{B}$.*

Proof. We must first show that for the purpose of proving this theorem we are allowed to increase $\mathscr{B}$. Suppose that $\mathscr{B}_0$ satisfies the conditions which were imposed on $\mathscr{B}$ after (22), and let $\mathfrak{p}$ be a prime of k not in $\mathscr{B}_0$. Suppose also that the hypotheses of the theorem hold for $\mathscr{B} = \mathscr{B}_0$ and $\mathfrak{a} = \mathfrak{a}_0$. Having chosen $\mathfrak{b}_\mathfrak{p}$ we can use weak approximation on $\mathbf{L}^1$ to find a positive 0-cycle $\mathfrak{a}'$ on $\mathbf{L}^1$ of degree N and defined over k which is close at every v in $\mathscr{B}_0$ to $\mathfrak{a}$ and close at $\mathfrak{p}$ to the projection of $\mathfrak{b}_\mathfrak{p}$. Now

$$L^*(\mathscr{B}_0 \cup \{\mathfrak{p}\}; -a_0a_1, c; \mathfrak{a}') = L^*(\mathscr{B}_0; -a_0a_1, c; \mathfrak{a}');$$

for writing both sides as products by means of (11), if there is a factor on the right hand side which is not present on the left, that factor must come from $\mathfrak{p}$ and is therefore equal to 1. But a continuous condition for $\mathscr{B}_0$ holds at $\mathfrak{a}'$ if and only if it holds at $\mathfrak{a}$, which it does by hypothesis. Hence the continuous conditions for $\mathscr{B}_0 \cup \{\mathfrak{p}\}$ hold at $\mathfrak{a}'$. Now suppose that the theorem holds for $\mathscr{B}_0 \cup \{\mathfrak{p}\}$; then there is a positive 0-cycle $\mathfrak{b}$ of degree N on W_0 defined over k whose projection on $\mathbf{L}^1$ is close to $\mathfrak{a}'$ in the topology induced by $\mathscr{B}_0 \cup \{\mathfrak{p}\}$. The same projection is close to $\mathfrak{a}$ in the topology induced by $\mathscr{B}_0$. So the theorem also holds for $\mathscr{B}_0$.

Note that if $\mathfrak{a}$ is actually the projection of a positive 0-cycle of degree N in W_0, then the continuous conditions certainly hold in view of (12); thus they are necessary conditions for the existence of such a 0-cycle. To simplify the notation, we assume henceforth that K is an algebraic number field; this will be true for the application in this paper because K will be constructed by means of Lemma 8. In view of the previous paragraph, we can assume that $\mathscr{B}$ is so large that it satisfies the conditions imposed on $\mathfrak{B}$ in the statement of Lemma 8 and it contains the additional place w which was adjoined to $\mathfrak{B}$ in the first paragraph of the proof of Lemma 8; and if b is as in Lemma 8 we also adjoin to $\mathscr{B}$ all the primes in k which divide b. By the analogue of Lemma 5, we can now choose $\mathfrak{a}''$ close to $\mathfrak{a}$ so that all the conditions like $L^*(\mathscr{B}; -a_0a_1, c; \mathfrak{a}'') = 1$ hold.

As was remarked in the last paragraph before (19), we can now increase $\mathscr{B}$ so that if $\lambda_0 = \alpha_0/\beta_0$ is a point of $\mathbf{L}^1(K)$ in $\mathfrak{a}''$, then α_0, β_0 are coprime and integral except perhaps at primes of K above a prime in $\mathscr{B}$. Now apply Lemma 8 with $M = 2$, where we take the $c(X, 1)$, normalized to be monic, to be the $P_i(X)$ and each U_v to be a small neighbourhood of the monic polynomial whose roots determine $\mathfrak{a}''$. Let $G(X)$ be given by Lemma 8; let $\mathfrak{a}'$ be the associated 0-cycle on $\mathbf{L}^1(k)$ and λ a point of $\mathbf{L}^1(K)$ in $\mathfrak{a}'$. For each v in $\mathscr{B}$, the cycle $\mathfrak{a}'$ is close to $\mathfrak{a}''$ in the v-adic topology; so (22) at λ is soluble in K_w for each w above v, by continuity. But $\lambda = \alpha/\beta$ with α, β coprime except at primes of K above a prime of $\mathscr{B}$. So

$$\prod_{\mathfrak{P}}(-a_0(\alpha,\beta)a_1(\alpha,\beta),c(\alpha,\beta))_{\mathfrak{P}} = L^*(\mathscr{B};-a_0a_1,c;\alpha,\beta) = 1,$$

where the product is taken over all primes $\mathfrak{P}$ not above a prime in $\mathscr{B}$ and such that $c(\alpha,\beta)$ is divisible to an odd power by $\mathfrak{P}$. Here the first equality holds by definition and the second one follows from the evaluation formula (8) by continuity. But if $c(X,1) = P_i(X)$, then the product on the left reduces to the single term for which $\mathfrak{P}$ is the prime of K above $\mathfrak{p}_i$ whose existence was proved by means of (16). Hence (22) at λ is locally soluble at this prime; and because these are the only primes not lying above a prime of $\mathscr{B}$ which divide any $c(\alpha,\beta)$ or any $a_r(\alpha,\beta)$ to an odd power, they are the only primes not lying above a prime of $\mathscr{B}$ at which local solubility might present any difficulty. Thus λ can be lifted to a point of that conic which is the fibre above λ, and the theorem now follows because weak approximation holds on conics. □

We now turn to the proof of Theorem 2(i), and therefore revert from W_0 to V_0. In principle, the idea of the proof is to construct a sequence of positive 0-cycles defined over k of decreasing degrees, each satisfying the appropriate local conditions, until we obtain a point P_0 in $V_0(k)$ satisfying the given local conditions; and indeed this is what we shall do in the last part of the proof. But it is not obvious how the local descriptions of successive elements of the sequence are related. So although the application of Theorem 3 to (21) shows that there is a positive 0-cycle of degree 8 satisfying any assigned local conditions, we do not yet know what local conditions to impose on it for the P_0 derived from it to be close in the topology induced by $\mathscr{B}$ to the adelic point which is our target. To cope with this, we first run the process backwards. We need only consider the continuous conditions $\mathscr{L}^* = 1$ introduced in Lemma 6.

From now on, any $\mathfrak{b}^r$ or $\mathfrak{b}^r_v$ will be a positive 0-cycle on V_0, defined over k or k_v respectively, and $\mathfrak{a}^r$ or $\mathfrak{a}^r_v$ will be its projection on $\mathbf{L}^1$. For each v in $\mathscr{B}$ we choose two distinct hyperplanes H'_v and H''_v, each defined over k_v and passing through A_v. Choose H', a hyperplane defined over k and close to each H'_v, and similarly for H''. The intersection $H' \cap H'' \cap V$ is a positive 0-cycle $\mathfrak{b}^1$ of degree 4 defined over k; and though $\mathfrak{b}^1$ may be irreducible over k it is reducible over k_v for each v in $\mathscr{B}$ because it has one point close to A_v. Thus we can write $\mathfrak{b}^1 = \mathfrak{b}^2_v \cup \mathfrak{b}^3_v$ where $\mathfrak{b}^2_v, \mathfrak{b}^3_v$ are positive 0-cycles of degrees 1 and 3 respectively defined over k_v and $\mathfrak{b}_v^2$ is close to A_v. Hence

$$1 = \mathscr{L}^*(\mathscr{B}; -a_0a_1, c; \mathfrak{a}^1) = \prod \ell^*(v; -a_0a_1, c; \mathfrak{a}_v^2 \cup \mathfrak{a}_v^3)$$
$$= \prod \ell^*(v; -a_0a_1, c; \mathfrak{a}_v^2) \prod \ell^*(v; -a_0a_1, c; \mathfrak{a}_v^3)$$

where the products are each taken over all v in $\mathscr{B}$. But the first product in the second line is 1, by continuity applied to (24); hence

$$\prod \ell^*(v; -a_0a_1, c; \mathfrak{a}_v^3) = 1. \tag{26}$$

Now let P_1 and P_2 be two points of $V_0(k)$; there is a linear system of ∞^6 curves on V which are the intersection of V with a quadric and have double points at P_1 and P_2. For each v in $\mathscr{B}$, let C'_v and C''_v be two such curves defined over k_v each of which also passes through the three points of $\mathfrak{b}_v^3$, and let Q'_v, Q''_v be quadrics defined over k_v which contain C'_v, C''_v, respectively, but neither of which contains the whole of V. Choose Q', a quadric defined over k, close to each Q'_v and touching V at P_1 and P_2, and similarly for Q''; since Q' is given by a single equation and the tangency conditions are linear in the coefficients, this is just a matter of weak approximation. The intersection

$$Q' \cap Q'' \cap V = 4\{P_1\} \cup 4\{P_2\} \cup \mathfrak{b}^4.$$

(This fails if Q' and Q'' have a common component; but we can ensure that this does not happen by requiring P_1, P_2 and $\mathfrak{b}^1$ to be in sufficiently general position. Similar remarks are needed at each stage of the proof.)

Much as before, $\mathfrak{b}^4 = \mathfrak{b}_v^5 \cup \mathfrak{b}_v^6$ over k_v for each v in $\mathscr{B}$, where each $\mathfrak{b}_v^5$ has degree 3 and is close to $\mathfrak{b}_v^3$, and each $\mathfrak{b}_v^6$ has degree 5; hence

$$\prod \ell^*(v; -a_0a_1, c; \mathfrak{a}_v^5) = 1$$

follows from (26) by continuity. But

$$\mathscr{L}(\mathscr{B}; -a_0a_1, c; \lambda_1) = \mathscr{L}(\mathscr{B}; -a_0a_1, c; \lambda_2) = 1$$

where λ_1, λ_2 are the projections of P_1, P_2 on $\mathbf{L}^1$; so

$$\prod \ell^*(v; -a_0a_1, c; \mathfrak{a}_v^6) = 1.$$

Now let P_3, P_4, P_5 be three further points of $V_0(k)$; then there is a linear system of ∞^9 curves on V which are the intersection of V with a quadric and pass through P_3, P_4, P_5. For each v in $\mathscr{B}$, let D'_v and D''_v be two such curves defined over k_v each of which also passes through the five points of $\mathfrak{b}_v^6$, and let R'_v, R''_v be quadrics defined over k_v which contain D'_v, D''_v respectively, but neither of which contains the whole of V. Choose R', a quadric defined over

k, close to each R'_v and passing through P_3, P_4, P_5, and similarly for R''. The intersection

$$R' \cap R'' \cap V = \{P_3\} \cup \{P_4\} \cup \{P_5\} \cup \mathfrak{b}^7,$$

where $\mathfrak{b}^7_v$ has degree 13. Much as before, $\mathfrak{b}^7 = \mathfrak{b}^8_v \cup \mathfrak{b}^9_v$ over k_v for each v in $\mathscr{B}$, where each $\mathfrak{b}^9_v$ is close to $\mathfrak{b}^6_v$, so that $\mathfrak{b}^8_v$ has degree 8 and

$$\prod \ell^*(v; -a_0a_1, c; \mathfrak{a}^8_v) = 1.$$

We now have the necessary road-map showing us how to go back. By Theorem 3, we can find a positive 0-cycle $\mathfrak{d}^8$ of degree 8 on V_0, defined over k and arbitrarily near to each $\mathfrak{b}^8_v$. With the same P_3, P_4, P_5 as before, there is a pencil of curves on V which are the intersections of V with a quadric and pass through P_3, P_4, P_5 and the points of $\mathfrak{d}^8$. Let $\mathfrak{d}^5$, of degree 5, be the residual intersection of the curves of this pencil; since the pencil contains a curve close to each D'_v and another close to each D''_v, it follows that $\mathfrak{d}^5$ is close to each $\mathfrak{b}^5_v$. (This time, the curves in the pencil do not all have a common component, because one of them is arbitrarily close to $R' \cap V$ and another to $R'' \cap V$.)

In the same way, we successively generate a 0-cycle $\mathfrak{d}^3$ on V_0 of degree 3 and arbitrarily close to each $\mathfrak{b}^3_v$, and then a point of $V_0(k)$ arbitrarily close to each A_v. This last is the point which we want. □

References

[1] J.-L. Colliot-Thélène & P. Swinnerton-Dyer – Hasse principle and weak approximation for pencils of Severi–Brauer and similar varieties, *J. Reine Angew. Math.* **453** (1994), 49–112.

[2] P. Salberger & A. N. Skorobogatov – Weak approximation for surfaces defined by two quadratic forms, *Duke Math. J.* **63** (1991), no. 2, 517–536.

[3] P. Swinnerton-Dyer – Rational points on pencils of conics and on pencils of quadrics, *J. London Math. Soc. (2)* **50** (1994), no. 2, 231–242.

[4] ———, Some applications of Schinzel's hypothesis to Diophantine equations, Number theory in progress, Vol. 1 (Zakopane-Kościelisko, 1997), de Gruyter, Berlin, 1999, 503–530.

Arithmetic of Higher-dimensional Algebraic Varieties
(B. Poonen, Yu. Tschinkel, eds.), p. 259–267
Progress in Mathematics, Vol. 226, © 2004 Birkhäuser Boston, Cambridge, MA

TRANSCENDENTAL BRAUER–MANIN OBSTRUCTION ON A PENCIL OF ELLIPTIC CURVES

Olivier Wittenberg
UMR 8628, Mathématiques, Bâtiment 425, Université de Paris-Sud, F-91405 Orsay, France • *E-mail :* olivier.wittenberg@ens.fr

Abstract. This short note gives an explicit example of transcendental Brauer–Manin obstruction to weak approximation. It has two features which the only previously known example of such obstruction did not have: the class in the Brauer group which is responsible for the obstruction is divisible, and the underlying algebraic variety is an elliptic surface.

1. Introduction

Let $\mathrm{Br}(X)$ denote the cohomological Brauer group $H^2_{\text{et}}(X, \mathbf{G}_m)$ of a scheme X. Let k be a number field and $\overline{k}$ be an algebraically closed extension of k. A class in the Brauer group of a projective smooth variety X over k is said to be *algebraic* if it belongs to the kernel of the restriction map $\mathrm{Br}(X) \to \mathrm{Br}(X_{\overline{k}})$, *transcendental* otherwise; this property does not depend on the choice of $\overline{k}$. For any prime number ℓ, the ℓ-primary part of the Brauer group over $\mathbf{C}$ fits into an exact sequence

$$0 \longrightarrow (\mathbf{Q}_\ell/\mathbf{Z}_\ell)^{b_2 - \rho} \longrightarrow \mathrm{Br}(X_{\mathbf{C}})\{\ell\} \longrightarrow H^3(X(\mathbf{C}), \mathbf{Z})\{\ell\} \longrightarrow 0,$$

Key words and phrases. Brauer–Manin obstruction, elliptic fibrations.

where b_2 and ρ respectively denote the second Betti number and the Picard number of $X_{\mathbf{C}}$, and $M\{\ell\}$ denotes the ℓ-primary part of M. Although this sequence does prove the non-triviality of $\mathrm{Br}(X_{\mathbf{C}})$ in many cases, e.g., when X is a $K3$ surface, transcendental classes are in general difficult to exhibit.

Almost all known instances of Brauer–Manin obstruction are thus explained by algebraic classes, the only exceptions being Harari's examples [**4**] with conic bundles over $\mathbf{P}^2_{\mathbf{Q}}$. Besides, in the particular case of pencils of curves of genus 1, results on the Hasse principle have been obtained only under the assumption that the 2-primary part of the Brauer group be "vertical", and therefore algebraic (see [**3**], § 4.7). The role of transcendental elements in the Brauer–Manin obstruction thus seems worthy of investigation. In this note we present an example of transcendental Brauer–Manin obstruction to weak approximation for an elliptic $K3$ surface over $\mathbf{Q}$, where "elliptic" means that it possesses a fibration in curves of genus 1, with a section, over $\mathbf{P}^1_{\mathbf{Q}}$. It should be noted that the class of order 2 which we will exhibit in $\mathrm{Br}(X_{\mathbf{C}})$ enjoys the property of being divisible (because $H^3(X(\mathbf{C}),\mathbf{Z}) = 0$ for a $K3$ surface), which was not the case in Harari's examples.

Acknowledgements: The author is most grateful to J.-L. Colliot-Thélène for sharing unpublished notes on the topic (which contain in particular the statement of Proposition 2.2), and would also like to thank him for his encouragements and many helpful conversations during the course of this research.

2. Preliminaries: 2-descent and the Brauer group of an elliptic curve

The subscript in $H^i_{\text{ét}}$ will be dropped, as we will only use étale cohomology. If G is an abelian group (resp. group scheme), ${}_nG$ will denote the n-torsion subgroup of G. Let k be a field of characteristic different from 2. The Hilbert symbol of a pair of elements $f, g \in k^\star$ will be denoted (f, g); it is the class of a quaternion algebra in ${}_2\mathrm{Br}(k)$. When X is a geometrically integral variety over k and L is an extension of k, $L(X)$ will denote the function field of X_L. The canonical morphism $\mathrm{Br}(X) \to \mathrm{Br}(k(X))$ is injective if in addition X is regular; this fact will be used without further mention. Let E be an elliptic curve over k whose 2-torsion points are rational. Fix an isomorphism of k-group schemes $(\mathbf{Z}/2\mathbf{Z})^2 \xrightarrow{\sim} {}_2E$. The kernel of the evaluation map at the zero section $\mathrm{Br}(E) \to \mathrm{Br}(k)$ will be denoted $\mathrm{Br}^0(E)$.

Lemma 2.1. *The group* $\mathrm{Br}^0(E)$ *is canonically isomorphic to* $H^1(k, E)$.

Proof. Let us write the Leray spectral sequence for the structure morphism $f\colon E \to \operatorname{Spec}(k)$ and the étale sheaf $\mathbf{G}_{\mathrm{m}}$. Since $f_\star\mathbf{G}_{\mathrm{m}} = \mathbf{G}_{\mathrm{m}}$, $R^1f_\star\mathbf{G}_{\mathrm{m}} = E \oplus \mathbf{Z}$ and $R^qf_\star\mathbf{G}_{\mathrm{m}} = 0$ for $q > 1$ by Tsen's theorem, we get an exact sequence

$$\operatorname{Br}(k) \longrightarrow \operatorname{Br}(E) \longrightarrow H^1(k,E) \longrightarrow H^3(k,\mathbf{G}_{\mathrm{m}}) \longrightarrow H^3(E,\mathbf{G}_{\mathrm{m}}).$$

The zero section induces retractions of $\operatorname{Br}(k) \to \operatorname{Br}(E)$ and of $H^3(k,\mathbf{G}_{\mathrm{m}}) \to H^3(E,\mathbf{G}_{\mathrm{m}})$, hence the lemma. □

The Kummer sequence

$$0 \longrightarrow {}_2E \longrightarrow E \xrightarrow{z\mapsto 2z} E \longrightarrow 0,$$

together with the previous lemma and the chosen isomorphism $(\mathbf{Z}/2\mathbf{Z})^2 \xrightarrow{\sim} {}_2E$, yields the exact sequence

$$0 \longrightarrow E(k)/2E(k) \xrightarrow{\delta} (k^\star/k^{\star 2})^2 \xrightarrow{\gamma} {}_2\operatorname{Br}^0(E) \longrightarrow 0. \tag{1}$$

We shall need explicit descriptions of the maps δ and γ. First choose distinct $p,q \in k^\star$ such that the Weierstrass equation

$$y^2 = x(x-p)(x-q) \tag{2}$$

defines E and the points $P = (p,0)$ and $Q = (q,0)$ are respectively sent to $(1,0)$ and $(0,1)$ via ${}_2E \xrightarrow{\sim} (\mathbf{Z}/2\mathbf{Z})^2$. It is well known (see e.g., **[9]**, p. 281) that $\delta(M) = (x(M)-q, x(M)-p)$ for $M \in E(k)$ if $M \notin {}_2E(k)$, that $\delta(P) = (p-q, p(p-q))$ and that $\delta(Q) = (q(q-p), q-p)$.

Proposition 2.2. *Let $f,g \in k^\star$. The classes of the quaternion algebras*

$$(x-p,f) \quad \textit{and} \quad (x-q,g) \in \operatorname{Br}(k(E))$$

actually belong to $\operatorname{Br}^0(E)$, and $\gamma(f,g) = (x-p,f) + (x-q,g)$.

Proof. By symmetry, it is enough to prove that $\gamma(f,1) = (x-p,f)$ in $\operatorname{Br}(k(E))$. Choose a separable closure $\overline{k}$ of k and let G_k be its Galois group over k. Likewise, choose a separable closure $\overline{k(E)}$ of $\overline{k}(E)$ and let $G_{k(E)}$ be its Galois group over $k(E)$. It follows from the Hochschild–Serre spectral sequence, Tsen's theorem and Hilbert's theorem 90 that the inflation map $H^2(k,\overline{k}(E)^\star) \to \operatorname{Br}(k(E))$ is an isomorphism. Let $\rho\colon H^1(k,E) \to H^2(k,\overline{k}(E)^\star/\overline{k}^\star)$ denote the composition of the canonical isomorphism $H^1(k,E) \xrightarrow{\sim} H^1(k,\operatorname{Pic}(E_{\overline{k}}))$ and the boundary of the exact sequence

$$0 \longrightarrow \overline{k}(E)^\star/\overline{k}^\star \longrightarrow \operatorname{Div}(E_{\overline{k}}) \longrightarrow \operatorname{Pic}(E_{\overline{k}}) \longrightarrow 0.$$

As shown in the annexe of [**2**], the diagram

$$\begin{array}{ccccc} \mathrm{Br}(k) & \longrightarrow & \mathrm{Br}(E) & \xrightarrow{\theta} & H^1(k,E) \\ \| & & \cap & & \\ \| & & \mathrm{Br}(k(E)) & & \Big\downarrow{-\rho} \\ \| & & \Big\downarrow\wr & & \\ \mathrm{Br}(k) & \longrightarrow & H^2(k,\overline{k}(E)^\star) & \longrightarrow & H^2(k,\overline{k}(E)^\star/\overline{k}^\star) \end{array}$$

commutes, where θ denotes the map which stems from the Leray spectral sequence (see Lemma 2.1). This enables us to carry out cocycle calculations for determining the image of $\gamma(f,1)$ in $H^2(k,\overline{k}(E)^\star/\overline{k}^\star)$. We shall use the standard cochain complexes. Let $\chi_f\colon G_k \to \mathbf{Z}$ be the map with image in $\{0,1\}$ whose composition with the projection $\mathbf{Z} \to \mathbf{Z}/2\mathbf{Z}$ is the quadratic character associated with $f \in k^\star/k^{\star 2} = H^1(G_k,\mathbf{Z}/2\mathbf{Z})$. The image of $(f,1)$ in $H^1(k,E)$ is represented by the 1-cocycle $a\colon \sigma \mapsto \chi_f(\sigma)P$. If $M \in E(k)$, let $[M]$ denote the corresponding divisor on $E_{\overline{k}}$. The 1-cochain with values in $\mathrm{Div}(E_{\overline{k}})$ defined by $\sigma \mapsto \chi_f(\sigma)([P]-[0])$ is a lifting of a. Its differential $(\sigma,\tau) \mapsto (\chi_f(\sigma)+\chi_f(\tau)-\chi_f(\sigma\tau))([P]-[0])$ is, as expected, a 2-cocycle with values in $\overline{k}(E)^\star/\overline{k}^\star$, which we may rewrite as $(\sigma,\tau) \mapsto (x-p)^{\chi_f(\sigma)\chi_f(\tau)}$; it represents the image of $\gamma(f,1)$ in $H^2(k,\overline{k}(E)^\star/\overline{k}^\star)$. Since $x-p$ is invariant under G_k, the same formula defines a 2-cocycle on G_k with values in $\overline{k}(E)^\star$. We thus end up with a 2-cocycle

$$\begin{array}{rcl} b\colon G_{k(E)} \times G_{k(E)} & \longrightarrow & \overline{k(E)}^\star \\ (\sigma,\tau) & \longmapsto & (x-p)^{\chi_f(\sigma)\chi_f(\tau)} \end{array}$$

which represents the image of $\gamma(f,1)$ in $\mathrm{Br}(k(E))$, at least modulo $\mathrm{Br}(k)$, where χ_m now denotes the lifting with values in $\{0,1\}$ of the quadratic character on $k(E)$ associated with $m \in k(E)^\star$. (Note that k is separably closed in $k(E)$, so that G_k identifies with a quotient of $G_{k(E)}$.) Choose a square root s of $x-p$ in $\overline{k(E)}$. Dividing b by the differential of the 1-cochain $\sigma \mapsto s^{\chi_f(\sigma)}$ gives the 2-cocycle $(\sigma,\tau) \mapsto (-1)^{\chi_{x-p}(\sigma)\chi_f(\tau)}$, which does represent the image of the cup-product $(x-p)\cup f$ by the composite map

$$H^1(k(E),\mathbf{Z}/2\mathbf{Z})^{\otimes 2} \to H^2(k(E),\mathbf{Z}/2\mathbf{Z}) \to \mathrm{Br}(k(E)).$$

We have now proved that $\gamma(f,1) = (x-p,f)$ in $\mathrm{Br}(k(E))/\mathrm{Br}(k)$, but the equality holds in $\mathrm{Br}(k(E))$ since $(x-p,f) = (y^2/(x-p)^3,f)$ evaluates to 0 at the zero section. $\square$

3. An actual example

The reader is referred to [4] for the definitions of weak approximation, Brauer–Manin obstruction, residue maps and unramified Brauer group.

Let Ω denote the set of places of $\mathbf{Q}$. Define the polynomials $p, q \in \mathbf{Q}[t]$ by $p(t) = 3(t-1)^3(t+3)$ and $q(t) = p(-t)$. It will be useful to notice that $p(t) - q(t) = 48t$. Let E be the elliptic curve over $\mathbf{Q}(t)$ defined by (2). Denote by $\mathscr{E}$ its minimal proper regular model over $\mathbf{P}^1_{\mathbf{Q}}$ (see [8]); it is a smooth surface over $\mathbf{Q}$ endowed with a proper flat morphism $f\colon \mathscr{E} \to \mathbf{P}^1_{\mathbf{Q}}$ whose generic fiber is isomorphic to E. A geometric fiber of f is either smooth or is a union of rational curves whose intersection numbers may be computed with Tate's algorithm [10]. One finds the following reduction types, in Kodaira's notation [5]: I_2 above $t = 0$, $t = 3$ and $t = -3$; I_6 above $t = 1$, $t = -1$ and $t = \infty$; the other fibers are smooth. Recall that a fiber of type I_n has n irreducible components $(C_i)_{1 \leqslant i \leqslant n}$, with $(C_i.C_{i+1}) = 1$, $(C_1.C_n) = 1$ and $(C_i.C_j) = 0$ if $|j - i| > 1$. Put

$$A = \gamma(6t(t+1), 6t(t-1)) = (x - p, 6t(t+1)) + (x - q, 6t(t-1)) \in \mathrm{Br}(E).$$

Proposition 3.1. *The class $A \in \mathrm{Br}(E)$ belongs to the subgroup $\mathrm{Br}(\mathscr{E})$.*

Proof. Let v be a discrete rank 1 valuation on $\mathbf{Q}(\mathscr{E})$ whose restriction to $\mathbf{Q}$ is trivial, and κ be its residue field. We shall prove that A has trivial residue at v. Let us choose a uniformiser π of v and put $\tilde{z} = z\pi^{-v(z)}$ for $z \in \mathbf{Q}(\mathscr{E})^\star$. It will be convenient to denote by $V\colon \mathbf{Q}(\mathscr{E})^\star \to \mathbf{Z} \times \kappa^\star$ the group homomorphism $z \mapsto (v(z), [\tilde{z}])$, where $[u]$ denotes the class in κ of $u \in \mathbf{Q}(\mathscr{E})$ if $v(u) = 0$. For $f, g \in \mathbf{Q}(\mathscr{E})^\star$, the residue of the quaternion algebra (f, g) at v is given by the tame symbol formula

$$\partial_v(f,g) = (-1)^{v(f)v(g)} \left[\frac{f^{v(g)}}{g^{v(f)}}\right] = (-1)^{v(f)v(g)} \left[\tilde{f}\right]^{v(g)} \left[\tilde{g}\right]^{v(f)} \in \kappa^\star/\kappa^{\star 2}.$$

Note that it only depends on $V(f)$ and $V(g)$. Furthermore, if $V(f)$ is a double, i.e., if $v(f)$ is even and $\tilde{f}$ is a square modulo π, then $\partial_v(f,g) = 1$. These remarks will be used implicitly throughout the proof.

Lemma 3.2. *The class $(-p, 6t(t+1)) + (-q, 6t(t-1)) \in \mathrm{Br}(\mathbf{Q}(t))$ is unramified over $\mathbf{P}^1_{\mathbf{Q}}$.*

Proof. The residue at a closed point of $\mathbf{P}^1_{\mathbf{Q}}$ other than $t = \alpha$ for

$$\alpha \in \{-3, -1, 0, 1, 3, \infty\}$$

is obviously trivial. It is straightforward to check that the remaining residues are also trivial. $\square$

Let us now turn to showing that $\partial_v(A) = 1$. As A is invariant under $t \mapsto -t$, we may assume $v(p) \leqslant v(q)$. If $v(x) < v(p)$, then $V(x-p) = V(x-q) = V(x)$, from which we deduce thanks to (2) that $V(x-p)$ and $V(x-q)$ are doubles. If $v(x) > v(q)$, then $V(x-p) = V(-p)$ and $V(x-q) = V(-q)$, hence the result by Lemma 3.2. From now on, we may and will therefore assume $v(p) \leqslant v(x) \leqslant v(q)$.

To begin with, suppose $v(p) < v(q)$. In this case, either $v(t-3) > 0$ or $v(t+1) > 0$. If $v(x) = v(q)$, then $V(x-p) = V(-p)$, hence

$$\partial_v(A) = \partial_v(-q(x-q), 6t(t-1))$$

by Lemma 3.2; but with a look at (2), one finds that both $v(-q(x-q))$ and $v(6t(t-1))$ are even. Suppose now $v(x) < v(q)$. It follows from (2) that $V(x-p)$ is a double, hence

$$\partial_v(A) = \partial_v(x-q, 6t(t-1)) = \partial_v(x, 6t(t-1)).$$

If $v(x)$ is even or if $[6t(t-1)]$ is a square in κ, which happens if $v(t-3) > 0$, we get $\partial_v(A) = 1$. If on the other hand $v(t+1) > 0$ and $v(x)$ is odd, then $[6t(t-1)] = 12$, which (2) shows to be a square in κ.

We are now left with the case $v(p) = v(q) = v(x)$. If $v(t) = 0$, then $v(t-3) = v(t-1) = v(t+1) = v(t+3) = 0$, so $v(6t(t+1)) = v(6t(t-1)) = 0$ and it suffices to prove that $v(x-p)$ and $v(x-q)$ are even, which follows from (2) and the equality $v(p) = v(x) = v(q) = v(p-q) = 0$. If $v(t) < 0$, then $V(6t(t+1)) = V(6t(t-1))$, so that $\partial_v(A) = \partial_v(x, 6t(t+1))$, which is trivial since both $v(x) = v(p) = 4v(t)$ and $v(6t(t+1))$ are even. Suppose finally that $v(t) > 0$. If $v(x-p) < v(t)$, then $V(x-p) = V(x-q)$ since $v(p-q) = v(t)$, and $\partial_v(A) = \partial_v(x-p, (t+1)(t-1)) = \partial_v(x-p, -1)$; if $v(x-p) = 0$, the residue is obviously trivial, and if $v(x-p) > 0$, which means that $[\tilde{x}] = [\tilde{p}] = -9$, (2) shows that -1 is a square in κ. We therefore assume $v(x-p) \geqslant v(t)$, which still leads to $[\tilde{x}] = [\tilde{p}] = -9$. As $v(p-q) = v(t)$, at least one of $v(x-p)$ and $v(x-q)$ is equal to $v(t)$. In either case, (2) implies that $v(x-p) + v(t)$ is even, so $(-9)^{v(t)}(-1)^{v(x-p)}$ is a square, hence $\partial_v(A) = \partial_v(x, 6t(t-1)) + \partial_v(x-p, (t+1)(t-1))$ is trivial. □

We shall now prove the following.

Theorem 3.3. *The class $A \in \mathrm{Br}(\mathscr{E})$ is transcendental and yields a Brauer–Manin obstruction to weak approximation on the projective smooth surface $\mathscr{E}$ over $\mathbf{Q}$.*

Proof. Let us first deal with the second part of the assertion. A glance at equation (2) shows that $\mathscr{E}$ has a $\mathbf{Q}_2$-point M_2 with coordinates $x = 1$ and $t = 2$. (Indeed, this equation defines an affine surface over $\mathbf{Q}$ endowed with a morphism to $\mathbf{P}^1_{\mathbf{Q}}$ whose smooth locus identifies with an open subset of $\mathscr{E}$.) Using the formula given in [**7**], Ch. XIV, §4, one easily checks that $A(M_2)$ is non-trivial. Now choose $N \in \mathscr{E}(\mathbf{Q})$ in the image of the zero section and let $M_v \in \mathscr{E}(\mathbf{Q}_v)$ be equal to N for any $v \in \Omega \setminus \{2\}$. This defines an adelic point $(M_v)_{v\in\Omega}$. The class $A(N) \in \mathrm{Br}(\mathbf{Q})$ is trivial since $A \in \mathrm{Br}^0(E)$; consequently, the evaluation of A at $(M_v)_{v\in\Omega}$ is non-trivial, which is an obstruction to weak approximation.

It remains to be shown that A is transcendental. The exact sequence (1) reduces this to the computation of $E(\mathbf{C}(t))/2E(\mathbf{C}(t))$.

Lemma 3.4. *The surface $\mathscr{E}$ is a $K3$ surface.*

Proof. The topological Euler–Poincaré characteristic $e(\mathscr{E}_{\mathbf{C}})$ of $\mathscr{E}_{\mathbf{C}}$ can be expressed in terms of that of the fibers and that of the base ([**1**], p. 97, prop. 11.4), which leads to $e(\mathscr{E}_{\mathbf{C}}) = 24$. Let $\chi(\mathscr{O}_{\mathscr{E}})$ denote the Euler–Poincaré characteristic of the coherent sheaf $\mathscr{O}_{\mathscr{E}}$. The canonical bundle $\mathscr{K}_{\mathscr{E}}$ of $\mathscr{E}$ is simply $f^\star\mathscr{O}(\chi(\mathscr{O}_{\mathscr{E}})-2)$ (see [**1**], p. 162, cor. 12.3); in particular it has self-intersection 0, hence $\chi(\mathscr{O}_{\mathscr{E}}) = 2$ by Noether's formula. We have now proved the triviality of $\mathscr{K}_{\mathscr{E}}$. That $H^1(\mathscr{E}, \mathscr{O}_{\mathscr{E}}) = 0$ follows from $\chi(\mathscr{O}_{\mathscr{E}}) = 2$ and Serre duality. □

Lemma 3.5. *The elliptic curve E has Mordell–Weil rank 0 over $\mathbf{C}(t)$.*

Proof. Let $\rho(\mathscr{E}_{\mathbf{C}})$ be the Picard number of $\mathscr{E}_{\mathbf{C}}$ and R be the subgroup of the Néron–Severi group $\mathrm{NS}(\mathscr{E}_{\mathbf{C}})$ spanned by the zero section and the irreducible components of the fibers. As follows from the output of Tate's algorithm, R has rank 20. On the other hand, $\rho(\mathscr{E}_{\mathbf{C}}) \leqslant 20$ since $\mathscr{E}$ is a $K3$ surface. The Shioda–Tate formula

$$\rho(\mathscr{E}_{\mathbf{C}}) = \mathrm{rank}(E(\mathbf{C}(t))) + \mathrm{rank}(R)$$

thus yields the result. □

This lemma shows that the $\mathbf{F}_2$-vector space $E(\mathbf{C}(t))/2E(\mathbf{C}(t))$ has dimension 2. Now the classes $\delta(P) = (t, t(t-1)(t+3))$ and $\delta(Q) = (t(t+1)(t-3), t)$ are independent over $\mathbf{F}_2$, hence span the whole kernel of γ. On the other hand $(t(t+1), t(t-1))$ is evidently not a combination of $\delta(P)$ and $\delta(Q)$, so that A has non-zero image in $\mathrm{Br}(\mathbf{C}(\mathscr{E}))$ and is therefore transcendental. □

Remark 3.6. It is actually true that $A(M) = 0$ in $\mathrm{Br}(\mathbf{Q})$ for *all* $M \in \mathscr{E}(\mathbf{Q})$. This is a consequence of the global reciprocity law and the fact that A vanishes on $\mathscr{E}(\mathbf{Q}_v)$ for all $v \in \Omega \setminus \{2\}$, which can be checked by a tedious computation.

Remark 3.7. It is possible to determine ${}_2\mathrm{Br}(\mathscr{E})$ completely if one is willing to compute explicit equations for $\mathscr{E}$. This involves blowing up the singular surface given by equation (2) a sufficient number of times. Alternately, one may observe that all fibers have type I_n (in other words, $\mathscr{E} \to \mathbf{P}^1_{\mathbf{Q}}$ is semi-stable), and then use the equations given by Néron in this case in [**6**], § III. Either way one finds that ${}_2\mathrm{Br}(\mathscr{E})$ is spanned by A modulo ${}_2\mathrm{Br}(\mathbf{Q})$ after writing out all possible residues of a general class $\gamma(f,g)$. On the other hand, the 2-torsion subgroup of the Brauer group of a complex $K3$ surface with Picard number 20 has rank 2 over $\mathbf{F}_2$, so ${}_2\mathrm{Br}(\mathscr{E}_{\mathbf{C}})$ is strictly larger than ${}_2\mathrm{Br}(\mathscr{E})/{}_2\mathrm{Br}(\mathbf{Q})$. It turns out that ${}_2\mathrm{Br}(\mathscr{E}_{\mathbf{C}})$ is spanned by A and the class of the quaternion algebra (x,t), which unexpectedly belongs to $\mathrm{Br}(\mathbf{Q}(\mathscr{E}))$ and only becomes unramified after extension of scalars to $\mathbf{Q}(\sqrt{-1},\sqrt{3})$.

Remark 3.8. In the semi-stable case, a computer program was written to carry out the calculations alluded to in the previous paragraph, since they often became quite lengthy. Its source code is available on request.

References

[1] W. Barth, C. Peters & A. V. de Ven – *Compact Complex Surfaces*, Ergebnisse der Mathematik und ihrer Grenzgebiete (3), vol. 4, Springer-Verlag, Berlin, 1984.

[2] J.-L. Colliot-Thélène & J.-J. Sansuc – La R-équivalence sur les tores, *Ann. Sci. École Norm. Sup. (4)* **10** (1977), no. 2, 175–229.

[3] J.-L. Colliot-Thélène, A. Skorobogatov & P. Swinnerton-Dyer – Hasse principle for pencils of curves of genus one whose Jacobians have rational 2-division points, *Invent. math.* **134** (1998), no. 3, 579–650.

[4] D. Harari – Obstructions de Manin transcendantes, Number theory (Paris, 1993–1994), London Math. Soc. Lecture Note Ser., vol. 235, Cambridge Univ. Press, Cambridge, 1996, 75–87.

[5] K. Kodaira – On compact analytic surfaces II, *Ann. of Math. (2)* **77** (1963), 563–626.

[6] A. Néron – Modèles minimaux des variétés abéliennes sur les corps locaux et globaux, *Inst. Hautes Études Sci. Publ. Math. No.* **21** (1964).

[7] J.-P. Serre – *Corps Locaux*, Hermann, Paris, 1968.

[8] I. R. Shafarevich – *Lectures on Minimal Models and Birational Transformations of Two Dimensional Schemes*, Notes by C. P. Ramanujam, Tata Institute of Fundamental Research Lectures on Mathematics and Physics, No. 37, Tata Institute of Fundamental Research, Bombay, 1966.

[9] J. H. Silverman – *The Arithmetic of Elliptic Curves*, Graduate Texts in Mathematics, vol. 106, Springer-Verlag, New York, 1992.

[10] J. Tate – Algorithm for determining the type of a singular fiber in an elliptic pencil, Modular functions of one variable, IV, Springer, Berlin, 1975, Lecture Notes in Math., Vol. 476, 33–52.

GLOSSARY

Abelian variety: A smooth projective geometrically integral group variety over a field. Elliptic curves are 1-dimensional abelian varieties.

Algebraic group: A smooth group scheme G over a field k. Some authors instead of insisting that G be smooth, assume only that G is of finite type.

Ample cone: See *cones of divisor classes.*

Automorphic form: Let G be a linear algebraic group over a global field k. Let $\mathbf{A}$ be the adele ring of k. An automorphic form is a function $G(k)\backslash G(\mathbf{A}) \to \mathbf{C}$ satisfying certain analytic conditions and behaving in a particular simple way under the action of some compact subgroup of $G(\mathbf{A})$.

Azumaya algebra: A sheaf of $\mathscr{O}_X$-algebras (on a scheme X) that is isomorphic étale locally on X to a finite-rank matrix algebra $\mathrm{M}_r(\mathscr{O}_X)$. If k is a field, an Azumaya algebra over $\operatorname{Spec} k$ is a finite-dimensional central simple algebra over k.

Big cone: See *cones of divisor classes.*

Birch and Swinnerton-Dyer conjecture: Let A be an abelian variety over a global field K and let $L(A,s)$ be the associated L-function. The Birch and Swinnerton-Dyer (BSD) conjecture asserts that $L(A,s)$ extends to an entire function and that $\operatorname{ord}_{s=1} L(A,s)$ equals the rank of $A(K)$. Moreover, the conjecture provides a formula for the leading coefficient of the Taylor expansion of $L(A,s)$ about $s=1$ in terms of other invariants of A.

Bombieri–Lang conjecture: See Lang's conjectures.

Brauer group: The cohomological Brauer group $\operatorname{Br}(X) = H^2_{\text{et}}(X, \mathbf{G}_m)$ of a scheme X. One can also define a Brauer group $\operatorname{Br}_{\text{Az}}(X)$ using Azumaya algebras, and it is conjectured that the two definitions agree when X is a smooth variety over a field.

Brauer–Manin obstruction: An obstruction to the Hasse principle that may exist for a smooth, geometrically integral variety X over a number field k. If the set of adelic points $X(\mathbf{A}_k)$ is nonempty, but the Brauer–Manin set $X(\mathbf{A}_k)^{\text{Br}}$ is empty, then one says that there is a Brauer–Manin obstruction to the Hasse principle. This obstruction was discovered by Manin. Skorobogatov found a surface for which $X(\mathbf{A}_k)^{\text{Br}}$ is nonempty, but $X(k)$ is empty: in order words, this is a counterexample to the Hasse principle not explained by the Brauer–Manin obstruction. If a class of varieties X satisfies

$$X(\mathbf{A}_k)^{\text{Br}} \neq \emptyset \implies X(k) \neq \emptyset,$$

one says that for that class, the Brauer–Manin obstruction to the Hasse principle is the only one. One can also formulate a Brauer–Manin obstruction to weak approximation.

Brauer–Manin set: A subset of the adelic points on a smooth geometrically integral variety X over a number field k: there is an evaluation pairing

$$\operatorname{Br}(X) \times X(\mathbf{A}_k) \to \mathbf{Q}/\mathbf{Z}$$

and the Brauer–Manin set $X(\mathbf{A}_k)^{\text{Br}}$ is defined as the set of $x \in X(\mathbf{A}_k)$ such that each $A \in \operatorname{Br}(X)$ pairs with x to give $0 \in \mathbf{Q}/\mathbf{Z}$. It contains the set $X(k)$ of rational points. (Some people also call $X(\mathbf{A}_k)^{\text{Br}}$ the set of Brauer points.)

Brauer–Severi variety: A twist of some projective space $\mathbf{P}^n$. Brauer–Severi varieties satisfy the Hasse principle. Some arithmetic properties of such varieties were first proved by Châtelet.

Calabi–Yau variety: A smooth projective integral variety X over $\mathbf{C}$ is a Calabi–Yau variety if $H^i(X, \mathscr{O}_X) = 0$ for $0 < i < \dim X$ and the canonical sheaf ω_X is trivial (i.e., $\omega_X \simeq \mathscr{O}_X$). Sometimes one insists in addition that $X(\mathbf{C})$ be simply connected.

Circle method: An analytic method for obtaining asymptotic formulas for the number of solutions to certain equations satisfying certain bounds.

Cones of divisor classes: Let X be a smooth projective integral variety over $\mathbf{C}$, let $\operatorname{NS}(X)$ be the Néron–Severi group of X, and let $\operatorname{NS}(X)_{\mathbf{R}}$ be the finite-dimensional $\mathbf{R}$-vector space $\operatorname{NS}(X) \otimes_{\mathbf{Z}} \mathbf{R}$. The cone generated by a set of classes in $\operatorname{NS}(X)_{\mathbf{R}}$ is the set of finite linear combinations with nonnegative coefficients of these classes. Important examples include the

- ample cone: generated by classes of ample divisors.
- nef cone: generated by nef classes (classes whose intersection number with any effective curve is nonnegative).
- big cone: generated by classes D such that $H^0(X, nD)/n^{\dim X}$ tends to a positive constant as $n \to \infty$.
- effective cone: generated by classes of effective divisors.

Each of these is contained in the next. They are not necessarily closed in $\mathrm{NS}(X)_{\mathbf{R}}$.

Cubic surface: A hypersurface of degree 3 in $\mathbf{P}^3$. A smooth cubic surface is a Del Pezzo surface of degree 3 (and vice versa).

Del Pezzo surface: A Del Pezzo surface is a Fano variety of dimension two. Over an algebraically closed field, the Del Pezzo surfaces are exactly the surfaces isomorphic to either $\mathbf{P}^1 \times \mathbf{P}^1$ or to a blowup of $\mathbf{P}^2$ at up to 8 points in general position. By *general position* we mean that no three points lie on a line, no six points lie on a conic, and no eight lie points lie on a singular cubic with one of the eight points on the singularity.

Descent: The word has two unrelated meanings in arithmetic geometry:

1. Fermat's "infinite descent," in modern terms, is a process of constructing, given a rational point on a variety, a point of smaller height; one then iterates the construction in order to contradict the finiteness of the set of rational points of bounded height, or to show (as in the Mordell–Weil theorem) that all rational points can be constructed from points of small height. The arguments involved can be reinterpreted as showing that, in some situations, the rational points on a variety can be expressed as a union of images of rational points from other varieties; hence the term "descent" is often used to signify any such construction.
2. The descent problem is as follows: Given a field extension L/K and a variety X over L, find a variety Y over K such that $X = Y \times_K L$. Weil gave a criterion for the existence of Y; his theory was vastly generalized in Grothendieck's theory of fpqc descent.

Diophantine set: Let R be a ring. A subset $A \subset R^n$ is diophantine over R if there exists a polynomial $f \in R[t_1, \ldots, t_n, x_1, \ldots, x_m]$ such that

$$A = \{\vec{t} \in R^n : \exists \vec{x} \in R^m \text{ such that } f(\vec{t}, \vec{x}) = 0\}.$$

The solution to Hilbert's tenth problem by Davis, Putnam, Robinson, and Matijaseviç proved that the diophantine sets over $\mathbf{Z}$ are the same as the recursively enumerable (listable) sets.

Effective cone: See *cones of divisor classes.*

Enriques Surface: A quotient of a K3 surface by a fixed-point free involution. Equivalently, the normalization of a singular surface of degree 6 in $\mathbf{P}^3$ whose singularities are double lines that form a general tetrahedron. Over $\mathbf{C}$ Enriques surfaces can be characterized cohomologically among smooth projective integral varieties by the following properties: $H^0(\Omega_X^2) = 0$ and $2K_X = 0$ but $K_X \neq 0$.

Faltings' theorem: If X is a smooth projective geometrically integral curve of genus > 1 over a number field k, then $X(k)$ is finite. This result had been conjectured by Mordell.

Fano variety: A smooth projective geometrically integral variety for which the anticanonical bundle $-K$ (that is, $\omega^{\otimes -1}$) is ample. This class of varieties is "simple" or "close to rational". For example, a deep theorem of Kollár–Miyaoka–Mori and independently Campana states that Fano varieties in characteristic 0 are rationally connected. One conjectures that for Fano varieties (and even for smooth projective geometrically integral rationally connected varieties in general) the Brauer–Manin obstruction to the Hasse principle is the only one. The Manin–Batyrev conjectures predict asymptotic estimates for points of bounded height on Fano varieties, but a counterexample of Batyrev–Tschinkel shows that the conjectures in their original form cannot hold in general.

See also Del Pezzo surface.

Fermat curve: The projective plane curve defined by $x^d + y^d = z^d$, for some positive integer d.

Fermat variety: A projective variety in $\mathbf{P}^n$ defined by the homogeneous equation $x_0^d + \cdots + x_n^d = 0$ for some positive integer d. (Sometimes one has some of the terms on the other side, or one generalizes by allowing coefficients other than 1.)

General type: A smooth projective geometrically integral variety X is of general type if there is a positive power of the canonical bundle whose global sections determine a rational map $f : X \dashrightarrow \mathbf{P}^n$ with $\dim f(X) = \dim X$. (If X is of general type then there exists some positive power of the canonical bundle such that the corresponding map is birational to its image.) One can show that this property depends only the birational class of X. Thus if Y is any geometrically integral variety birational to a smooth projective geometrically integral variety X, one says that Y is general type if and only if X is of general type. Because of Hironaka's resolution of singularities for varieties in characteristic 0, this definition applies to any geometrically integral variety Y in characteristic 0. There is a different definition of general type that gives the same answer in characteristic 0, but does not depend on resolution of singularities, and hence works even in characteristic p.

Geometric invariant theory: A theory that studies the quotient of a variety by the action of an algebraic group (when such a quotient exists).

Grassmannian variety: A smooth projective geometrically integral variety $G(n,k)$ whose points correspond to linear subspaces of a fixed dimension k in a fixed n-dimensional vector space. Sometimes instead one parameterizes linear subvarieties of dimension k in $\mathbf{P}^n$ (this is equivalent to parameterizing $(k+1)$-dimensional subspaces of an $(n+1)$-dimensional vector space).

Group cohomology: The functors $H^i(G,-)$ are defined as the right derived functors of the functor taking a G-module A to its subgroup of G-invariant elements. When G is a profinite group, such as a Galois group, one usually restricts the functor to continuous G-modules.

Hardy–Littlewood circle method: See circle method.

Hasse principle: A variety X over a global field k satisfies the Hasse principle if the existence of points on X over every completion k_v of k implies the existence of a k-rational point on X. Hasse proved that any projective variety defined by a single nondegenerate quadratic form satisfies this.

Height function: The (global) height of a point $(x_0 : \cdots : x_n) \in \mathbf{P}^n(\mathbf{Q})$ with $x_i \in \mathbf{Z}$ satisfying $\gcd(x_0, \ldots, x_n) = 1$ is $\max|x_i|$. It is a measure of the arithmetic complexity of the point. More generally, for any smooth projective geometrically integral variety X over a number field k with a metrized line bundle $\overline{\mathscr{L}} := (\mathscr{L}, \{\|\cdot\|_{P,v}\})$, one can define a local height by $H_{\overline{\mathscr{L}},s,v}(P) := \|s\|_{P,v}^{-1}$ where $s \in H^0(X, \mathscr{L})$ is nonvanishing at $P \in X(k_v)$, and then define a global height $H_{\overline{\mathscr{L}},s}(P) := \prod_v H_{s,v}(P)$ for any adelic point P on X outside the support of s. If moreover $P \in X(k)$, then $H_{\overline{\mathscr{L}},s}(P)$ is independent of the choice of s. Sometimes, for instance when studying Mordell–Weil groups of abelian varieties, it is convenient to define a logarithmic height function $h(P) = \log H(P)$, where H is a height function as above.

Heights can also be interpreted as self-intersection numbers in arithmetic intersection theory.

When a smooth projective geometrically integral variety X over a number field k has infinitely many rational points, one can fix a height function $H = H_{\overline{\mathscr{L}}}$ associated to an ample line bundle $\mathscr{L}$ and study the size of the finite set $\{\, P \in X(k) : H(P) \leq B \,\}$ as $B \to \infty$.

Hilbert's tenth problem: Let R be a commutative ring. Hilbert's tenth problem for R is to determine if there is an algorithm that decides whether or not a given system of polynomial equations with coefficients in R has a solution over R.

Homogeneous space: A homogeneous space under an algebraic group G over a field k is a k-variety X with an action of G such that $G(\overline{k})$ acts transitively on $X(\overline{k})$. For example, if G is a linear algebraic group, and H is a Zariski closed subgroup, then G/H is a homogeneous space under G.

Jacobian: The Jacobian of a nonsingular projective curve X is an abelian variety J of dimension equal to genus of X whose points, loosely speaking, correspond to line bundles of degree 0 on X. More precisely, if $X(k)$ is nonempty, the group $J(k)$ is isomorphic to the group $\operatorname{Pic}^0(X)$ of isomorphism classes of line bundles (or divisor classes) of degree 0 on X.

K3 surface: A smooth projective geometrically integral surface with trivial canonical bundle and trivial fundamental group (or equivalently, a Calabi–Yau variety of dimension 2).

Lang's conjectures (about rational points):

1. Suppose k is a number field and X is a variety over k of general type. Then $X(k)$ is not Zariski dense in X. (This was independently conjectured by Bombieri.)
2. Suppose k is a number field and X is a variety over k. All but finitely many k-rational points on X lie in the special set.
3. Let X be a variety over a number field k. Choose an embedding of k into $\mathbf{C}$, and suppose that $X(\mathbf{C})$ is hyperbolic: this means that every holomorphic map $\mathbf{C} \to X(\mathbf{C})$ is constant. Then $X(k)$ is finite.

Linear algebraic group: An algebraic group that is affine as a variety. Equivalently, a closed algebraic subgroup of the algebraic group GL_n for some n.

Local-global principle: Another name for the Hasse principle.

Metrized line bundle: Let k be a global field. Let Ω be the set of all places of k. Let X be a smooth projective geometrically integral variety over k, and $\mathscr{L}$ a line bundle on X. An adelic metrization of $\mathscr{L}$ is a family of v-adic metrics on $\mathscr{L} \otimes_k k_v$ subject to some continuity and compatibility conditions.

For example, let $X = \mathbf{P}^n$ over $\mathbf{Q}$ and $\mathscr{L} = \mathscr{O}(1)$. Let $x_0, \ldots, x_n$ be the standard basis for $H^0(X, \mathscr{L})$. For each place v of k, and each $P \in X(\mathbf{Q})$, define $\| \, \|_{P,v}$ by the formula

$$\|s\|_{P,v} := \left(\max_j \left| \frac{s}{x_j}(P) \right| \right)^{-1}$$

for any $s \in H^0(X, \mathscr{L})$ not vanishing at P. Then the family $\{\| \cdot \|_{P,v}\}$ is an adelic metrization of $\mathscr{L}$.

Adelic metrizations are used to define height functions.

Minimal model program: A research program to find, within each birational equivalence class of varieties, a particularly simple representative variety, and then to study the properties of these simple varieties. Over an algebraically closed field k, the following is known: In dimension 1, each birational class of irreducible curves contains a unique smooth projective curve. In dimension 2, each birational class that does not consist of rational or ruled surfaces contains a unique smooth projective surface containing no smooth rational curve with self-intersection -1. In dimension 3 in the characteristic 0 case, a minimal model program was completed by Mori. Partial progress has been made in some other cases.

Mordell–Weil theorem: The statement "If A is an abelian variety over a number field k, then the abelian group $A(k)$ is finitely generated." The result can be generalized to any field k that is finitely generated over its minimal subfield. The group $A(k)$ is called the Mordell–Weil group, and its rank is called the Mordell–Weil rank. All known proofs of this theorem are variants of the following strategy: One first proves the "weak Mordell–Weil theorem", which is the statement that $A(k)/mA(k)$ is finite for any $m \geq 1$. One then chooses a set of coset representatives for $mA(k)$ in $A(k)$ and shows that for all but finitely many points in $A(k)$, it is possible to subtract the appropriate representative and divide by m to make the height go down. This process is known as descent.

Mori theory: The minimal model program for 3-folds over algebraically closed fields of characteristic 0, completed by Mori in 1988. In contrast with the case of surfaces, it was necessary to allow 3-folds with mild singularities, and introduce new birational transformations in addition to blow-ups, called flips and flops.

Nef cone: See *cones of divisor classes*.

Néron–Severi group: The Néron–Severi group $\mathrm{NS}(X)$ of a smooth projective geometrically integral variety X is the quotient of the Picard group $\mathrm{Pic}(X)$ by the subgroup of line bundles that after base extension to an algebraic closure are algebraically equivalent to 0. It is a finitely generated abelian group.

Picard group: The Picard group $\mathrm{Pic}(X)$ of a variety or scheme X is the group of isomorphism classes of line bundles on X.

Picard number: The rank of the Néron–Severi group of a smooth projective geometrically integral variety.

Principal homogeneous space: Another name for torsor.

Rationally connected variety: A smooth projective integral variety X over an uncountable algebraically closed field k of characterstic 0 satisfying one of the following equivalent conditions:

1. For any two points $x, y \in X(k)$ there exists a rational map $\varphi : \mathbf{P}^1 \to X$ such that $\varphi(0) = x$ and $\varphi(\infty) = y$.
2. For any n points $x_1, \ldots, x_n \in X(k)$ there exists a rational map $\varphi : \mathbf{P}^1 \to X$ such that $\{x_1, \ldots, x_n\}$ is a subset of $\varphi(\mathbf{P}^1)$.
3. For any two points $x, y \in X(k)$ there exist rational maps $\varphi_i : \mathbf{P}^1 \to X$ for $i = 1, \ldots, r$ such that $\varphi_1(0) = x$, $\varphi_r(0) = y$, and for each $i = 1, \ldots, r-1$ the images of φ_i and φ_{i+1} have nontrivial intersection.

(When k has characteristic p, these definitions are no longer equivalent.) If X is over an arbitrary field k of characteristic 0, one says that X is rationally connected if $X_{\overline{k}}$ is.

It is conjectured that for a smooth projective geometrically integral variety X over a number field, if it is rationally connected, then the Brauer–Manin obstruction to the Hasse principle is the only one.

Restriction of scalars: Let L/K be a finite extension of fields. Let X be an L-variety. Then the restriction of scalars $R_{L/K}X$ is a K-variety $\mathscr{X}$ such that for every k-algebra A, one has $\mathscr{X}(A) = X(A \otimes_K L)$, functorially in A. (Also called *Weil restriction.*) If X is an algebraic group, then so is $\mathscr{X}$.

Schinzel's hypothesis: The conjecture that if $f_1, \ldots, f_r \in \mathbf{Z}[x]$ are irreducible and no prime divides

$$f_1(n)f_2(n)\cdots f_r(n)$$

for all $n \in \mathbf{Z}$, then there are infinitely many $n \in \mathbf{Z}$ such that $|f_1(n)|, \ldots, |f_r(n)|$ are simultaneously prime.

Selmer group: A group constructed out of Galois cohomology, whose purpose is to give an upper bound for the Mordell–Weil rank of an abelian variety over a global field. In theory it can be computed.

Shimura variety: A smooth projective variety V over $\mathbf{C}$ having a Zariski open subset whose set of complex points is analytically isomorphic to a quotient of a bounded symmetric domain X by a congruence subgroup of an algebraic group G that acts transitively on X. One can show that V descends to a variety over a number field contained in $\mathbf{C}$. Examples include moduli spaces $X_0(N)$ of elliptic curves with extra structure and Shimura curves which parametrize quaternionic-multiplication abelian surfaces with extra structure.

Special set: The algebraic special set of a variety X is the Zariski closure of the union of all positive-dimensional images of morphisms from abelian varieties to X over an algebraic closure. This set contains all rational curves in X, since $\mathbf{P}^1$ is dominated by any elliptic curve.

Toric variety: Over an algebraically closed field, a toric variety is a normal integral variety X containing a dense open subvariety T such that T has the structure of a torus and the left translation action of T on T extends to an action of T on X.

Torsor: Let B be a variety over a field k and let G be a (smooth) algebraic group over k. A right B-torsor under G is a B-scheme X with right G-action given by a B-morphism $X \times_k G \to X$ such that for some étale covering

$$\{U_i \to B\}$$

there is a G-equivariant isomorphism of U_i-schemes from $X \times_B U_i$ to $G \times_k U_i$ for all i. These are also called principal homogeneous spaces.

One can show that a torsor $X \to B$ is trivial (that is, isomorphic to $G \times_k B \to B$) if and only if it admits a section.

Torus: An algebraic group G over a field k that, after base extension to a separable closure of k, becomes isomorphic to a power of the multiplicative group $\mathbf{G}_m$.

Twist: Given a variety X over a field k, a twist Y is another k-variety such that X and Y become isomorphic after base extension to a separable closure k^s of k. The set of isomorphism classes of twists of X/k is in bijection with the cohomology set $H^1(\mathrm{Gal}(k^s/k), \mathrm{Aut}\, X_{k^s})$, where X_{k^s} is the base extension of X to k^s.

Universal torsor: Let X be a smooth projective geometrically integral variety over a field such that $\mathrm{Pic}(X_{\overline{k}})$ is free of finite rank over $\mathbf{Z}$. Let T be an algebraic torus over k with character group $\mathfrak{X}^*(T)$. A T-torsor over X gives rise to a homomorphism $\chi : \mathfrak{X}^*(T) \to \mathrm{Pic}(X_{\overline{k}})$ of $\mathrm{Gal}(\overline{k}/k)$-modules. In the case where χ is an isomorphism, the torsor is called universal.

Waring's problem: Given k, find the smallest number g_k such that every positive integer is a sum of g_k positive kth powers. The "easier Waring's problem" refers to the analogous problem where the kth powers are permitted to be either positive or negative. Variant: Given k, find the smallest number G_k such that every sufficiently large positive integer is a sum of G_k positive kth powers.

Weak approximation: For a smooth geometrically integral variety X over a global field k, one says that weak approximation holds if $X(k)$ is dense in the set of adelic points $X(\mathbf{A}_k)$.

Zeta function: The zeta function of a scheme X of finite type over $\mathbf{Z}$ is the analytic function

$$\zeta_X(s) := \prod_x \left(1 - N(x)^{-s}\right)^{-1},$$

where the product is taken over all closed points $x \in X$, and $N(x)$ denote the size of the residue field of x. The product converges for $\mathrm{Re}(s) > \dim X$, and the function is extended to other parts of the complex plane by analytic continuation, if possible. It is conjectured to have a meromorphic continuation to the whole plane. If $X = \operatorname{Spec} \mathbf{Z}$, then $\zeta_X(s)$ is the Riemann zeta function

$$\prod_{\text{prime } p} \left(1 - p^{-s}\right)^{-1} = \sum_{n=1}^{\infty} n^{-s}.$$

If X is of finite type over a finite field $\mathbf{F}_q$, then $\zeta_X(s)$ is a rational function of q^{-s}.

INDEX